Advances in 3G Enhanced Technologies for Wireless Communications

For a listing of recent titles in the *Artech House Mobile Communications Series,*
turn to the back of this book.

Advances in 3G Enhanced Technologies for Wireless Communications

Jiangzhou Wang
Tung-Sang Ng
Editors

Artech House
Boston • London
www.artechhouse.com

Library of Congress Cataloging-in-Publication Data
Advances in 3G enhanced technologies for wireless communications/Jiangzhou Wang, Tung-Sang Ng, editors.
p. cm. — (Artech House mobile communications series)
Based on selected papers from the Fourth IEEE Workshop on Emerging Technologies in Circuits and Systems—Third Generation (3G) Mobile Technologies and Applications, held Nov. 29–Dec. 1, 2000, in Hong Kong.
Includes bibliographical references and index.
ISBN 1-58053-302-7 (alk. paper)
1. Wireless communication systems—Technological innovations. I. Wang, Jiangzhou. II. Ng, Tung-Sang. III. Series.
TK5103.2 .A388 2002
621.382—dc21 2001056654

British Library Cataloguing in Publication Data
Advances in 3G enhanced technologies for wireless communications. — (Artech House mobile communications series)
1. Wireless communication systems 2. Mobile communication systems 3. Code division multiple access
I. Wang, Jiangzhou II. Ng, Tung-Sang
621.3'8456

ISBN 1-58053-302-7

Cover design by Igor Valdman

685 Canton Street
Norwood, MA 02062

International Standard Book Number: 1-58053-302-7
Library of Congress Catalog Card Number: 2001056654

10 9 8 7 6 5 4 3 2 1

Contents

Preface

The Fourth IEEE Workshop on Emerging Technologies in Circuits and Systems—Third Generation (3G) Mobile Technologies and Applications was held from November 29 to December 1, 2000, in Hong Kong. It was a very successful workshop with more than 100 participants attending, mainly from industry. Speakers who are working on the development of 3G products in industry were invited. Their presentations, such as WCDMA, EDGE, CDMA2000, wireless Internet, and software radio, were comprehensive and covered different aspects of this important emerging technology. To benefit a larger group of people, we managed to persuade some of the presenters to take time from their busy schedules to write up their presentations. Their efforts enabled us to compile this informative book, which we consider most suitable for both research engineers and postgraduate students.

This book contains seven chapters. In Chapter 1, Dr. Mamoru Sawahashi of NTT DoCoMo comprehensively describes features of the WCDMA air-interface and the essential WCDMA technologies associated with its performance. The chapter describes physical channel and spreading code assignment, transport channel multiplexing (including rate matching), synchronization techniques focusing on the radio interface synchronization, SIR measurement-based fast transmitting power control, and diversity techniques (including rake-time diversity, intercell diversity, and transmitting diversity) in the forward link. Moreover, adaptive antenna array processing, which can further improve system capacity, is also presented.

In Chapter 2, Dr. Hiroshi Furukawa of NEC discusses in detail downlink power control and enhancement of downlink performance. Two proposals to enhance downlink performance are presented: optimum allocation of downlink orthogonal code and site selection diversity transmission power control.

In Chapter 3, Benoist Sébire, Janne Parantainen, and Guillaume Sébire of Nokia provide an overview of GERAN with a focus on release 5 and harmonization with UTRAN. The chapter covers different aspects of GERAN, including 3G services, the GERAN reference architecture, and protocol architectures. Emphasis is placed on the radio protocols that constitute the main change from GSM/EDGE to the Iu-capable GERAN.

In Chapter 4, Dr. Qiang Wu and Dr. Eduardo Esteves of Qualcomm describe the technical details of the CDMA2000 high-rate packet data system, including hybrid ARQ, turbo coding, scheduling, adaptive modulation, and performance simulations.

In Chapter 5, Dr. Vincent Lau of Lucent Technologies analyzes the peak-to-average ratio of a CDMA signal, which determines the backoff factor of a power amplifier allowing for reductions in the clipping of input signals and hence spectral regrowth.

In Chapter 6, Haseeb Akhtar, Dr. Emad Abdel-Lateef Qaddoura, Dr. Abdel-Ghani Daraiseh, and Russ Coffin of Nortel Networks discuss the wireless Internet. This chapter describes the framework of IP mobility, which enables users to gain access to the Web while roaming freely. The framework is designed to extend the mobility management (with centralized directory management), end-to-end security, device independence, and application independence that allow users to use Internet applications on any device and any network.

Finally, in Chapter 7, Yik-Chung Wu, Professor Tung-Sang Ng, and Dr. Kun-Wah Yip of the University of Hong Kong discuss the software radio, which is a promising technique for implementing 3G and future mobile systems incorporating multiple standards. This chapter covers fundamental issues in software radio such as bandpass sampling, decimation filtering, and fractional sampling-rate conversion.

Acknowledgments

We appreciate all authors for providing these important contributions. They spent much time preparing their chapters in addition to their normal busy schedules. The editors also would also like to thank Yik-Chung Wu of the University of Hong Kong for his help in formatting the final manuscripts.

Jiangzhou Wang and Tung-Sang Ng
Hong Kong
February 2002

1

WCDMA Enhanced Technologies

Mamoru Sawahashi

After the global-level standardization in the Third Generation Partnership Project (3GPP) and development with enthusiastic efforts of *wideband code-division multiple access* (WCDMA) [1, 2] the commercial service was launched in October 2001 in Japan. Along with the start of IMT-2000 commercial service, the dawn of the genuine era of wireless Internet is upon us. *Direct-sequence CDMA* (DS-CDMA) wireless access, on which WCDMA is based, has numerous advantages over *time-division multiple access* (TDMA) or *frequency-division multiple access* (FDMA) including single-frequency reuse, soft handoff (or site diversity), enhanced radio transmission through RAKE combining, and direct capacity increase through sectored antennas.

A list of the key features of the WCDMA physical layer follows:

- Intercell asynchronous operation and three-step fast cell search;
- Flexible realization of various levels of *quality of service* (QoS) for various transport channels by rate matching associated with channel coding;
- *Signal-to-interference power ratio* (SIR)–based fast *transmission power control* (TPC) to satisfy the required quality level for a physical channel with minimum transmission power;
- Significant gains in link capacity and coverage through the use of many diversity techniques, such as coherent RAKE time diversity using pilot symbol-assisted channel estimation, space diversity,

intercell (sector) diversity, and transmit diversity (only in the forward link);

- A high level of flexibility in offering different multirate services (up to 2 Mbps) through orthogonal variable spreading factor multiplexing and orthogonal multicode transmission;
- Capacity enhanced techniques such as interference cancellation and adaptive antenna array diversity.

This chapter comprehensively describes features of the WCDMA air-interface and the essential WCDMA technologies associated with its performance. This chapter is organized as follows. Section 1.1 explains the physical channel and spreading code assignment. Then, transport channel multiplexing including rate matching is described in Section 1.2. After synchronization techniques focusing on radio interface synchronization are explained in Section 1.3, SIR measurement-based fast TPC is described in Section 1.4. Section 1.5 discusses the various diversity techniques, including RAKE time diversity, intercell (sector) diversity, and transmit diversity in the forward link. Finally, the adaptive antenna array processing technique, which further improves system capacity, is presented.

1.1 Physical Channel and Spreading Code Assignment

1.1.1 Physical Channel

WCDMA has a three-layered channel structure: physical [3, 4], transport, and logical. The physical channels provide several transport channels to the *medium access control* (MAC) layer, which is a sublayer of the data link layer (Layer 2). The MAC layer provides several different logical channels to a higher layer, the *radio link control* (RLC) layer. The physical channels are classified by spreading codes, carrier frequency, and in-phase (I)/quadrature-phase (Q) assignment. One radio frame of a physical channel has a frame length of 10 ms and comprises 15 slots. Thus, the slot length is equal to a basic updating unit of adaptive fast TPC and channel estimation of coherent RAKE combining and is optimized to the value of 0.667 ms taking into account a tradeoff between the frame efficiency and the tracking ability of fast TPC and channel estimation against fast fading variation. The number of channel-coded information bits conveyed by each physical channel differs according to the type of physical channel and spreading factor. The features of the major physical channels are described next:

1. *Primary-common control physical channel (P-CCPCH):* One P-CCPCH is defined for each sector in the forward link. The P-CCPCH has a fixed spreading factor of 256 (15 Ksps) and carries the broadcast channel transport channel. The P-CCPCH is not transmitted during the first 256-chip duration; instead, the P-SCH and S-SCH are transmitted during that period at each slot.
2. *Secondary-common control physical channel (S-CCPCH):* Multiple S-CCPCHs, which are common channels in the forward link, are defined in each cell (sector) and carry paging information and lower data information from a higher layer.
3. *Physical random-access channel (PRACH):* Multiple PRACHs, which are common channels in the reverse link, are defined and used to carry the RACH transport channel comprising lower information data from a higher layer.
4. *Dedicated physical channel (DPCH):* A DPCH is assigned to each *mobile station* (MS) in both the forward and reverse links. It comprises a *dedicated physical control channel* (DPCCH) and a *dedicated physical data channel* (DPDCH). A DPDCH consists of a channel-coded data sequence, and more than one DPDCH can be assigned to one DPCH. A DPCCH is used for Layer 1 control of DPCH and one DPCCH is defined for one DPCH. A DPCCH comprises pilot bits for coherent channel estimation, TPC bits, *transport format combination indicator* (TFCI) bits, and *feedback information* (FBI) bits designating the control information for transmit diversity in the forward link (thus, FBI bits are defined only in the reverse link).
5. *Common pilot channel (CPICH):* A CPICH is the common pilot channel used for channel estimation, path search for RAKE combining (generation of power delay profile), and the third step, that is, scrambling code identification in the three-step cell search method. Two kinds of CPICHs are defined: primary CPICH and secondary CPICH. The primary CPICH has two-symbol data sequences associated with two antennas. Without transmit diversity all symbol sequences with all 1's are transmitted from antenna 1; with transmit diversity, the second primary CPICH with different symbol sequences from those of the first primary CPICH are also transmitted from antenna 2 in addition to the first primary CPICH. In future applications of smart antennas for spot beam transmis-

sion, the secondary CPICH will be defined, which will be spread by the primary or secondary scrambling code.

6. *Synchronization channel (SCH):* The SCH is a common channel in the forward link, which is used for cell search. Primary and secondary SCHs are used for the first and second steps of the three-step cell search method. They are transmitted only during the 256-chip period at the beginning of each slot.
7. *Acquisition indication channel (AICH):* The AICH is a common channel in the forward link used for random access control. It is used as a pair comprising a PRACH and PCPCH.
8. *Page indication channel (PICH):* The PICH is a common channel in the forward link and is associated with S-CCPCH, in which the PCH transport channel is mapped.
9. *Physical downlink shared channel (PDSCH):* The PDSCH is a common channel in the forward link, which carries the DSCH transport channel and is used for high-rate packet data transmission.
10. *Physical common packet channel (PCPCH):* The PCPCH is a common channel in the reverse link, which carries the CPCH transport channel and is used for high-rate packet data transmission.

The frame structure of the DPCH in the reverse and forward links is illustrated in Figure 1.1(a) and 1.1(b), respectively. The DPDCH and DPCCH are code multiplexed into I and Q channels, respectively, in the reverse link. Because the DPCCH with a fixed rate (i.e., spreading factor) and DPDCH with variable rate transmission are separated from each other in the orthogonal phase, fluctuation of the amplitude during variable transmission can be decreased. Meanwhile, the DPCCH and DPDCH are alternatively time multiplexed within a slot in the forward link.

1.1.2 Spreading Code Assignment

WCDMA adopts a two-layered spreading code assignment [5], which combines a channelization code with the repetition period of the corresponding symbol rate and a scrambling code with the repetition of the frame interval. The orthogonal variable spreading factor code [5, 6] is used as the channelization code. The spreading code assignment for each physical channel is given

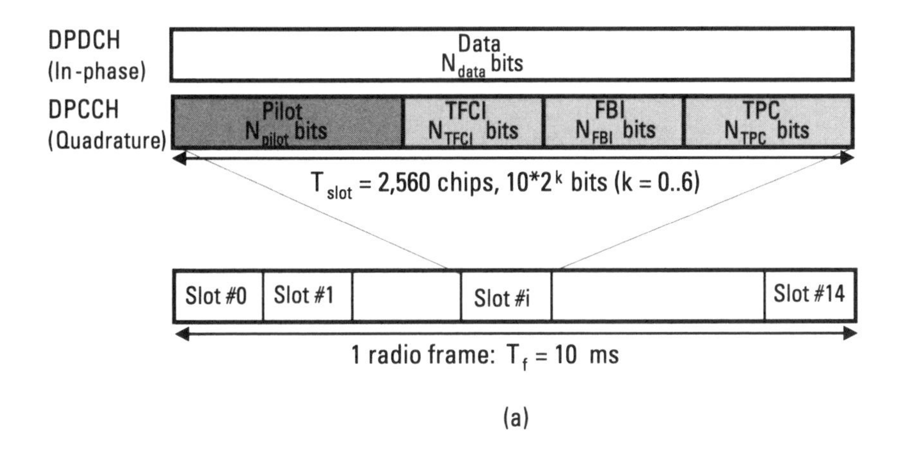

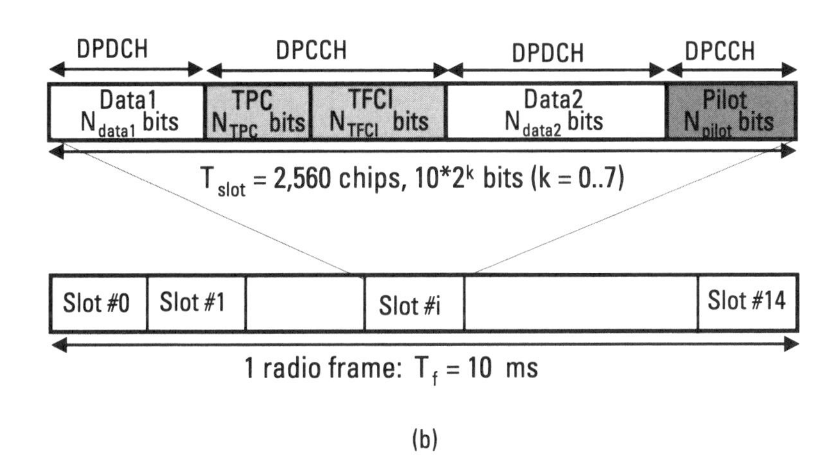

Figure 1.1 Frame structure of DPCH: (a) reverse link and (b) forward link.

in Table 1.1. The *spreading factor* (SF) of 4 to 256 is used for S-CCPCH and DPCH.

1.1.2.1 Channelization Code

Starting from $C_{ch,1,0}$ (1) (SF = 1), the orthogonal variable spreading factor code that has a length of 2^{k-1} chips in the kth layer, is recursively generated based on the formula given next, resulting in the tree-structured code generation as shown in Figure 1.2 [6].

Table 1.1
Spreading Code Allocation

	Channelization Code Repetition Period = Data Symbol Period		**Scrambling Code Repetition Period = 10-ms Frame**
Forward Link	User identification (4–512 chips)		Cell (sector) identification (38,400 chips)
CPICH	#0	SF = 256	Primary
P-CCPCH	#1	SF = 256	Primary
S-CCPCH	Arbitrary	SF = 4–256	Primary (secondary)
DPCH	Arbitrary	SF = 4–256	Primary (secondary)
AICH	Arbitrary	SF = 256	Primary (secondary)
PICH	Arbitrary	SF = 256	Primary (secondary)
Reverse Link	Code channel identification in multicode transmission (4–256 chips)		User identification (38,400 chips)
DPCH	Arbitrary	SF = 4–256	Primary (secondary)

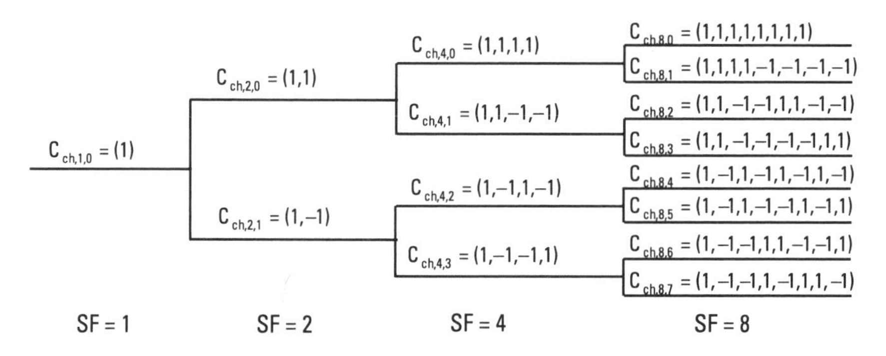

Figure 1.2 Generation method of orthogonal variable spreading factor (SF) codes.

$$\begin{bmatrix} C_{ch,2^{(k+1)},0} \\ C_{ch,2^{(k+1)},1} \\ C_{ch,2^{(k+1)},2} \\ C_{ch,2^{(k+1)},3} \\ \vdots \\ C_{ch,2^{(k+1)},2^{(k+1)}-2} \\ C_{ch,2^{(k+1)},2^{(k+1)}-1} \end{bmatrix} = \begin{bmatrix} C_{ch,2^k,0} & C_{ch,2^k,0} \\ C_{ch,2^k,0} & -C_{ch,2^k,0} \\ C_{ch,2^k,1} & C_{ch,2^k,1} \\ C_{ch,2^k,1} & -C_{ch,2^k,1} \\ \vdots & \vdots \\ C_{ch,2^k,2^k-1} & C_{ch,2^k,2^k-1} \\ C_{ch,2^k,2^k-1} & -C_{ch,2^k,2^k-1} \end{bmatrix} \tag{1.1}$$

The k orthogonal variable spreading factor codes of the kth layer are orthogonal to each other. Furthermore, any two codes belonging to different layers are orthogonal except for when one code is not the mother code of the other. For example, $C_{ch,2,0}$ and $C_{ch,4,2}$ are orthogonal to each other. When $C_{ch,2,0}$ is already assigned, any code below this code on the code tree cannot be used; this is a restriction of the code assignment. The codes of $C_{ch,256,0}$ and $C_{ch,256,1}$ are commonly used for all cells for the primary CPICH and P-CCPCH in the forward link, respectively. The channelization codes of other physical channels are assigned from a higher layer.

1.1.2.2 Scrambling Code

Cell (sector)-specific and user-specific scrambling codes are assigned in the forward and reverse links, respectively. In the reverse link, the repetition period of the scrambling code is 10 ms and that with the repetition period of 256 chips is optionally defined for future application of multiuser detection. The long scrambling code is truncated by a duration of 38,400 chips from the beginning of the Gold sequence with a repetition period of 2^{24} chips. There are 2^{24} long scrambling codes.

The scrambling code in the forward link is generated by truncating the 38,400 chips from the beginning of the Gold sequence with the repetition period of 2^{18} and its shifted version by 131,072 chips. The 8,192 scrambling codes are grouped into 512 scrambling-code groups, where each group comprises one primary scrambling code with 15 corresponding secondary scrambling codes. The primary scrambling code is first used, and then the secondary scrambling codes are used to cover any shortage in the channelization code set associated with the primary scrambling code. Then 512 primary scrambling codes are divided into 64 primary scrambling code groups (hereafter simply denoted as group), each including 8 primary scrambling codes. This group-wise divided primary scrambling code structure is used for the three-step cell search algorithm, which is described in Section 1.3.

1.1.2.3 Synchronization Code

A synchronization code is used to spread an SCH and comprises a *primary synchronization code* (PSC) and *secondary synchronization code* (SSC) both with a length of 256 chips, which are used for the *primary SCH* (P-SCH) and *secondary SCH* (S-SCH), respectively. Let PSC be denoted as C_{PSC}, which is a complex-value code sequence with the same sequence for real and imaginary parts expressed as follows:

$$C_{\text{PSC}} = (1 + j) \times \{a, a, a, -a, -a, a, -a, -a, a, a, a, -a, a, -a, a, a\}$$

where

$$
\begin{aligned}
a &= \{x_1, x_2, x_3, \ldots, x_{16}\} \\
&= \{1, 1, 1, 1, 1, 1, -1, -1, 1, -1, 1, -1, 1, -1, -1, 1\}
\end{aligned}
\tag{1.2}
$$

Let 16 SSCs be denoted as $C_{SSC,k}$ (k = 1, 1, 2, . . . , 16). Then, $C_{SSC,k}$ is generated by multiplying the jth component ($1 \le j \le 256$) of vector Z of a common sequence with the length of 256 chips and the jth component of the nth column of H_8 of the Hadamard matrix, where $n = 16 \times (k - 1)$. Let $h_n(j)$ and $z(j)$ be the jth symbol of nth column of the Hadamard matrix and the jth symbol of a common sequence, respectively. By selecting 16 columns from 256 columns at every 16 columns, the 16 $C_{SSC,k}$ are generated as

$$
\begin{aligned}
C_{SSC,k} &= (1 + j) \times \\
&\quad \langle h_n(0) \times z(0), h_n(1) \times z(1), h_n(2) \times z(2), \ldots, h_n(255) \times z(255) \rangle
\end{aligned}
$$

where

$$
\begin{aligned}
Z &= \{-b, b, b, -b, b, b, -b, -b, b, -b, b, -b, -b, -b, -b, -b\} \\
b &= \{x_1, x_2, x_3, x_4, x_5, x_6, x_7, x_8, -x_9, -x_{10}, -x_{11}, -x_{12}, -x_{13}, -x_{14}, \\
&\qquad -x_{15}, -x_{16}\}
\end{aligned}
\tag{1.3}
$$

1.1.2.4 Spreading

In the reverse link, the channelization codes are independently spread into I/Q channels by using different orthogonal variable spreading factor codes and are weighted by weighting factor G, which denotes the transmitted amplitude (power) ratio of DPDCH to DPCCH. Complex spreading is applied to the physical channel: One is a code truncated by 38,400 chips from the beginning of the Gold sequence with a repetition period of 2^{24}, and the other is truncated by 38,400 chips of the shifted first Gold sequence by 16,777,233 chips. Thus, the spreading using channelization codes and the scrambling codes are expressed as

$$
\begin{aligned}
S_I &= D_I C_I - D_Q C_Q \\
S_Q &= D_I C_Q + D_Q C_I
\end{aligned}
\tag{1.4}
$$

where $D_{I(Q)}$ denotes the I/Q components of the chip data sequence spread by channelization codes and $C_{I(Q)}$ represents the I/Q components of a long scrambling code. In this *quadrature phase-shift keying* (QPSK) spreading, the carrier-phase transition by π-occurs across the zero point, thus incurring an increasing nonlinear distortion of the power amplifier. Therefore, in the 3GPP standard, the *hybrid phase-shift keying* (HPSK) scheme was adopted, which decreased the possibility of the phase transmission crossing the zero point [5, 7]. The long scrambling code sequence used for spreading is generated from the two original scrambling codes based on the following equation:

$$C_{\text{long},n}(i) = c_{\text{long},1,n}(i)(1 + j(-1)^i c_{\text{long},2,n}(2\lfloor i/2 \rfloor)) \tag{1.5}$$

In the forward link, P-SCH and S-SCH are spread by only PSC and SSC, respectively, commonly used for both I/Q channels. The other physical channels, except for SCH, are first spread by an identical channelization code with SF = m for both the I/Q channels and then complex-scrambled by the two scrambling code sequences.

1.2 Transport Channel Multiplexing

1.2.1 Explanation of Data Format for Layer 1

This section first briefly explains the terminology used for data transfer between the MAC layer and Layer 1 [8]. A transport block, which corresponds to an RLC-PDU (protocol data unit), is a basic unit for data transfer between the MAC layer and Layer 1. A cyclic redundancy check calculation result for error detection in Layer 1 is added to every transport block. A set of transport blocks simultaneously transferred between the MAC layer and Layer 1 in the same transport channel is called a *transport block set.* The size of the transport block is the length of the transport block defined in bit form. The size of each transport block belonging to one transport block set is uniform and is a fixed value. The number of bits within a transport block set is called the transport block set size.

The arrival time interval of transport block sets between the MAC layer and Layer 1 is called the *transmission time interval* (TTI), which is equal to the channel interleaving length. The TTI is some integer times the radio frame length (= 10 ms) and is defined as 10, 20, 40, or 80 ms in the 3GPP standard. The transport format is a format in which a transport block set is transferred between the MAC layer and Layer 1 on a transport channel every TTI. The transport format comprises two attributes: the dynamic part and the semistatic part. Attributes of the dynamic part are the transport block size, transport block set size, and TTI, and those for the semistatic part are error correction (channel coding) scheme, such as the type of error correction, coding rate, and the size of the cyclic redundancy check.

The *transport format set* (TFS) is defined as a set of transport formats used for the transport channels. Within one TFS, the semistatic parts of all transport channels are identical; however, the dynamic parts may be changed every TTI in order to achieve variable-rate transmission. The transport channels are simultaneously multiplexed into Layer 1 as a coded composite transport channel. Each transport channel in the coded composite transport channel has an available TFS; however, only one transport format is used at each TTI. Thus, the combination of possible transport formats of all transport channels transferred in the same Layer 1 at each TTI is defined as a *transport format combination* (TFC). Furthermore, a set of TFCs applied to the coded composite transport channel is called a *transport format combination set* (TFCS). The indicator designating the TFC is called the *transport format combination indicator* (TFCI). TFCI bits are multiplexed into the DPCCH of each DPCH. In the receiver, the TFCI bits are employed to decode Layer 1 data sequences and demultiplex transport blocks transferred on one physical channel.

In addition to the explicit TFCI detection method, the blind transport format detection method using cyclic redundancy check to trace the surviving trellis path ending at the zero state among the possible transport formats is also specified in the 3GPP standard (note that blind detection is used only for the forward link) [9].

1.2.2 Transport Channel

A transport channel [3, 8] is defined as a channel that is used to transfer various kinds of data to the MAC layer. The major transport channels are described next. The mapping relationships between the major physical channels and transport channels are given in Figure 1.3.

1. *Broadcast channel (BCH):* The BCH is a forward link transport channel that is employed for broadcasting system- and cell-specific information. The BCH is always transmitted over the entire cell and has a single transport format.
2. *Forward access channel (FACH):* The FACH is a forward link transport channel that is commonly used for multiple MSs and for transmitting low-rate user information from a higher layer.

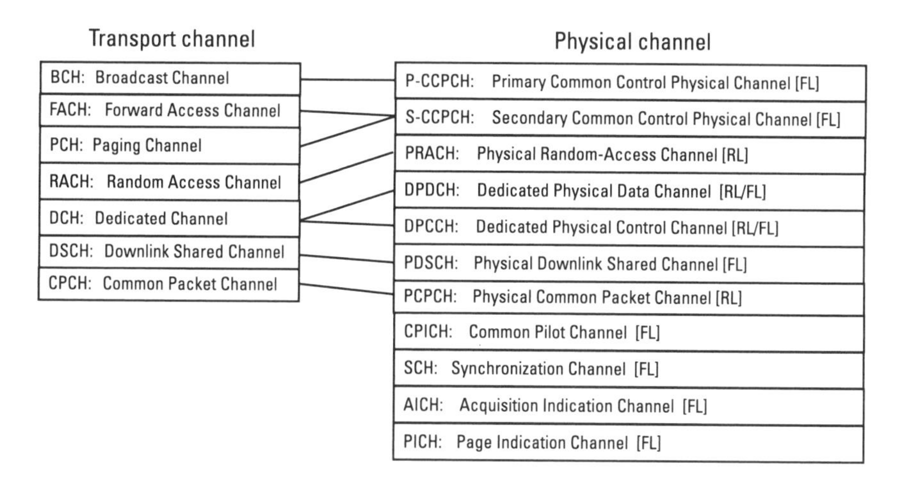

Figure 1.3 Relationship between the physical channel and the transport channel.

3. *Paging channel (PCH):* The PCH is a forward link transport channel that is transmitted over the entire cell and is used to transmit paging information.
4. *Random-access channel (RACH):* The RACH is a reverse link transport channel, which is received from the entire cell. The RACH is characterized by collision risk and by being transmitted using open-loop-type TPC.
5. *Dedicated channel (DCH):* The DCH is a forward link and reverse link transport channel, which is transmitted either throughout the entire cell or over only a part of the cell using a beam-forming transmission with an adaptive antenna. The DPCH is used for the transmission of user data and is assigned to each MS. Variable-rate transmission and fast TPC are applied to the DPCH.
6. *Downlink shared channel (DSCH):* The DSCH is a forward link transport channel shared by several MSs. The DSCH is used primarily for high-rate packet data transmission and is transmitted either throughout the entire cell or over only a part of the cell using beam-forming transmission.
7. *Common packet channel (CPCH):* The CPCH is a reverse link transport channel and is associated with a dedicated channel on the forward link, which provides power control and CPCH control commands. The CPCH is used for high-rate data transmission in random access channels.

1.2.3 Multiplexing and Rate Matching

We now discuss transport channel multiplexing and its associated rate matching [9]. The flow of the transport channel multiplexed into a physical channel in the reverse link is depicted in Figure 1.4. First, cyclic redundancy check parity bits required for block error detection at the receiver are calculated for the original data sequence per transport block of each transport channel. Then, the calculated cyclic redundancy check bits are attached to each transport block. All transport blocks with cyclic redundancy check bits within one TTI are serially concatenated followed by channel coding.

For channel coding, convolutional coding or turbo coding is used in the 3GPP specification. For the common transport channels such as BCH, PCH, and RACH, convolutional coding with a rate of 1/2 and a constraint length of 9 bits is used. Convolutional coding with a rate of 1/3 (1/2) is also used for FACH and DPCH with a lower channel bit rate, and turbo coding [10] with a rate of one-third and a constraint length of 4 bits is used for FACH and DPCH with higher channel bit rates.

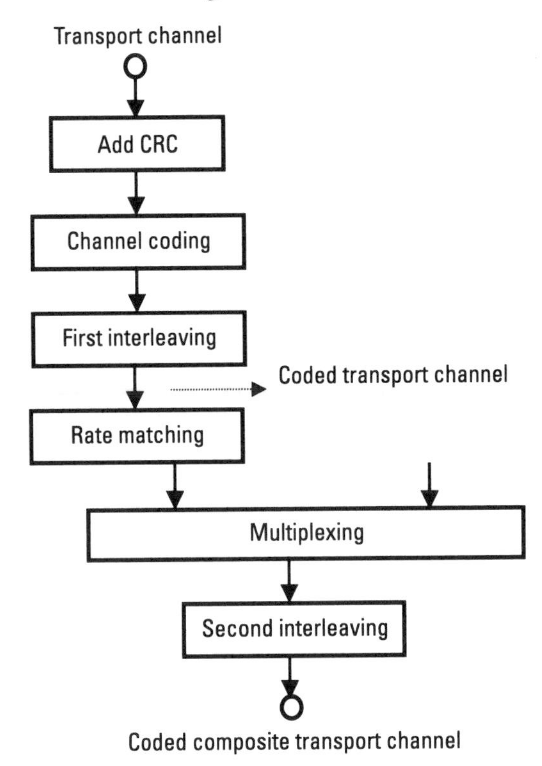

Figure 1.4 Transport channel multiplexing (reverse link).

After the coded data sequence of each transport channel is interleaved over the length of the TTI (first interleaving), rate matching is performed according to the required QoS and the number of bits. The data sequence of each transport channel after rate matching is segmented and interleaved over one radio frame length (second interleaving). Finally, the coded composite transport channel containing all transport channels is multiplexed into a physical channel. As described previously, the first channel interleaving is performed before rate matching of each transport channel in the reverse link. Meanwhile, discontinuous transmission is allowed when there is no transmitted data sequence in the radio frame of a certain transport channel in the forward link. Thus, the rate matching is performed independently for each transport channel before the first interleaving.

As shown in Figure 1.5, transport channels with different bit rates and QoS levels are multiplexed and transferred into one physical channel. A transport block is a basic unit for data transfer between the MAC layer and Layer 1 [in Figure 1.5 of the transport channel, $a(b)$ represents the bth block of transport channel a]. The required QoS, that is, the *block error rate* (BLER) or *bit error rate* (BER) of the physical channel, is achieved by changing the transmit power or data modulation scheme according to the fading variation. In general, the QoS level of one physical channel can be controlled by changing the target SIR in fast TPC employing outer loop control so that the output BLER or BER is equal to the required value as explained later.

However, the average received signal energy per bit-to-interference and background noise spectrum density ratio (E_b/I_0), thus, the received signal

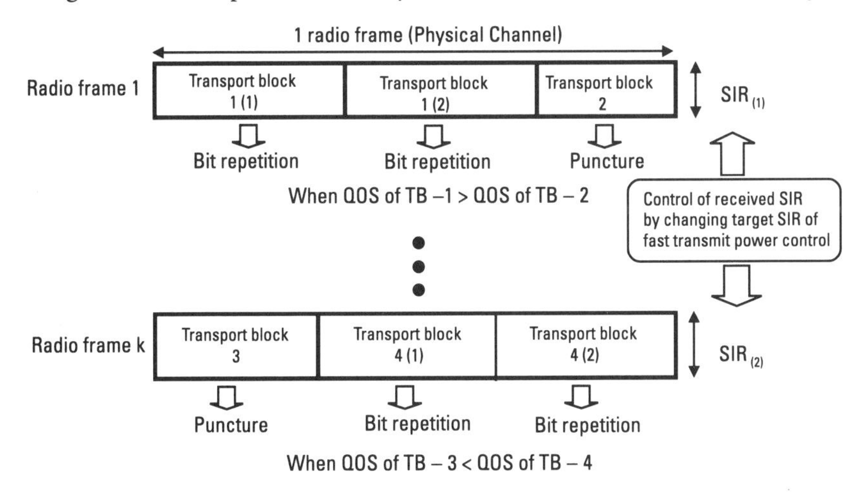

Figure 1.5 Operation principle of rate matching. QOS: quality of service.

power, is an almost constant value during one radio frame interval. Therefore, to bundle various transport channels with different QoSs into one physical channel, the required QoSs of various transport channels are simultaneously satisfied with respect to the identical average received signal power by changing the number of coded bits of each transport channel after channel decoding (this process is called *rate matching*). That is, by repeatedly transmitting some coded bits at a regular interval, the BLER or BER is improved. In contrast, if encoded bit sequences are punctured at a regular interval, the received quality is degraded. In this way, the number of bits of each transport channel multiplexed into the physical channel is flexibly changed for every radio frame by rate matching as described hereafter.

In the reverse link, rate matching is performed for the coded data sequence of each transport channel after the first interleaving. The number of bits of each transport channel to be repeated or punctured is calculated based on the rate-matching attributes signaled from a higher layer. The discontinuous transmission, when there is no coded transmitted data sequence of a certain transport channel multiplexed into a physical channel, is not permitted. Thus, the spreading factor, that is, the symbol data rate, of a physical channel is first determined according to the total number of bits per radio frame of all transport channels multiplexed into the physical channel. Then, rate matching is performed so that the sum of the bits of all transport channels per radio frame after rate matching should be equal to the bits per radio frame accommodated into the physical channel having the assigned spreading factor. Let $N_{i,j}$ and $\Delta N_{i,j}$ be the number of coded data bits of transport channel i per radio frame with TFC j before rate matching and the number of bits per radio frame to be bit-repeated or punctured (the positive and negative values of Δ denote the bit repetition and puncture), respectively. The value of $Z_{i,j}$ needed for the calculation of $\Delta N_{i,j}$ is recursively computed from the following equations using the rate-matching attribute value, RM_i.

$$Z_{0,j} = 0$$

$$Z_{i,j} = \left\lfloor \frac{\left\{ \left(\sum_{m=1}^{i} \mathrm{RM}_m x N_{m,j} \right) x N_{\mathrm{data},j} \right\}}{\sum_{m=1}^{I} \mathrm{RM}_m x N_{m,j}} \right\rfloor \quad \text{for all } i = 1, \ldots, I \tag{1.6}$$

where $N_{\text{data},j}$ is the total number of bits per radio frame to be assigned to code the composite transport channel with TFC j and $\lfloor x \rfloor$ denotes the integer value defined as $x - 1 \le \lfloor x \rfloor \le x$. Using the value of $Z_{i,j}$ recursively calculated from (1.6), $\Delta N_{i,j}$ is derived from the following equation:

$$\Delta N_{i,j} = Z_{i,j} - Z_{i-1,j} - N_{i,j} \qquad \text{for all } i = 1, \ldots, I \qquad (1.7)$$

In the reverse link, rate matching is performed per radio frame based on (1.7). Meanwhile, in the forward link, unlike the reverse link, discontinuous transmission is applied when there are no transmitted coded data bits of a certain transport channel. Thus, the rate-matching pattern does not necessarily change for each radio frame.

Rate matching in the reverse link is performed as follows. The number of bits per TTI of transport channel i before rate matching, $N_{i,h}^{\text{TTI}}$, is first calculated for the corresponding TFC h belonging to TFCS. Then, from the value of $N_{i,h}^{\text{TTI}}$, and the number of radio frames of transport channel i over TTI, F_i, the corresponding number of bits per radio frame is derived for all TFC belonging to TFCS. Thus, rate matching is performed such that the number of total bits per radio frame for TFC $h_{\max}$, when the summation of bits per radio frame of all transport channels is maximized, is equal to the number of bits per radio frame accommodated by a physical channel, that is to say, the number of bits per radio frame. Then the number of bits per TTI to be bit-repeated or punctured is computed for each transport channel. Based on the rate-matching pattern obtained, the number of bits per radio frame of each transport channel is updated every TTI.

Consequently, when transport channels having different TTI are multiplexed, the number of total bits belonging to a radio frame is changed at the shortest TTI at every TTI. If the number of bits per radio frame of transport channel i after rate matching is lower than the maximum number of bits assigned to that transport channel, discontinuous transmission is performed during an interval corresponding to the number of bits to be shortened.

1.3 Asynchronous Cell Sites and Synchronization

1.3.1 Synchronization in UTRAN

In asynchronous cell site operation, which is the most prominent feature in WCDMA, flexible system deployment from outdoors to indoors is possible, since no external timing source such as the *Global Positioning System* (GPS) is required. Synchronization techniques [11] for flexibly supporting asynchronous cell site operation are roughly classified into three processes as shown in Figure 1.6.

The first is the node synchronization, which has two aspects: node synchronization between *radio network controller* (RNC) and node B [defined as a logical node that conducts radio transmissions with MSs using one or multiple cells and that physically corresponds to the *base station* (BS)] and that between several node B's. In UTRAN, each node does not necessarily have a common timing reference. Thus, respective node counters, that is, the RNC frame number counter and node B frame number counter, may not be phase aligned. The node synchronization between the RNC and node B is used for estimating the timing differences between the RNC and node B without compensating the phase difference between the RNC and B frame number counters.

Node synchronization between node B's can be achieved via the node synchronization between the RNC and node B in order to determine internode B timing reference relations. This node synchronization is used for determining the intercell timing relationship with neighboring cell lists, which is needed for achieving fast cell search by an MS at handover. Second is the transport channel synchronization between UTRAN and MS. This synchronization is established by using a common frame number called a connection frame number, which is associated with Layer 2 according to

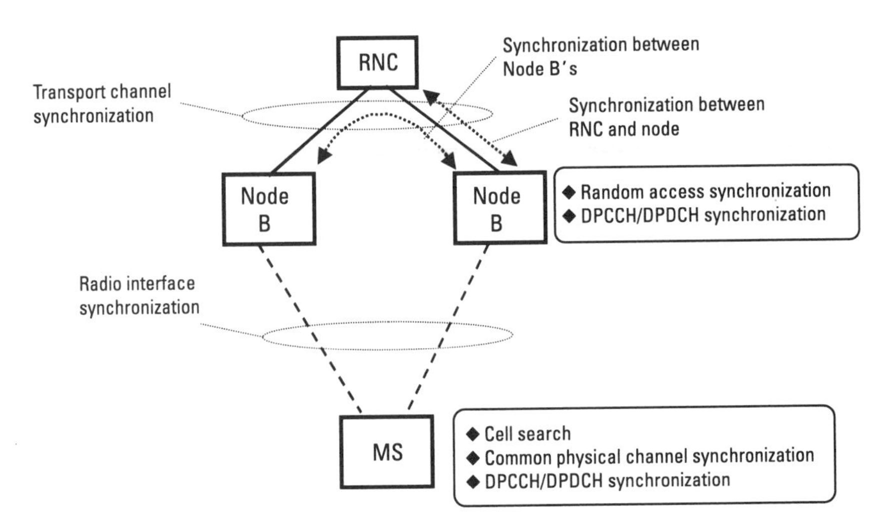

Figure 1.6 Synchronization in UTRAN.

the transport block set. The third is the radio interface synchronization, that is, synchronization of the physical channel between node B and the MS. This is further categorized into three processes. In the first process, an MS must quickly acquire forward link scrambling code synchronization and measure the average received signal power of the best cell site (having the least propagation path loss) at the beginning of communication. We call this process cell search. The second and third processes are the synchronizations of a common channel and a dedicated channel, which follow after a cell search process.

1.3.2 Cell Search

Cell search is categorized into three cases. The first case is the initial cell search when the MS is switched on; the second is the target cell search (i.e., in which the number of candidate cells is limited) in the active mode for searching for *soft handover* (SHO) candidates; and the third is the target cell search in the idle mode for searching for a temporary perching cell. For all of these cases, in general, accomplishing these tasks takes longer in an intercell asynchronous system than it does in a synchronous system because different cell sites use different scrambling codes. Thus, the fast cell search algorithm was proposed and refined in the standardization process.

1.3.2.1 Three-Step Cell Search Algorithm

To reduce the initial cell search time in asynchronous cell site operation, a three-step cell search method using scrambling code masking was originally proposed [12]. Subsequently, the original cell search method was refined in the standardization process. The forward link frame structure in the 3GPP standard required for the three-step cell search is illustrated in Figure 1.7.

The BS transmits a continuous CPICH, primary SCH, and secondary SCH over the 256-chip duration at the beginning of each slot. The spreading codes for the CPICH and the DPCHs are taken from a set of OVSF codes, thereby maintaining mutual orthogonality between the CPICH and DPCHs. These channels are further scrambled by a cell-specific scrambling code with a 10-ms repetition period (= 38,400-chip duration), which is equal to the data frame length. The PSC for the primary SCH is common to all cell sites, and the SSC for the secondary SCH denotes the group index into which each of the scrambling codes is grouped beforehand. As described in Section 1.1.2.2, the total number of scrambling codes to be searched is 512, which is divided into 64 groups of eight codes each.

The operational flow of the three-step cell search algorithm is illustrated in Figure 1.8. The three-step cell search is performed as follows.

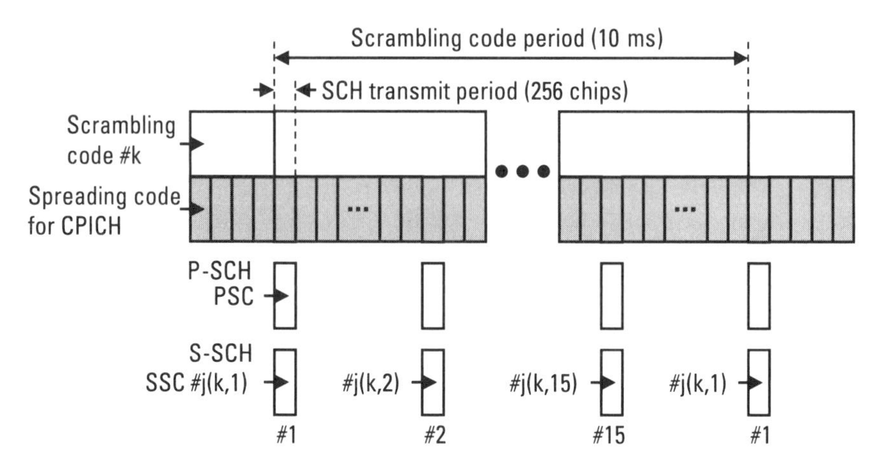

Figure 1.7 Forward link frame structure of CPICH and SCH.

First Step

The MS first calculates the correlation between the received sampled chip data sequence (actually an oversampled chip data sequence) and a PSC replica using the PSC-matched filter in order to detect the P-SCH time position. The correlation of the matched filter output at time t is represented as

$$\psi_1(t) = \frac{1}{256 \cdot T_c} \int_0^{256T_c} r(t - \mu) c_{\text{PSC}}(256T_c - \mu) d\mu \tag{1.8}$$

Then, let $\psi_1(\tau, m, n)$ be the value of $\psi_1(t)$ at $t = \tau + mT_{\text{slot}} + nT_{\text{frame}}$. The matched filter output was averaged over period T_1 (= $N_1 T_{\text{frame}}$) to decrease the influence of fading variations and the impact of interference and background noise. The averaged value $\psi_1(\tau, m, n)$ is expressed as

$$\overline{\Psi}_1(\tau) = \frac{1}{15N_1} \sum_{n=1}^{N_1} \sum_{m=1}^{15} \left| \psi_1(\tau, m, n) \right|^2 \tag{1.9}$$

Based on the results of (1.9), we regard the time at which the maximum average correlation of $\overline{\Psi}_1(t)$ is obtained as the received timing of the primary SCH of the target cell, that is, $\max_{\tau} \overline{\Psi}_1(\tau) \Rightarrow \hat{\tau}$. However, at the moment, frame synchronization has not yet been established. Thus, there are 15 ambiguities for the parameter m.

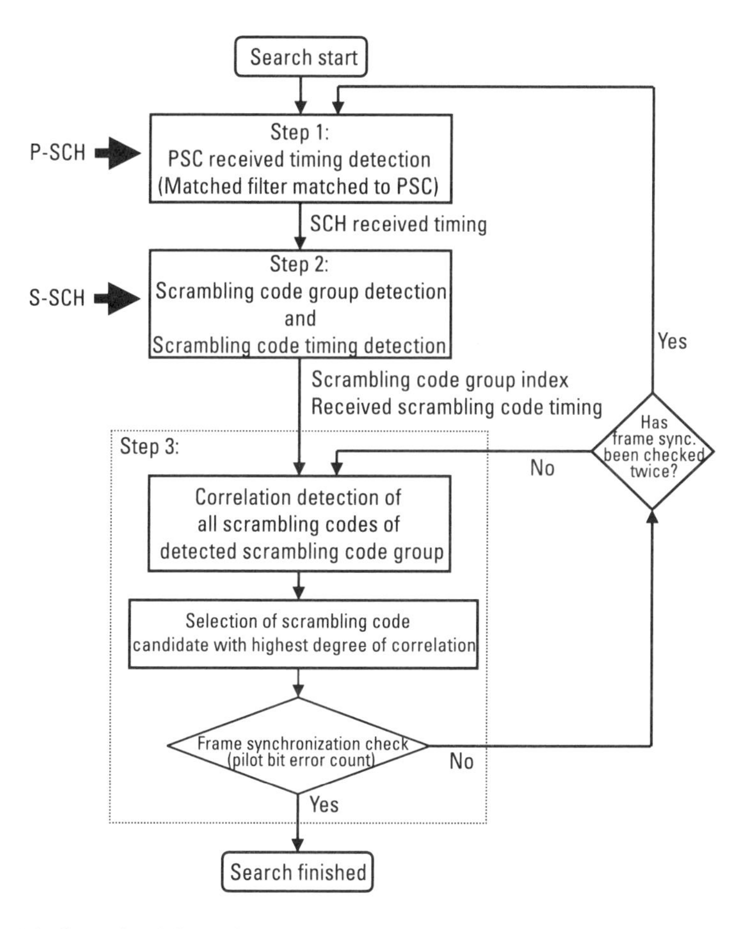

Figure 1.8 Operational flow of three-step cell search method.

Second Step

In the second step, by using the received timing of secondary (primary) SCH, $\hat{\tau}$ detected in the first step, the correlation between the received sampled chip data sequence and 16 SSC patterns at $\hat{t} = \hat{\tau} + mT_{\text{slot}} + nT_{\text{frame}}$ are calculated as

$$\psi_2(x, \hat{\tau}, m, n) = \frac{1}{256 \cdot T_c} \int_0^{256T_c} r(\hat{t} - \mu) c_{\text{SSC},x}(256T_c - \mu) d\mu \tag{1.10}$$

As in the first step, instantaneous correlation $\psi_2(x, \hat{\tau}, m, n)$ is also averaged over the T_2 interval ($= N_2 T_{\text{frame}}$). Two methods are used for averaging the correlation peak in the second step: coherent summation and power summation. The averaged correlation value in the second step using the power summation method is represented as

$$\overline{\Psi}_2(s, m) = \frac{1}{15N_2} \sum_{n=1}^{N_2} \sum_{k=1+m}^{15+m} \left| \psi_2[i(s, k \bmod 15 + 1), \hat{\tau}, m \bmod 15 + 1, n] \right|^2 \quad (1.11)$$

Because the primary SCH has already been detected, coherent averaging can be conducted by compensating for the phase variation due to fading using the correlation output of the primary SCH as a phase reference. By using the coherent summation method, the impact of instantaneous interference and background noise can be decreased further compared to the averaging done with the power summation method. The averaged correlation value in the second step by coherent summation is represented as

$$\overline{\Psi}_2(s, m) = \frac{1}{15N_2} \left| \sum_{n=1}^{N_2} \sum_{k=1+m}^{15+m} \psi_2[i(s, m \bmod 15 + 1, \hat{\tau}, m) \xi^*(\hat{\tau}, m \bmod 15 + 1, n)] \right|^2 \quad (1.12)$$

where $\xi(\hat{\tau}, m \bmod 15 + 1, n)$ is the value of $\Psi_1(t)$ at $t = \hat{\tau} + mT_{\text{slot}} + nT_{\text{frame}}$.

In the current 3GPP standard, the transmit timings for SCHs of sectors within the same cell are adjusted so that the SCHs from different sectors are not slot-aligned within one chip duration (T_c). Thus, although the possibility that the SCHs from different cells are slot-aligned within T_c is very low, it is not zero in an intercell asynchronous system. When SCHs from different cells are completely slot-aligned, the correlation output of PSC-matched filter in the first step is increased because PSC is common to all cell sites. Therefore, this occurrence may increase the detection accuracy of the first step. However, because coherent averaging with correlation output of the primary SCH as a phase reference is conducted in the second step, the detection error may increase due to the degradation of the reference signal in that case. Therefore, to reduce the detection error in the second step when the primary SCHs from different cell sites are slot-aligned, we proposed the use of a combined averaging scheme of coherent averaging and noncoherent averaging (power averaging) for the second step.

In our proposed scheme, for the first N_{Av2} iterative loops of the cell search, the second step used coherent averaging and afterward used noncoherent detection from the simulation results of the detection probability. (We set N_{Av2} equal to 3 in the experiments.) Then, from the SSC set and frame timing, which provide the maximum correlation peak, scrambling code group S and frame timing of the target cell is obtained: $\max_{s,m} \overline{\Psi}_2(s, m) \Rightarrow \hat{s}, \hat{m}$.

Third Step

Finally, the scrambling code is identified by taking a partial correlation between the received sampled chip data sequence and each of the candidate scrambling codes. Let $L(\hat{s}, i)$ be the index of the ith scrambling code belonging to group $\hat{s}$. The correlation for the scrambling code $L(\hat{s}, i)$ is computed over the H_3 symbol duration using CPICH. The correlation output is expressed as

$$\Psi_3(L(\hat{s}, i), \hat{\tau}, \hat{m}) = \frac{1}{H_3}\sum_{j=1}^{H_3}\left|\frac{1}{256T_c}\int_{256jT_c+\hat{\tau}}^{256(j+1)T_c+\hat{\tau}} r(t+\eta)c_{L(\hat{s},i),0}(\eta + \hat{m}T_{\text{slot}} - \hat{\tau})d\eta\right|^2 \tag{1.13}$$

where we assumed a fixed data modulation phase for the channelization code of CPICH. The scrambling code that provides the maximum correlation in (1.13) is determined as the scrambling code to be searched, that is, when $\Psi_3(L(\hat{s}, i), \hat{m}, \hat{\tau}) \geq \alpha\overline{\psi}_1(\hat{\tau})$, then $c_{L(\hat{s},i),0} \Rightarrow c_{\hat{k},0}$. To reduce false detection, a verification mode was added by using a frame synchronization check. When the synchronization verification failed two consecutive times, the cell search process was restarted from the first step.

As we explained, the correlations of the primary and secondary SCHs are averaged over intervals T_1 and T_2, respectively, so that the influence of fading variations should be reduced. However, in slow fading environments (i.e., those with low mobility), the false detection in the first and second step is increased, especially when the received signal power is low, since the influence of fading variation is not sufficiently eliminated. Therefore, *time-switched transmit diversity* (TSTD) was proposed and the application of

TSTD to SCH was determined in 3GPP. In TSTD, the primary and secondary SCH pair is alternatively transmitted slot by slot from two antennas. Thus, when the fading correlation between two antennas is small, variations of the received signal power due to fading are mitigated. False detection is thus decreased due to the transmit diversity effect without changing the cell search processing at an MS compared to that without TSTD.

Figure 1.9 shows the measured laboratory experimental results of the probability distribution of the cell search time with a fading maximum Doppler frequency, f_D, as a parameter using the 4.096-Mcps WCDMA experimental system with TSTD [13, 14]. In addition to CPICH and SCHs, 10 DPCHs without fast TPC were transmitted as a channel load. An L = 2 path Rayleigh fading channel with average equal power was assumed because we confirmed that field experimental results conducted near Tokyo could be well approximated using this model where two to three paths with unequal average received signal power were observed. The transmit power

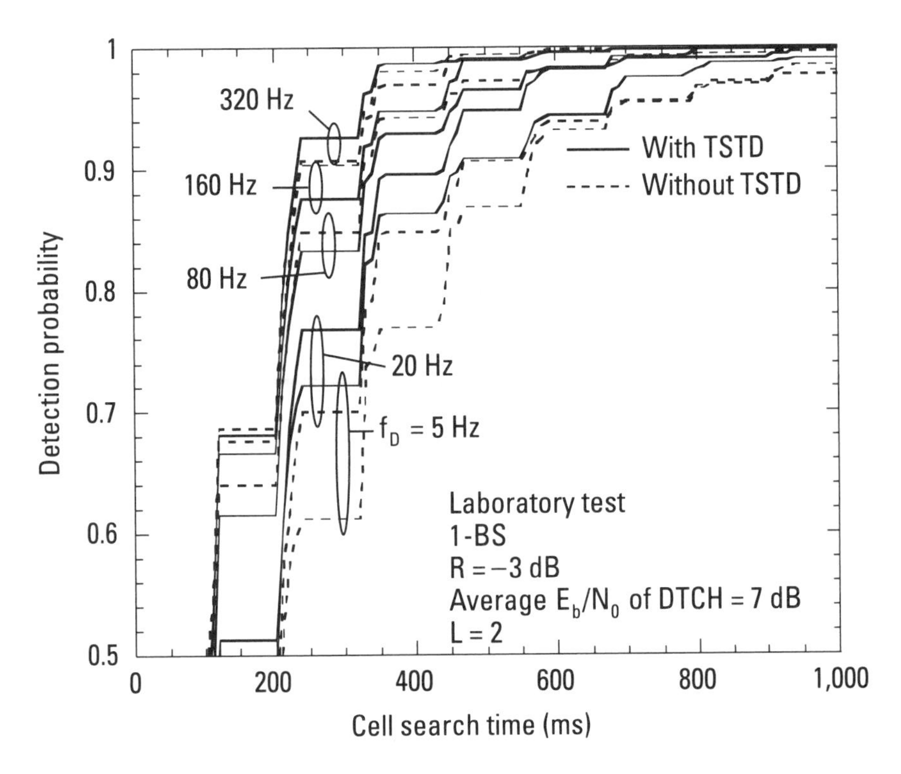

Figure 1.9 Probability distribution of cell search time using TSTD.

ratio of CPICH to DPCH and average received E_b/N_0 of DPCH were set to $R = -3$ and 7 dB, respectively. We set T_1, T_2, and T_3 to 40, 30, and 10 ms, respectively. Figure 1.9 shows that as f_D becomes larger, the cell search time becomes shorter because false detections are decreased. The figure also shows that by using TSTD, the cell search time when f_D is low (such as 5 and 20 Hz) is decreased because false detections are mitigated when the received signal level drops. As a result, the cell search time at the detection probability of 90% with TSTD is decreased by approximately 100 ms compared to that without TSTD. The cell search can be completed within approximately 250 ms at a probability of 90% with TSTD, when $R = -3$ dB and $f_D = 5$ Hz.

Figure 1.10 shows the measured field experimental results of the probability distribution of the cell search time with the transmit power ratio of CPICH to DPCH, R, as a parameter using the 4.096-Mcps WCDMA experimental system in asynchronous two-cell site environments with simultaneous 20 DPCHs without fast TPC [13]. Averaging times T_1, T_2, and T_3 were set to identical values in Figure 1.9. For comparison, the laboratory

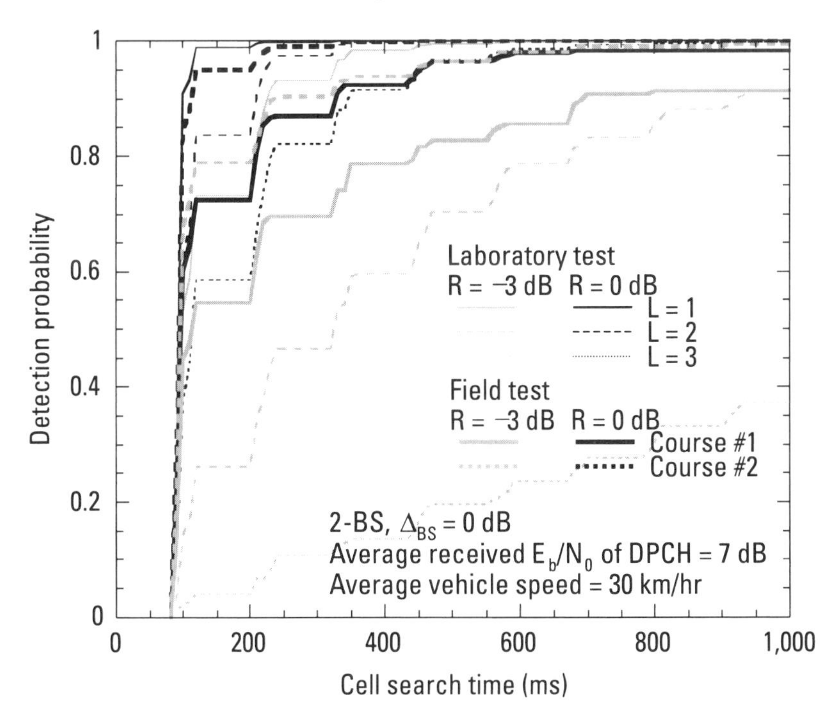

Figure 1.10 Probability distribution of the cell search time with R as a parameter (field test).

experimental results with L = 1, 2, 3 path Rayleigh fading channel are also given. The average received E_b/N_0 per DTCH at BS1 over the measurement test courses were both 7 dB. Figure 1.10 shows that the cell search time performance in the two-cell case, which was slightly better than that in the single-cell case, was different from the laboratory results. This is explained later in the chapter. In the two-cell case in the field experiments, although the average received E_b/N_0 values of SCH and CPICH over the test course from two cell sites were equal, the instantaneous received signal powers of the two-cell sites were different due to the independent shadowing variations. Therefore, the MS can search the scrambling code of the cell site with a larger receive signal power utilizing the compensation effect of the received level degradation, that is, the site diversity effect. This site diversity effect is considered larger than the increase in interference from the other cell. In the two-cell case, cell search can be completed within approximately 700 and 240 ms at a probability of 90%, when $R = -3$ dB, for course 1 and course 2, respectively.

1.3.2.2 Cell Search Algorithm Using the Mobile-Assisted Measuring and Informing Method

For the target cell search in the active mode, the number of target cells is much smaller than that in the initial cell search. Furthermore, the power consumption required for cell search operation matters little since the transmit circuitry including the power amplifier is active. Thus, this three-step fast cell search can also be applied to the target cell search in the active mode. Meanwhile, in the idle mode, when the three-step cell search algorithm is used in the same way as in the initial search, the cell search time becomes longer than that in a synchronous system because the MS must perform its search during a full span of a scrambling code repetition interval. This brings an increase in the intermittent reception time ratio, which is defined as the ratio of waking-up time to sleeping time of an MS in the idle mode. Therefore, the power consumption of the MS increases. (Note that long battery life is an inevitable requirement for a good portable mobile terminal.) Therefore, a fast target cell search algorithm used during intermittent reception in the idle mode of an MS was proposed [15]. The operational flow of the proposed cell search algorithm at the MS is illustrated in Figure 1.11.

The relative transmitting timing of scrambling codes in the forward link is shown in Figure 1.12. Let $BS^{(k)}$ be the temporary perching BS during intermittent reception from which the MS receives control information from the CCPCH. In the conventional target cell search method, $BS^{(k)}$ informs the MS of scrambling codes $L_j^{(k)}$ of the target BS candidates, $BS_j^{(k)}$ $(1 \leq j$

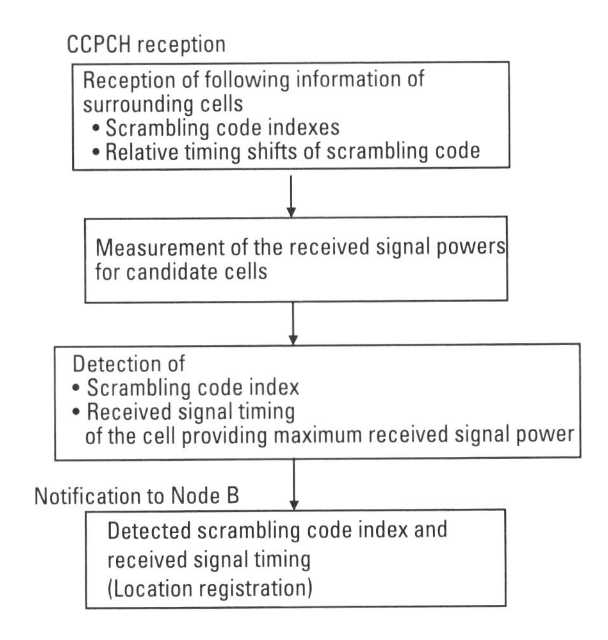

Figure 1.11 Fast cell search algorithm in idle mode.

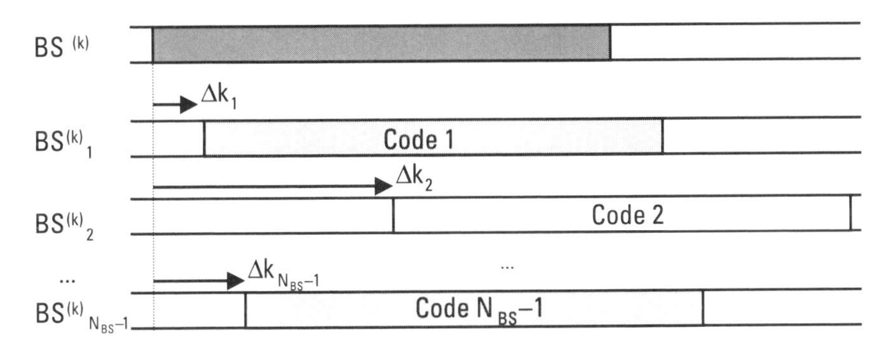

Figure 1.12 Relative transmitting timing of scrambling codes in the forward link.

$< N_{\mathrm{BS}}$), where N_{BS} is the number of BS candidates to be searched. However, in the proposed method [15], $\mathrm{BS}^{(k)}$ informs the MS of the average relative time shift of the scrambling codes of $\mathrm{BS}_j^{(k)}$ to that of $\mathrm{BS}^{(k)}$, Δk_j $(= \overline{\tau_j^{(k)} - \tau^{(k)}})$ $(1 \le j < N_{\mathrm{BS}})$, where $\tau_j^{(k)}$ and $\tau^{(k)}$ are the time delays of $\mathrm{BS}_j^{(k)}$ and $\mathrm{BS}^{(k)}$, respectively, in addition to the scrambling codes of $\mathrm{BS}_j^{(k)}$. Since MS knows $\tau^{(k)}$ beforehand, the MS can estimate the average time delay of $\mathrm{BS}_j^{(k)}$, $\bar{\tau}_j^{(k)}$, using the relation denoted as $\bar{\tau}_j^{(k)} = \Delta k_j + \tau^{(k)}$.

Thus, the MS can know the scrambling codes and the average received time delay for each $BS_j^{(k)}$, which corresponds to the situation in which the MS knows the relative time shift of a single scrambling code of the target cell sites in the intercell synchronous system. Therefore, the target cell search (i.e., in which the number of candidate cells is limited) can be achieved as fast as in intercell synchronous systems.

To achieve the proposed algorithm, each $BS^{(k)}$ must know the relative time shifts of the different scrambling codes of the surrounding $BS_j^{(k)}$. We use the mobile-assisted measuring and informing method for this. Before SHO operation, the MS searches for the target cell to which an MS performs SHO in the active mode. When the average received power difference between the CPICH of the original $BS^{(k)}$ and that of the surrounding $BS_j^{(k)}$ is smaller than the predetermined SHO threshold, the MS sends to the original $BS^{(k)}$ a request message for SHO operation. At the same time, the MS measures the reception time difference (time shift) between the scrambling code of the original BS and that of the target $BS_j^{(k)}$ and sends this time shift information to the $BS^{(k)}$ using the dedicated control channel associated with *dedicated data channel* (DTCH). Therefore, if each $BS^{(k)}$ restores the time shift information of the scrambling codes of surrounding cell sites, our algorithm can be realized. In general, the location of the MS when it measures the received scrambling code timing difference between the scrambling code of $BS^{(k)}$ and that of $BS_j^{(k)}$ is different. Since the propagation delays from these BSs are different, the measured time shift values at the MS are different according to the location of the MS to measure. Therefore, each BS should average the accumulated time shift values of the scrambling codes reported by each MS, and update this averaged time-shift information.

Reference [15] reported that the target cell search time every 720 ms at the cell detection probability of 95% is accomplished within 5.9 ms (this corresponds to the intermittent time ratio required for the target cell search to become 0.82%), when the transmit power ratios of the CPICH and CCPCH required for cell search to a DPCH are 3 and 6 dB, respectively. In this simulation, the instantaneous power delay profile was first generated by coherently accumulating the despreading signals of pilot symbols over a 512-chip duration (= 125 μs) using four correlators over a period of 3 × 720 ms for 19 target cell-site candidates using the search window with a 10-chip duration (= 2.4 μs).

1.3.3 Random Access

MS transmits a random-access channel [4, 16] based on the higher layer control signaling by BCH such as transmitting timing, spreading code, and preamble signature after receiving BCH carried in P-CCPCH in the forward link. The corresponding physical channel is PRACH. The slotted ALOHA protocol is used for PRACH transmission to avoid collision of simultaneously transmitted PRACHs with an identical spreading code sequence. A simplified version of random-access operation is illustrated in Figure 1.13. As shown in the figure, MS starts a PRACH transmission at the beginning of one of the 15 predefined time slots, called access slots, during a two radio-frame interval. The forward link AICH is divided into forward link access slots, and the reverse link PRACH is divided into reverse link access slots. Transmission of the acquisition indicator of the preamble part only starts at the beginning of the forward link access. Similarly, the transmission of the RACH preamble part and a message part only starts at the beginning of a reverse link access slot.

PRACH comprises several preambles with 4,096 chips each and a message part with 10- or 20-ms length. A preamble is a short-interval signal for achieving fast code acquisition of the succeeding subsequent message part. The coded data sequence and pilot and TFCI symbols are quadrature multiplexed into an I channel and a Q channel, respectively. The spreading code for the preamble part is a combination of the preamble signature and the scrambling code. The preamble signature is a code sequence of 256 repetitions of a signature with 16 chips. The scrambling code for the preamble part is a truncated code sequence with 4,096 chips from the beginning of the long scrambling code, which is assigned by BCH signaling. Meanwhile, the message part is first spread with dual-channel spreading using a different channelization code; then, it is spread via complex spreading using a long scrambling code with lengths of 38,400 chips. For the channelization code, orthogonal variable spreading factor code with SF = 256 and with SF = 32-256 is used for the control part in the Q channel and for the coded data part in the I channel, respectively.

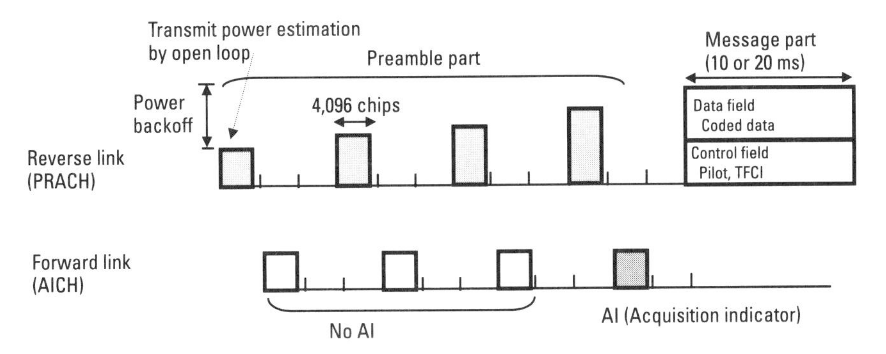

Figure 1.13 Simplified random-access operation.

The correspondence between the signature of a preamble part s ($1 \leq s \leq 16$) and channelization code of a message part is predefined. Therefore, by detecting a signature s in a preamble part, BTS can simplify code synchronization processing in a message part. That is, for the channelization code of a control part, one orthogonal variable spreading factor code with a 256-chip length for every 16 code, $C_{ch,256,m}$ is selected in accordance with the 16 signatures s based on the relation of $m = 16(s - 1) + 15$. However, orthogonal variable spreading factor code $C_{ch,\mathrm{SF},m}$ is obtained according to the spreading factor from 32 to 256 and s using the corresponding relation $m = \mathrm{SF} \times (s - 1)/16$ for the channelization code of a data part. In this way, because a scrambling code is known to node B in advance and the relationship between the preamble signature and channelization code in a message part is negotiated, node B can easily find the spreading code and received timing of a message part by detecting the preamble part. Furthermore, the transmit power of a preamble part is gradually increased from the initial value until the preamble is detected in order to decrease the amount of interference to other channels (via a process called *power ramping*) [2].

1.4 SIR Measurement-Based Fast TPC

Fast TPC based on SIR measurement of RAKE-combined signals is used to minimize the transmission power according to the traffic load in both the reverse and forward links. This results in increased capacity by reducing the interference to other users in other cells and the user's own cell. Fast TPC comprises two loops as shown in Figure 1.14: the inner loop and outer loop.

Inner loop operation is performed as follows. In the RAKE combiner, the despread signals associated with resolved paths are multiplied by the complex conjugate of their channel gain estimates and summed. Therefore, if the SIR measurement is done after RAKE combining, it is affected by the channel estimation error. Then, instead of directly measuring the SIR after RAKE combining, a more accurate SIR measurement method was proposed [17, 18], in which, first, the SIR on each resolved path is measured, and then the SIRs of all the resolved paths are summed to obtain the SIR (which is equivalent to the one at the output of the RAKE combiner). By doing so, we obtain an SIR measurement that has less influence on the channel estimation error. The SIR measurement is summarized next.

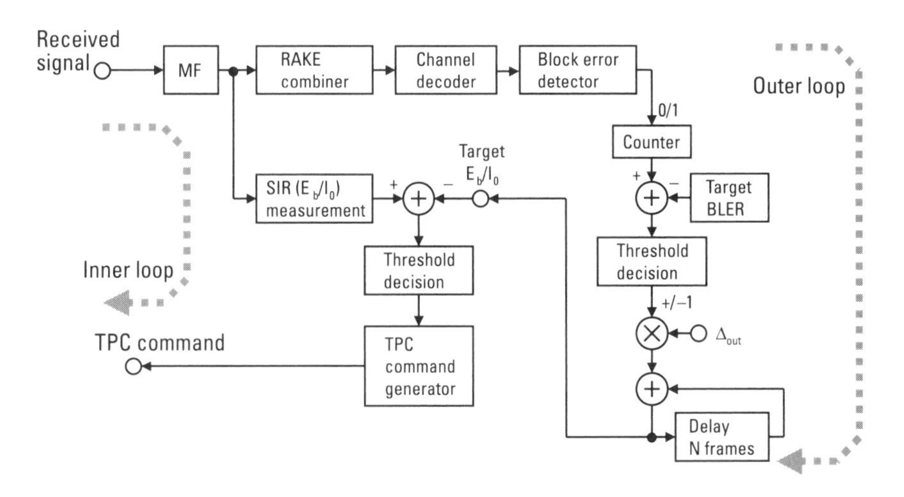

Figure 1.14 SIR-based adaptive TPC with outer control loop. MF: matched filter.

First, signal power $\tilde{S}_l(k)$ of the kth slot associated with the lth path is computed using the received N_p pilot symbols belonging to one slot terminal. The signal power $\tilde{S}_l(k)$ is given by

$$\tilde{S}_l(k) = \left|\tilde{r}_l(k)\right|^2 \tag{1.14}$$

where

$$\tilde{r}_l(k) = \frac{1}{N_p}\sum_{n=0}^{N_p-1} r_l(n, k)\exp\left(-\frac{j\pi}{4}\right) \tag{1.15}$$

since we assume that the modulation phase of N_p pilot symbols is $\pi/4$ radians. The instantaneous interference plus background noise power of the lth path, $\tilde{I}_l(k)$, is computed as the squared error of the received N_p pilot samples:

$$\tilde{I}_l(k) = \frac{1}{N_p}\sum_{n=0}^{N_p-1}\left|r_l(n, k)\exp(-j\pi/4) - \tilde{r}_l(k)\right|^2 \tag{1.16}$$

Then, $\tilde{I}_l(k)$ is averaged using a first-order filter with a forgetting factor $\mu(<1)$ over the multiple-slot terminal to obtain

$$\bar{I}_l(k) = \mu\bar{I}_l(k-1) + (1-\mu)\tilde{I}_l(k) \tag{1.17}$$

The SIR at the kth slot associated with the lth path $\lambda_l(k)$ is given by

$$\tilde{\lambda}_l(k) = \tilde{S}_l(k)/\tilde{I}_l(k) \tag{1.18}$$

Finally, the SIR at the kth slot, $\overline{\lambda}(k)$, is obtained as

$$\overline{\lambda}(k) = \sum_{l=0}^{L-1} \tilde{\lambda}_l(k) \tag{1.19}$$

The measured SIR was compared to the target SIR and the TPC command was generated, which was transmitted to raise or lower the mobile transmit power by ±1 dB every 0.667 ms. Even if the received SIRs are the same, the received quality (BLER) is not the same because the BLERs are affected by the number of paths, maximum Doppler frequency (which depends on the speed of the vehicle), SIR measurement, and so on. Therefore, the outer loop controls the target SIR with a more gradual updating interval compared to the inner loop so that the measured BLER or BER is equal to the target value. In general, a BLER-based outer loop is used. BLER is measured by calculating the number of cyclic redundancy check results that coincide with the value attached to every transport block. Because the required BLER becomes a very small value for high-speed and high-quality data transmission (e.g., with the required BER of 10^{-6}), it takes a much longer time to calculate the BLER. As a result, outer loop control cannot track changes in the propagation conditions. Therefore, in these cases, outer loop control based on BER measurement of the tentative decision data symbols before channel decoding (i.e., after RAKE combining) with decision data symbols after channel decoding as a reference can be applied. The reference data symbols are generated by reencoding and interleaving binary decision data symbols after channel decoding. Although a data decision error occurs in the decoded data sequence, its impact on the reference symbols is considered to be very small.

1.5 Diversity

1.5.1 Coherent RAKE Combining (RAKE Time Diversity)

Pilot symbol-assisted coherent detection is used for both the reverse and forward links [19, 20]. The block diagram of the pilot symbol-assisted coherent RAKE combiner is illustrated in Figure 1.15(a). The received multipath signals are despread by the matched filter and resolved into L-multipath components of transmitted modulated data that are received via different propagation paths with different delay times. The coherent RAKE combiner output is expressed at the nth symbol position of the kth slot associated with the lth path ($l = 0, 1, \ldots, L-1$) using despread signal $r_l(n, k)$ as (QPSK data modulation is assumed here)

$$\tilde{d}(n, k) = \sum_{l=0}^{L-1} r_l(n, k)\tilde{\xi}_l^*(k) \tag{1.20}$$

where $\tilde{\xi}_l^*(k)$ represents the channel estimates. The output data sequence, $\tilde{d}(n, k)$, is deinterleaved and channel decoded to recover the transmitted binary data sequence. To achieve accurate channel estimation that works satisfactorily in a fast fading environment, we presented an improved channel estimation filter called a *weighted multislot averaging* (WMSA) channel estimation filter [20] as shown in Figure 1.15(b). After obtaining the instantaneous channel estimates of each slot, the channel estimates, $\hat{\xi}_l(j+i)$s, of

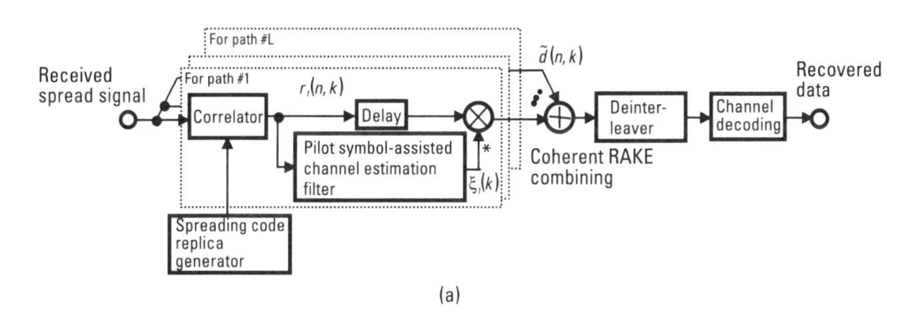

(a)

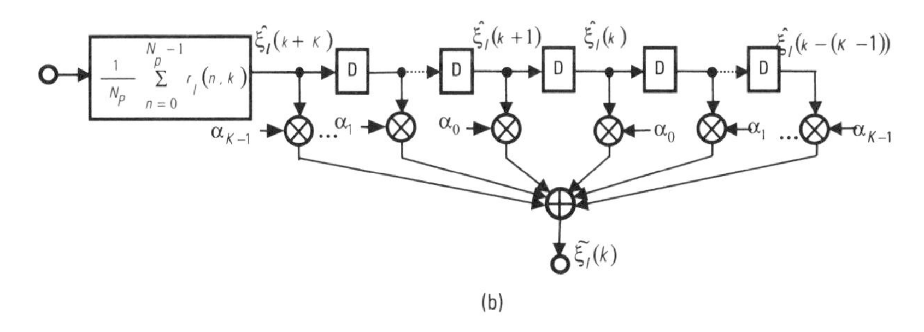

(b)

Figure 1.15 Coherent RAKE receiver: (a) receiver structure and (b) WMSA channel estimation filter.

$2J$-multiple slots ($I = -J + 1, \ldots, 0, 1, \ldots, J$) are then weighted and summed to obtain the final channel estimate, $\bar{\xi}_l(k)$, as

$$\bar{\xi}_l(k) = \sum_{i=0}^{J-1} \alpha_i \hat{\xi}_l(k - i) + \sum_{i=1}^{J} \alpha_{i-1} \hat{\xi}_l(k + i) \tag{1.21}$$

where α_i is the real-valued weight. Using the WMSA channel estimation filter, accurate channel estimation is possible, particularly in slow fading environments. The optimum value of α_i varies according to the fading correlation between succeeding slots in a real fading channel. Therefore, we proposed [21] an adaptive WMSA channel estimation filter, in which a weighting factor is adaptively controlled by measuring an inner product of the averaged despread pilot signals of successive slots.

The SIR measurement interval M is a design parameter and is closely related to the TPC delay. In general, as M increases, the SIR measurement becomes more accurate. However, a TPC delay of more than one slot is required. As the TPC delay increases, the required E_b/I_0 increases because the ability of fast TPC to track fading degrades. Figure 1.16 shows the required average E_b/I_0 for the average BER of 10^{-3} as a function of the TPC delay with M (signal power measurement interval in symbols for E_b/I_0 measurement) as a parameter in the 32-Kbps data rate channel.

A single-user case was evaluated since the interference from other power-controlled users can be approximated well by Gaussian noise and can be combined with the background noise. We assumed the number of paths $L = 2$ and $f_D = 80$ Hz. However, the forward link channel was not a faded channel and its received signal power was set to be sufficiently high such that the average BER was 0. For comparison, the results without fast TPC are also plotted in Figure 1.16. For the given TPC delay (less than three slots), as M increases, the required E_b/I_0 can be reduced because accuracy of the SIR measurement improves.

However, the impact of TPC delay must be considered. Using the frame structure, the TPC delay can be one slot if $M = 10$ symbols or less, but it is two slots if $M = 20$ symbols or more. As the TPC delay increases, the tracking ability of TPC against fading tends to be lost. Hence, a trade-off occurs between improved accuracy of the SIR measurement and increased TPC delay. As a result, $M = 10$ symbols (one-slot TPC delay) yields slightly lower required E_b/I_0 than $M = 40$ symbols (two-slot TPC delay).

We evaluated the BER performance of coherent RAKE combining with SIR-based fast TPC in field experiments conducted in an area near Tokyo. The cell site and mobile transmitter/receiver antennas were located

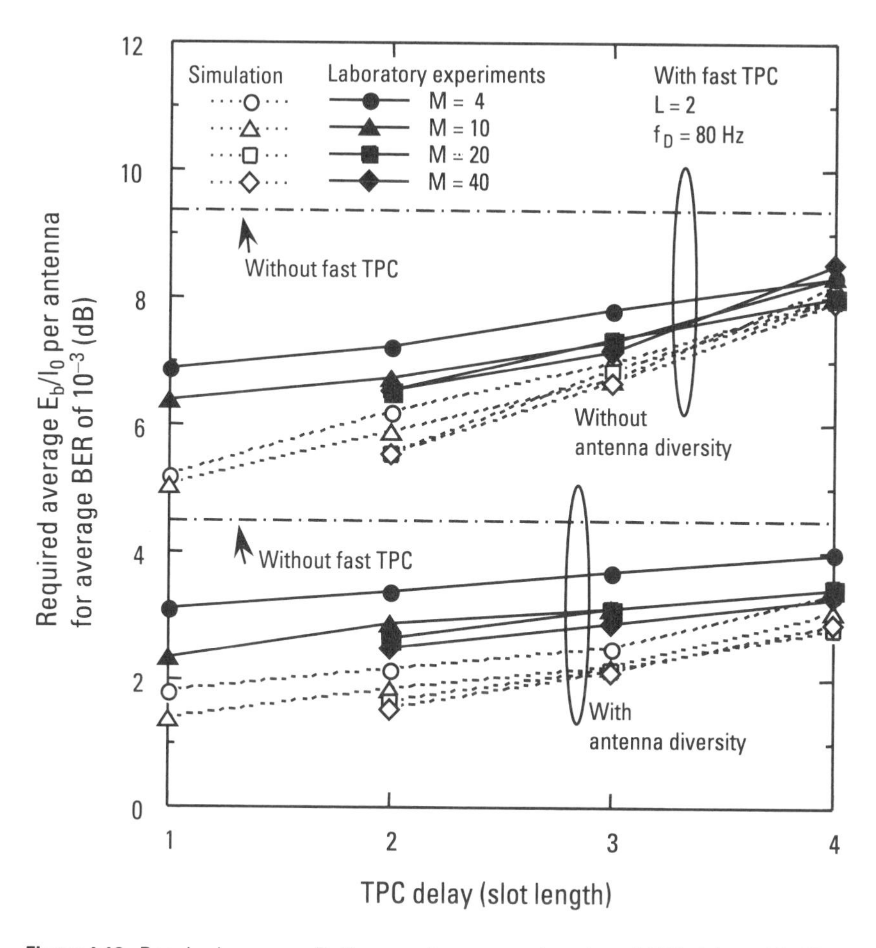

Figure 1.16 Required average E_b/I_0 per antenna as a function of TPC delay with M as a parameter.

59 and 2.9m off the ground, respectively. A measurement vehicle equipped with the mobile receiver was driven along roads at distances of 0.75–1.35 km from the cell site at the average speed of approximately 30 km/hr. The measurement course passes through a business zone that is lined with office buildings and factories. Other conditions are given in detail in [22]. The average delay spread of the test course was approximately 1 μs. The test course first experienced clear two-path and single-path fading at the middle of the course. Then, three-path fading with unequal average power was observed at the end of the course.

Figure 1.17 plots the measured average BER performance of the 32-Kbps data rate user in the single-user and two-user cases (one interfering

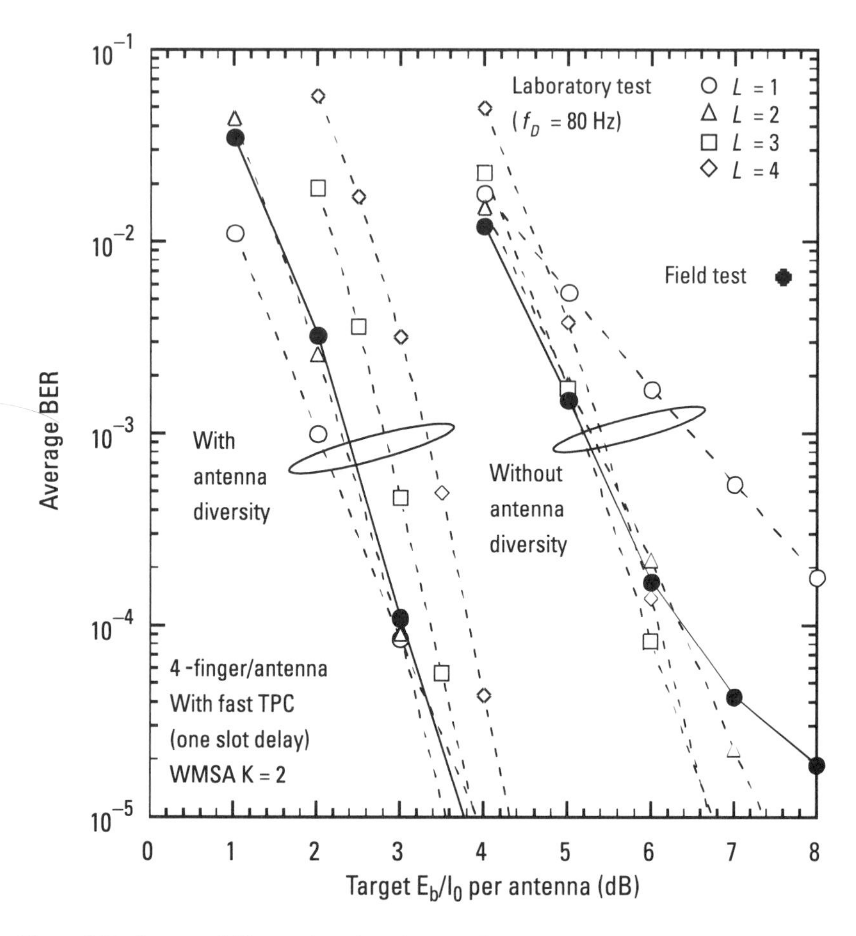

Figure 1.17 Average BER as a function of target E_b/I_0 per antenna for field experiments.

user with a 64-Kbps data rate assuming the same BER independently employing fast TPC), as a function of the TPC target E_b/I_0 value. (Note that E_b/I_0 is calculated as E_b/I_0 = SIR + 10 log(3/2) dB, since convolutional coding with the rate of 1/3 and QPSK data modulation were used in the experiments [21].) Two MSs established radio links with BS 1. A WMSA channel estimation filter with $J = 2$ was used. Laboratory experimental results of the single-user case using the L-path model with f_D = 80 Hz are also plotted for comparison. The results clearly show that the target E_b/I_0 when an interfering user exists becomes almost the same in order to achieve the same BER as that of the single-user case, implying that fast TPC worked satisfactorily in a real fading channel. The measured numbers of active RAKE fingers per antenna along test courses #1 and #2 are 2.0 and 1.6, respectively.

Figure 1.17 shows that the measured BER performance is almost the same as the laboratory-measured BER performance when $L = 2$. The field-measured BER performance results are in good agreement with those estimated from the laboratory experiment. The figure also shows that two-branch space diversity (antenna diversity) reception can reduce the target E_b/I_0 by approximately 3 dB at the average BER of 10^{-3}. With space diversity reception, the average BER of 10^{-3} can be achieved at the required E_b/I_0 of approximately 3 dB per antenna.

The measured average BER performance of the 64-Kbps channel using turbo coding is plotted in Figure 1.18 as a function of the MS relative

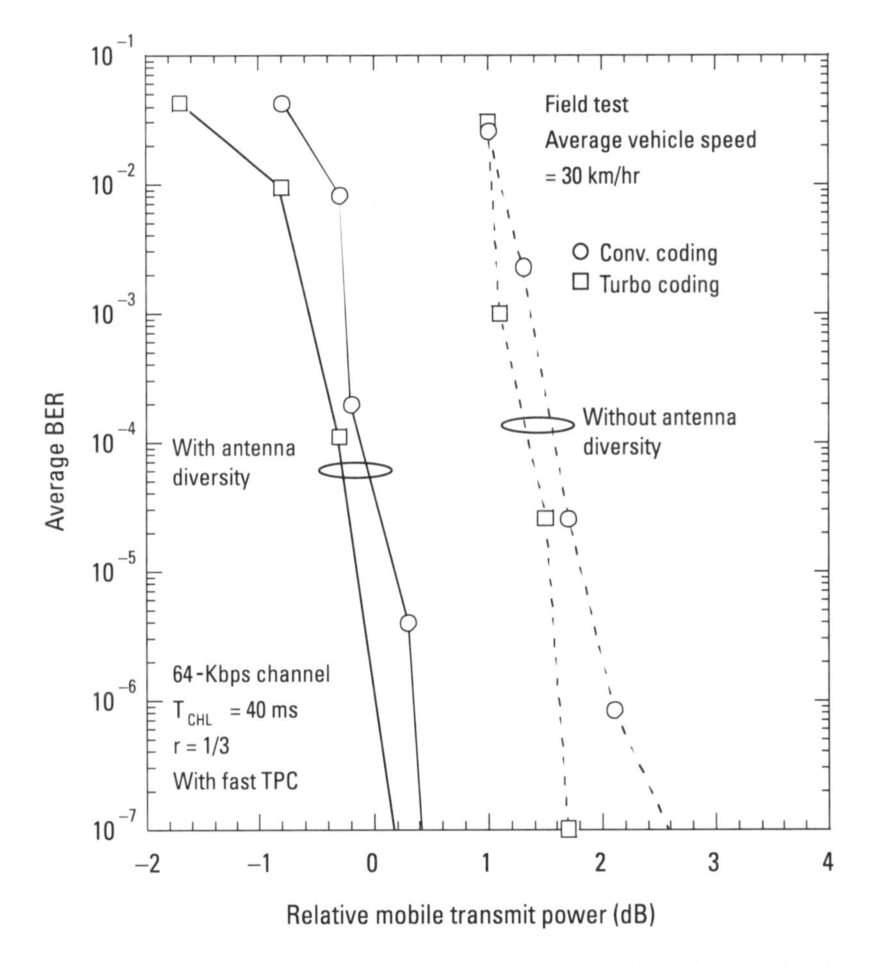

Figure 1.18 Average BER of 64-Kbps data transmission with turbo coding as a function of mobile transmit power for field experiments.

transmit power with the channel interleaving length of T_{CHL} = 40 ms [23]. Turbo coding with a rate of r = 1/3 and a constraint length of K = 4 bits (generator polynomials are 13, 15, and 15 in octal notation) was used, while the rate and constraint of convolutional coding, for reference, were r = 1/3 and K = 9 bits, respectively. Primary interleaving [9, 24] and multistage interleaving [9, 25], which offer a greater capability for randomization compared to the block interleaving method, were used as turbo interleaving and channel interleaving methods, respectively.

In the experimental system, Max-log-Map decoding was used as the soft-in/soft-out decoder and the number of iterations, m, was assumed to be eight, which was sufficiently large. From Figure 1.18, the MS average transmit power for the average BER of 10^{-6} using turbo coding can be decreased by approximately 0.6 (0.3) dB compared to that using convolutional coding without (with) antenna diversity reception. Although the superiority of turbo coding to convolutional coding was confirmed in an actual multipath fading channel, this difference was decreased compared to the laboratory experiments assuming a fixed delay time for each path using a fading simulator, that is, superiority was confirmed to be above 1.0 dB. This abatement in the improvement with antenna diversity reception indicated that in an actual fading channel in the field experiments, the impact of path search for RAKE combing and SIR measurement for fast TPC diminished the improvement in performance of the turbo coding due to a very low received signal power level.

1.5.2 Site Diversity (Soft/Softer Handover)

Soft handoff or site diversity ("site diversity" hereafter) [26, 27], which was first implemented in the IS-95 CDMA standard [28], is an essential technique together with fast TPC for improving transmission impairment due to multipath fading and shadowing near the cell edge. The simplified configuration of site diversity is illustrated in Figure 1.19. In the forward link, the same original information sequences before channel coding are transferred to N BSs (N is the number of BSs with which the MS is associated) through the back-haul, that is, the wired transmission line between the BS and RNC, from an RNC and transmitted from two BSs using different scrambling codes. The received signals after RAKE combining at the MS are combined symbol by symbol with *maximum ratio combining* (MRC) followed by channel decoding such as soft-decision Viterbi decoding or turbo decoding.

With intersector diversity in the reverse link, the RAKE-combined signal of each sector is combined with MRC in the same way as in the

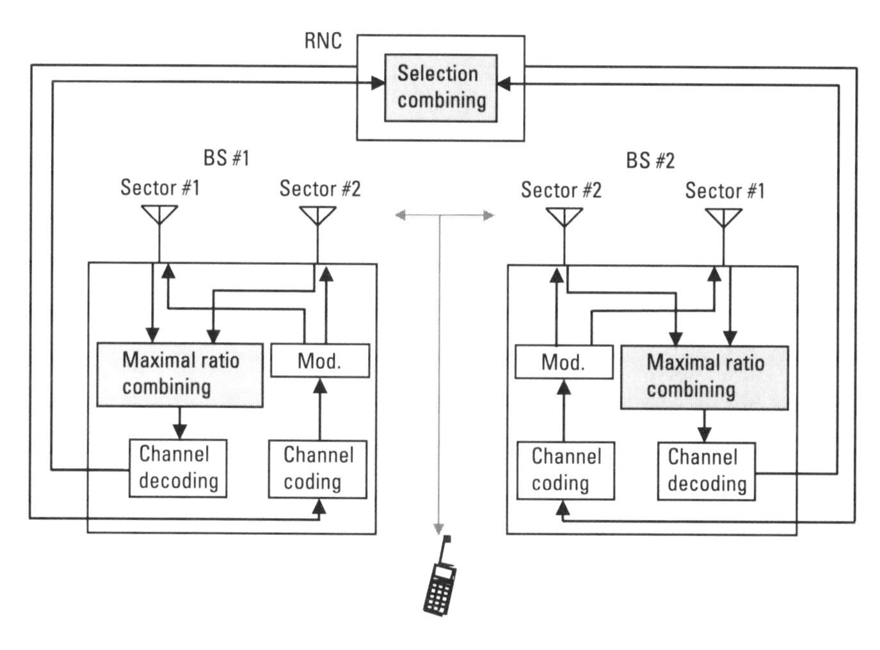

Figure 1.19 Simplified configuration of site diversity.

forward link. In contrast, in the reverse link, a hard-decision data sequence after channel decoding at each BS is transferred to the RNC via the back-haul with the reliability information associated with each traffic channel. The transferred data sequences are selection combined every selection period, according to the reliability information.

1.5.2.1 Reverse Link

The performance of reverse link intercell site diversity depends on the type of reliability information that is used. Therefore, we present a two-step selection combining scheme using two types of reliability information [29, 30]: cyclic redundancy check results calculated over selection interval T_{SEL} and the average received SIR measured over interleaving interval T_{ILV}. In our scheme, we use the number of slots with a measured SIR value greater than the target value of fast TPC, N_{SIR}, which is the number of TPC command bits to lower the transmit power during T_{ILV}, instead of the actual measured SIR value. This is because the transfer capacity in the back-haul required for the reliability information of intercell site diversity can be significantly decreased (note that only 4 bits/frame are required for denoting the SIR average over one frame). The selection combining at the RNC was performed in two steps.

1. When multiple decoded data sequences transferred from *N* cell sites (BSs) indicate no cyclic redundancy check error, then the one data sequence over T_{SEL} among the data sequences yielding the successful cyclic redundancy check result is selected.
2. When all cyclic redundancy check results transferred from *N* BSs indicate frame errors have occurred, the data sequence during T_{SEL} with the larger N_{SIR} over T_{ILV} is selected.

The field experiments using intercell site diversity were conducted in an area near Tokyo in order to measure the BER performance in a 32-Kbps data rate channel. The measurement course is a road running north and south, which passes through the middle of two BSs. The distance between BS 1 and BS 2 is approximately 2.5 km. The middle point of the measurement course is approximately 1,300 and 1,200m apart from BS 1 and BS 2, respectively. On either side of the measurement course is a low-rise factory area. The view from BS 1 was *line of sight* (LOS) except at the end of the course, where it was *non-line of sight* (NLOS) from BS 2 due to the tall buildings.

We set the SHO threshold to 3 dB. The difference in the measured average received signal powers from the two BSs was approximately 1 dB. Thus, the measurement course is an SHO area within the prescribed threshold. The power delay profiles with 1 (2) and 2 (1) paths were observed in the first half and the latter half of the course from BS 1 (BS 2). We set T_{SEL} = 10 ms. In the experiments, fast TPC was used only in the reverse link. The received signal power was set to be sufficiently high so that there was no TPC command bit error.

The measured time variations of the instantaneous BER and received E_b/I_0 at BS 1 and BS 2, after intercell site diversity, and the mobile transmit power averaged over one radio frame length are plotted in Figure 1.20. The target E_b/I_0 at each BS was set to 7 dB so that the average BER after intercell site diversity was approximately 10^{-3}. The figure shows that bit errors occurred when the received E_b/I_0 at each BS dropped; however, the instantaneous received E_b/I_0 after intercell site diversity was maintained at an almost constant level. Therefore, the measured BER after intercell site diversity significantly improved; nevertheless, the BER measured at each BS was remarkably degraded due to the reduced signal level caused by shadowing and fading variations. Because the target E_b/I_0 was set to satisfy the average BER of 10^{-3} after selection combining, burst errors rarely occurred since convolutional coding was used.

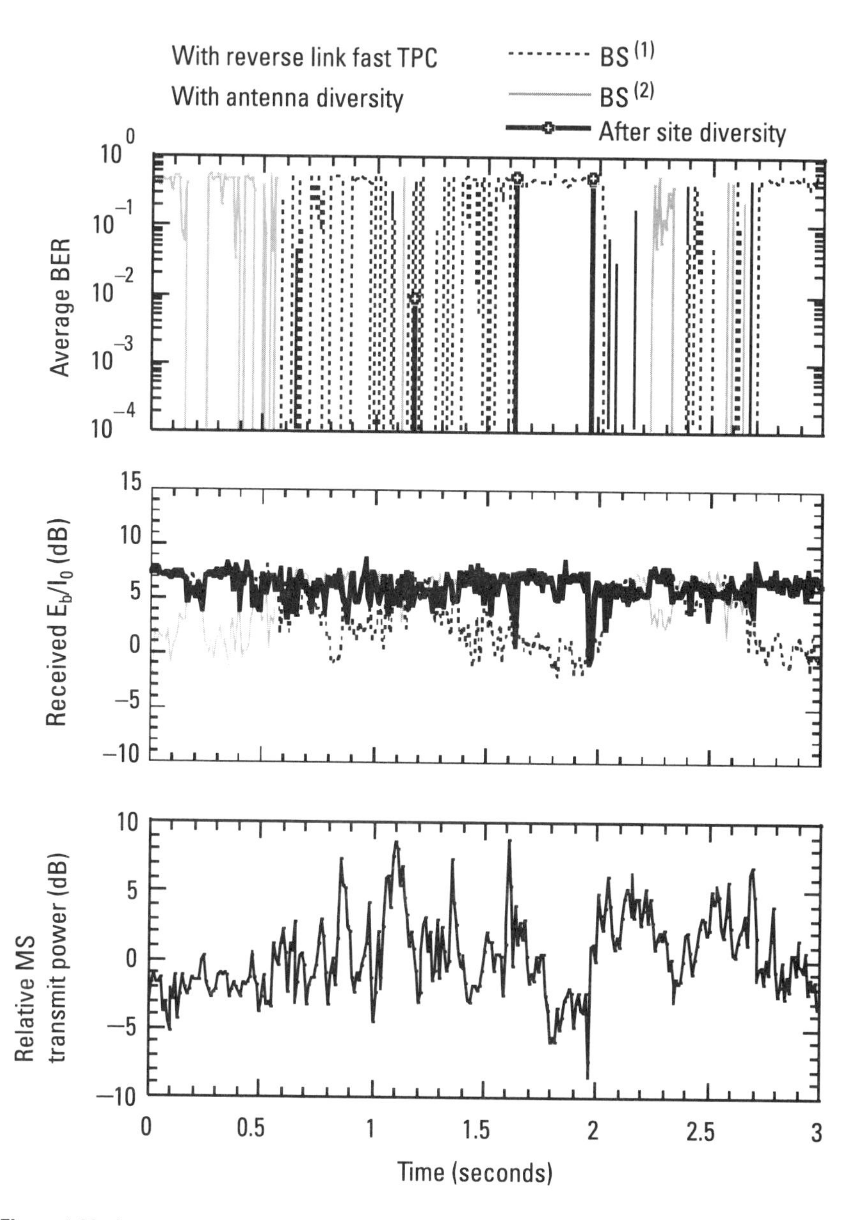

Figure 1.20 Instantaneous time variations in the reverse link intercell site diversity for field experiments.

The measured average BER performance during intercell site diversity is shown in Figure 1.21 as a function of the relative transmit power of the MS, which was normalized for when the average BER of 10^{-3} was achieved during intercell site diversity. The performance is also plotted when only a one-site connection with BS 1 or BS 2 was established. Although the received signal powers from BS 1 and BS 2 are almost equal, the BER performance when the radio link is connected to BS 2 is better than with BS 1 since the received signal power from BS 2 is greater than that from BS 1 in the longer part of the measurement course. Figure 1.21 shows that the required transmit power of the MS satisfying the average BER of 10^{-3} in intercell site diversity can be decreased by approximately 2.0 dB compared to that in the one-site connection.

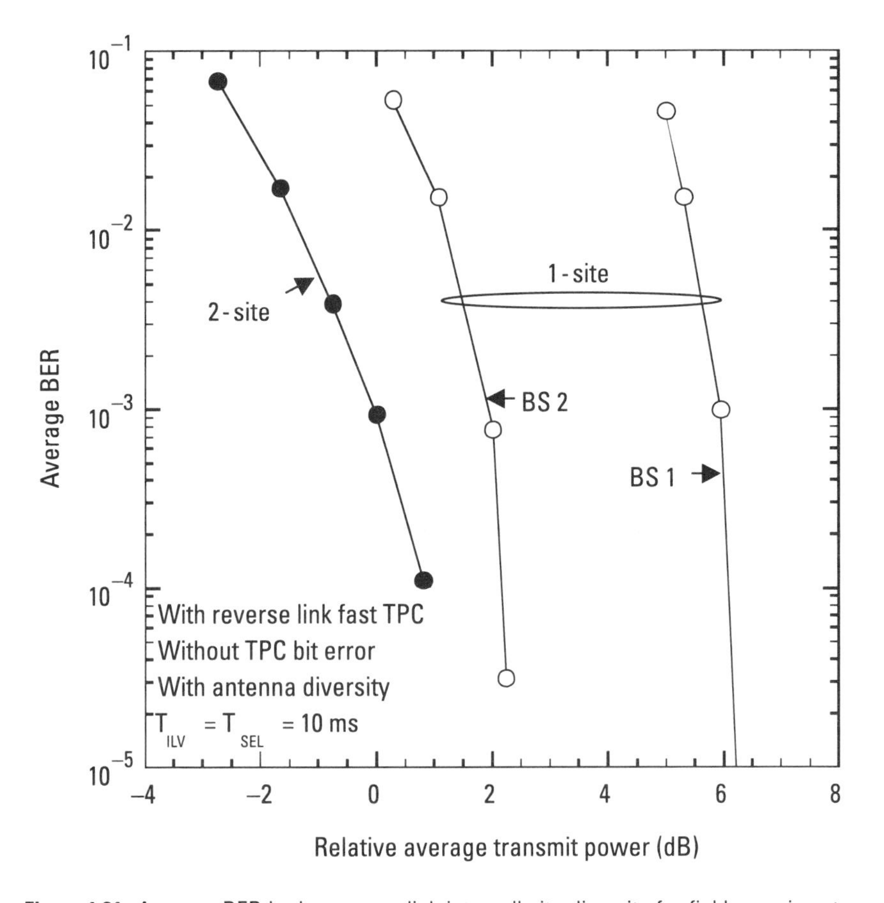

Figure 1.21 Average BER in the reverse link intercell site diversity for field experiments.

1.5.2.2 Forward Link

When fast TPC is applied in the forward link intercell site diversity mode, each BS independently follows the TPC command bit sent from the MS via the reverse link. Therefore, the transmit power of each BS differs when a TPC command bit error occurs in the reverse link. An increase in the difference between the transmit powers of the BSs causes a reduction in site diversity gain and an increase in the interference to other users.

To overcome this problem, several schemes that compensate for the BS transmit power were proposed [31, 32]. In the method proposed in [31], each BS controls its instantaneous transmit power by using a forgetting factor so that the difference between the instantaneous transmit power and the BS-specific reference transmit power calculated by averaging the instantaneous values does not become large. However, it is difficult to quickly track variations in path loss including shadowing due to the movement of the MS. The method in [32] reduces TPC bit error by sending the same TPC bit over several slots in the site diversity mode; this prevents the transmit power difference between BSs from becoming too large. However, in addition to the problem described in [32], the TPC delay increases. Therefore, we proposed the following two-step algorithm to reduce the impact of TPC errors and keep the transmit power of the BSs the same as that shown in Figure 1.22 [29].

1. *First loop:* The standard transmit powers, $P_{\mathrm{REF}}^{(k)}$, of all BS_k are compensated by $\Delta P^{(k)}$ (in decibels) according to the dedicated control channel from a MS based on the average SIR measurement at an MS.

$$\Delta P^{(k)} = \text{Target\_}E_b/I_0 - \text{Measured\_total\_}E_b/I_0 \qquad \text{(dB)} \qquad (1.22)$$

   where the Measured_total_E_b/I_0 and Target_E_b/I_0 are the measured E_b/I_0 after RAKE combining and the target E_b/I_0 at an MS, respectively. The $P_{\mathrm{REF}}^{(k)}$ is constant during the length of the G slot and its value of n $(= g \times G)$th slot $P_{\mathrm{REF}}^{(k)}(n)$ is updated every G slot as $P_{\mathrm{REF}}^{(k)}(g \times G) = P_{\mathrm{REF}}^{(k)}((g-1) \times G) + \Delta P^{(k)}$.

2. *Second loop:* The instantaneous transmit power, $P_{\mathrm{CL}}^{(k)}(n)$, is controlled according to the TPC command bits (Δ_{TPC}) by introducing

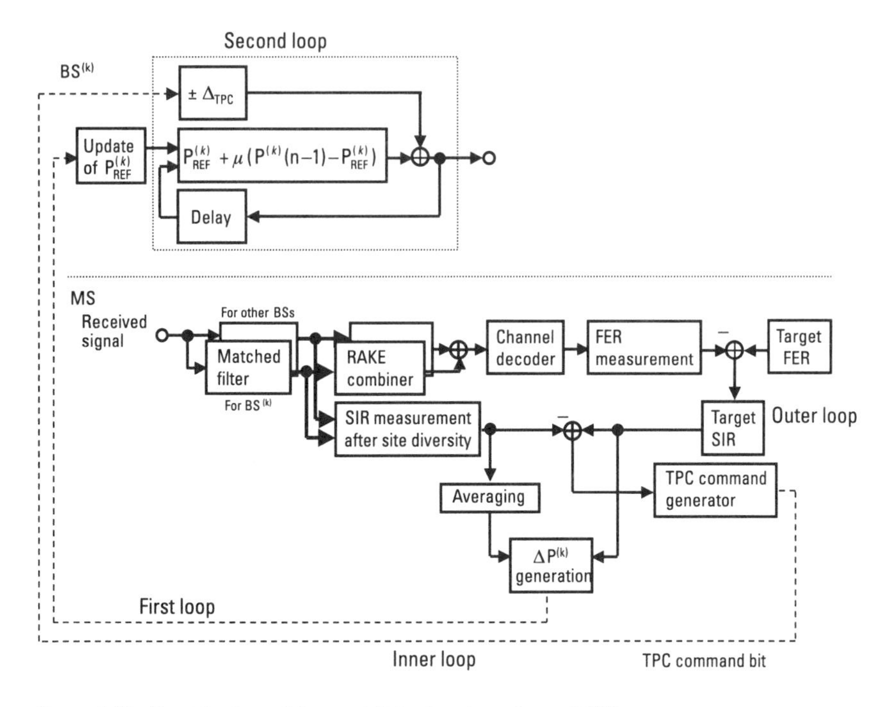

Figure 1.22 Combination of forward link site diversity and TPC.

forgetting factor μ using the standard transmit power compensated in the first loop:

$$P_{\mathrm{CL}}^{(k)}(n) = P_{\mathrm{REF}}^{(k)} + \mu(P_{\mathrm{CL}}^{(k)}(n-1) - P_{\mathrm{REF}}^{(k)}) + \Delta_{\mathrm{TPC}} \tag{1.23}$$

The measured average BER performance in the forward link when intercell site diversity is applied is plotted in Figure 1.23 as a function of the total BS average transmit power. The measurement course and experimental conditions were the same as those in Figure 1.21. Fast TPC was used for the reverse and forward links. The number of maximum RAKE fingers for the BS and MS was 4. The forgetting factor was set to $\mu = 0.8$. In the course of making our measurements, we observed that the instantaneous transmit power is controlled around the standard transmit power without dispersing to the maximum output during the measuring process. Figure 1.23 shows that the total transmit power of the two BSs at the average BER of 10^{-3} in the intercell site diversity is decreased by approximately 0.3 dB compared to a one-cell site connection. This improvement is small compared to that in the reverse link because the increase in interference due to transmis-

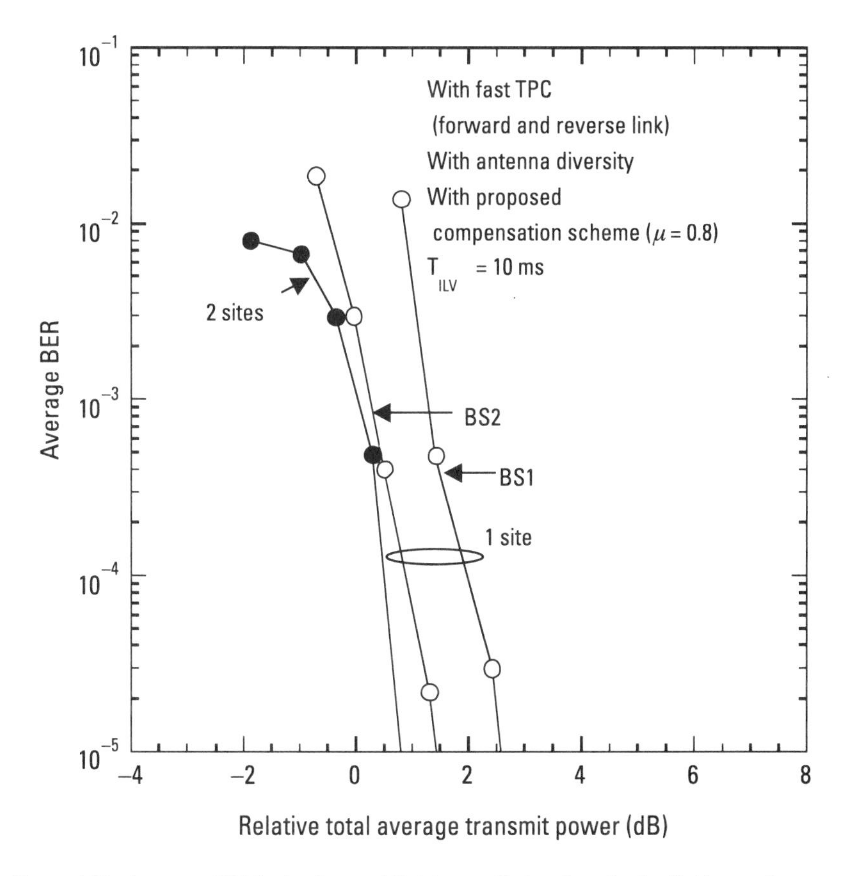

Figure 1.23 Average BER in the forward link intercell site diversity for field experiments.

sions from two BSs diminished the diversity effect. Thus, the *site selection diversity transmit* (SSDT) power control [33], in which only the primary BS transmits control bits to decrease the interference, was proposed.

1.5.3 Transmit Diversity

Transmit diversity employing several antennas at a BS can improve the forward link transmission performance without increasing the complexity of the MS [34, 35]. Therefore, several transmit diversity schemes were adopted in the 3GPP standard [4, 16]. The CPCHs with the same spreading code, but different data modulation patterns, are transmitted from two antennas in the same carrier phase. Two open-loop-type transmit diversity schemes were adopted in the 3GPP standardization: TSTD [4] and *space-time transmit diversity* (STTD) [4, 36].

STTD, which is used for the CCPCH, transmits two data sequences in parallel after coding from two antennas using the identical channelization code. Because the fading correlation between the two antennas is low, the fluctuation in the received signal level due to fading is mitigated. The operational principle and coding scheme of STTD are illustrated in Figure 1.24(a) and 1.24(b), respectively. Let $S(m)$ be the QPSK symbol data sequence of a DPCH denoted as $S(m) = \exp j\phi(m)$, where $\phi(m) \in \{h\pi/2 + \pi/4;\ h = 0\text{–}3\}$ is the QPSK modulation phase. Then, two successive symbols, $S(m)$ and $S(m+1)$, are treated as a pair, where m denotes an even number. The two symbol sequences, $d^1(m)$ and $d^2(m)$, for antennas 1 and 2 generated in the STTD encoder are expressed, respectively, as

$$d^1(m) = S(m) \quad \text{and} \quad d^1(m+1) = S(m+1) \tag{1.24a}$$

$$d^2(m) = -S^*(m+1) \quad \text{and} \quad d^2(m+1) = S^*(m) \tag{1.24b}$$

It is clear that the orthogonality between the two data sequences is maintained irrespective of the spreading code sequence.

The measured average BER performance with STTD is plotted in Figure 1.25 when fast TPC was not applied in the forward link as a function of the average received E_b/N_0, where N_0 is the multipath interference plus background noise power density. The measurement course was course 1 described in [37]. The performance with and without antenna diversity

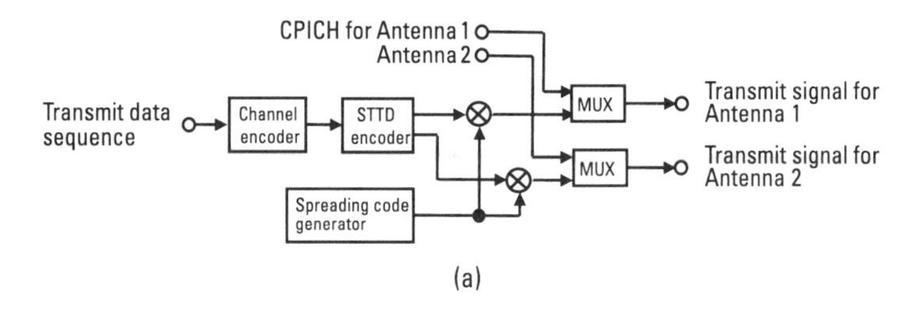

(a)

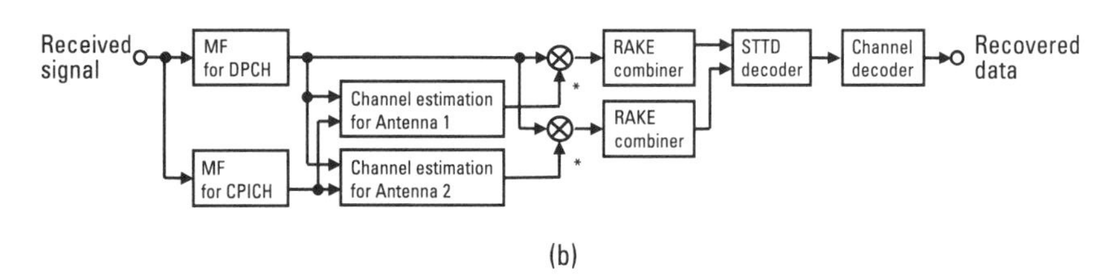

(b)

Figure 1.24 Block diagram of transceivers using STTD: (a) node B transmitter and (b) MS receiver.

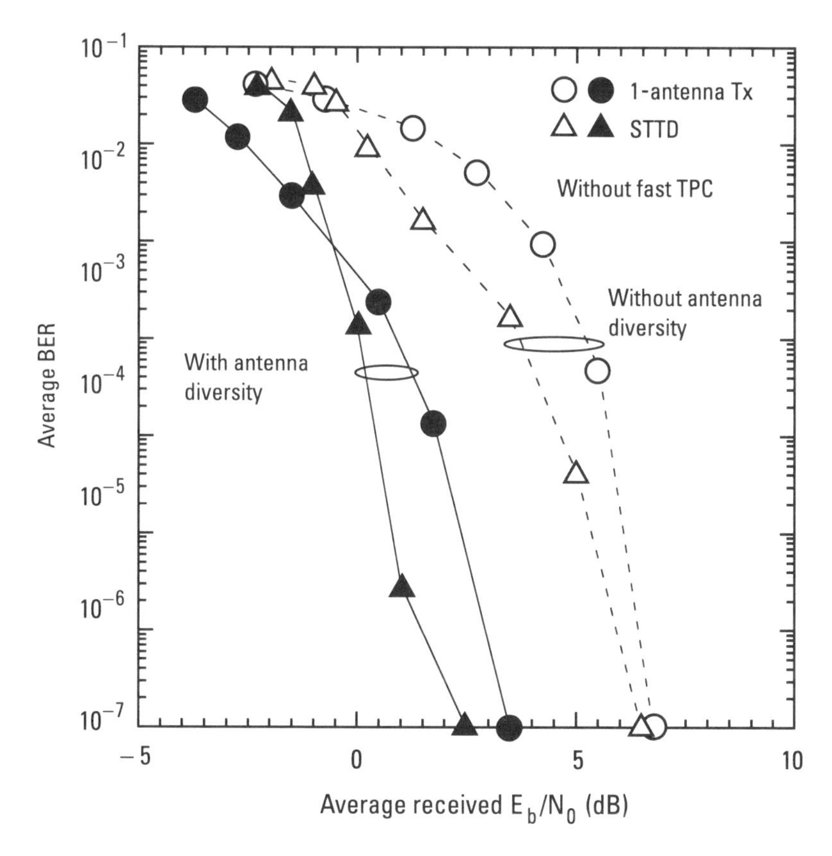

Figure 1.25 Average BER performance as a function of the average received E_b/I_0 using STTD for field experiments.

reception at an MS is shown in the figure. The BER performance with signal antenna transmission is also depicted for comparison. Figure 1.25 shows that the average required received E_b/N_0 at the average BER 10^{-3} with STTD was decreased by approximately 1.5 dB without antenna diversity reception and 1.0 dB with antenna diversity reception. The improvement using STTD with antenna diversity reception became smaller than that without antenna diversity because the degradation of the channel estimation due to a lower received level offset the additional diversity effect by STTD when using RAKE path diversity and antenna diversity reception. From the figure, the effectiveness of STTD for a channel without TPC, such as a common control channel, was elucidated in a real multipath-fading channel.

Meanwhile, closed-loop-type (two modes were standardized) transmit diversity is used for DPCHs, in which the transmit antenna weights are controlled by the FBI generated at the MS [16]. The simplified block diagram of the BS transmitter when closed-loop transmit diversity is used is illustrated in Figure 1.26. Let $W_1 = A_1 e^{j\phi 1}$ and $W_2 = A_2 e^{j\phi 2}$ be the transmit antenna weights. Thus, in mode 1, the transmitted phase of the second antenna, ϕ_2, is changed with the accuracy of $\pi/2$ according to the FBI from the MS so that the received SIR after combining is maximum. This is expressed as $\phi_1 = 0$, $\phi_2 = \{\pm\pi/4, \pm 3\pi/4\}, A_1 = A_2 = \sqrt{1/2}$. Meanwhile, the transmitted amplitudes of two data sequences are also controlled by FBI bits as well as the transmitted carrier phase in mode 2.

Here, we briefly explain the operation of mode 1. The MS generates the FBI bit b_n at slot n based on the received carrier phases transmitted from two antennas of CPICH, $\theta_{1,n}^{\mathrm{CP}}$ and $\theta_{2,n}^{\mathrm{CP}}$. Let $\hat{\theta}_{1,n}^{\mathrm{CP}}$ and $\hat{\theta}_{2,n}^{\mathrm{CP}}$ be the estimated values of $\theta_{1,n}^{\mathrm{CP}}$ and $\theta_{2,n}^{\mathrm{CP}}$, respectively. Then, the FBI bit b_n is estimated as

$$\text{If } -\pi/2 \leq \left(\hat{\theta}_{1,n}^{\mathrm{CP}} - \hat{\theta}_{2,n}^{\mathrm{CP}}\right) < \pi/2, \text{ then } b_n = 0; \text{ otherwise } b_n = 1, \quad \text{for even slot } n \tag{1.25a}$$

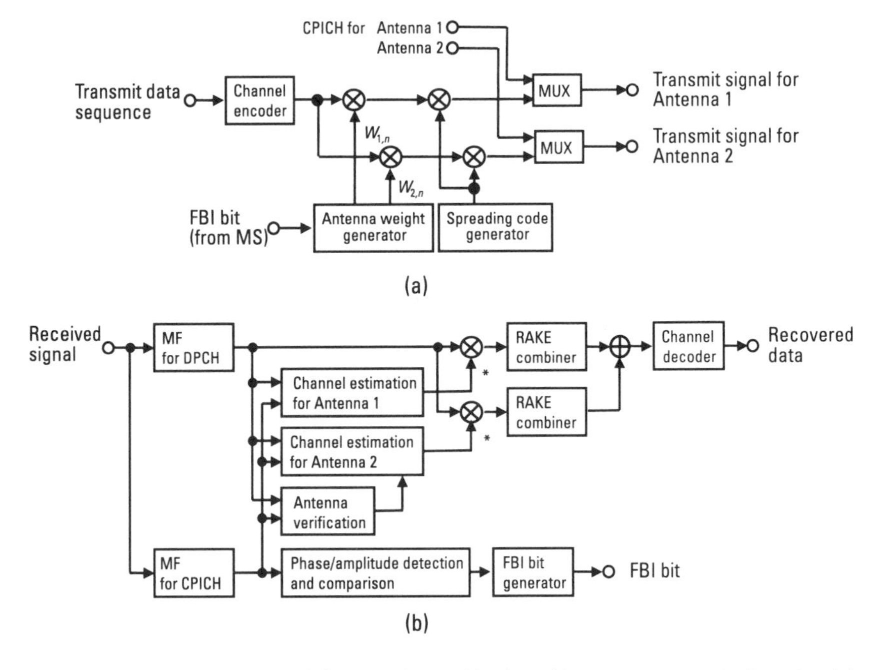

Figure 1.26 Block diagram of BS transmitter with closed-loop-type transmit diversity: (a) node B transmitter and (b) MS receiver.

$$\text{If } 0 \le (\hat{\theta}_{1,n}^{\text{CP}} - \hat{\theta}_{2,n}^{\text{CP}}) < \pi, \text{ then } b_n = 0; \text{ otherwise } b_n = 1, \quad \text{for odd slot } n \tag{1.25b}$$

According to the decoded results of FBI bit $\hat{b}_n$, the BS determines the tentative transmit carrier phase of DPCH at the $\varphi_{2,(n+1)}$ slot on the second antenna as

$$\begin{aligned} &\text{If } \hat{b}_n = 0, \text{ then } \varphi_{2,(n+1)} = 0; \text{ otherwise } \varphi_{2,(n+1)} = \pi, \\ &\qquad \text{for even slot } n \\ &\text{If } \hat{b}_n = 0, \text{ then } \varphi_{2,(n+1)} = \pi/2; \text{ otherwise } \varphi_{2,(n+1)} = -\pi/2, \\ &\qquad \text{for odd slot } n \end{aligned} \tag{1.26}$$

Finally, the transmit carrier phase at the $(n + 1)$ slot on the second antenna is derived from estimated values $\varphi_{2,n}$ and $\varphi_{2,(n+1)}$ as

$$\phi_{2,(n+1)} = (\varphi_{2,n} + \varphi_{2,(n+1)})/2 \tag{1.27}$$

Because the actual transmit power-controlled reverse link channel, FBI bit errors occur. In that case, the BS transmits a DPCH signal with the different carrier phase from the proper one that the MS estimated and informed. Therefore, the MS estimates the transmitted antenna weights, namely, the transmitted carrier phase for mode 1 at each slot of the DPCH to use the valid reference channel estimate of CPICH or DPCH of other slots (this process is called *antenna verification*) [16].

The measured average BER performance using closed-loop modes 1–3 is plotted in Figure 1.27 as a function of the total transmit power of two antennas when fast TPC was applied to both the forward and reverse links with the same target E_b/I_0 value (the fast TPC in the reverse link generates the FBI bit errors) [38]. In addition to mode 1 [called *phase diversity* (PD)-$\pi/2$ hereafter], two other transmit antenna weight generation methods are evaluated for comparison. The second method is also adaptive transmitting phase diversity with the accuracy of π such as $\phi_1 = 0$, $\phi_2 = \{0, \pm\pi\}$, $A_1 = A_2 = \sqrt{1/2}$ (which we call PD-π hereafter), and the third method is a *selection diversity* (STD) [29], which adaptively changes the amplitude of the transmitting antenna as $\phi_1 = \phi_2 = 0$, $(A_1, A_2) = \{(1,0),(0,1)\}$.

The transmit power was normalized by the average transmit power at the average BER of 10^{-3} without transmit diversity and without antenna diversity reception at the MS. The antenna verification was applied in order

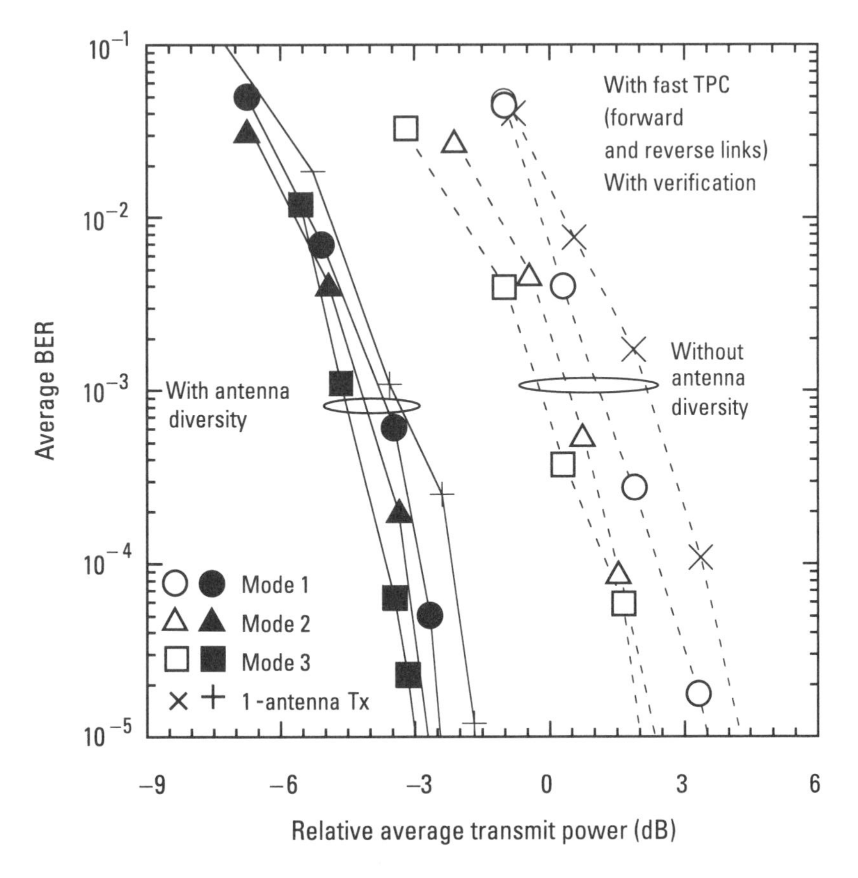

Figure 1.27 Average BER performance as a function of the total BS transmit power with a closed-loop-type transmit diversity for field experiments (without FBI bit error or antenna verification).

to decrease the influence of the FBI bit error under fast transmit power-controlled reverse link. However, we confirmed that the performance improvement using closed-loop mode transmit diversity with antenna verification when the FBI bit error occurs is smaller than the results without FBI bit error. The total required transmit power at the average BER of 10^{-3} using PD-$\pi/2$ and PD-π was decreased by approximately 1.0–1.5 (0.3–0.7) dB without (with) antenna diversity reception compared to the case without transmit diversity. The improvement using selection diversity was the largest, and the total required transmit power was decreased by approximately 2.0 dB without and 1.0 dB with MS antenna diversity reception compared to the case without transmit diversity. The reason that the effect of transmit diversity is small with antenna diversity reception is considered in a later discussion.

Because performance degradation due to multipath fading is sufficiently mitigated by the use of RAKE path diversity and antenna diversity reception in addition to fast TPC, further improvement by transmit diversity is small. Furthermore, the accuracy of antenna verification is degraded since the received signal power becomes small, especially with PD-$\pi/2$ having the smaller resolution of the controlled phase difference between the two antennas. As a result, the total required transmit power with PD-$\pi/2$ is greater than that with PD-π.

1.6 WCDMA Capacity Enhanced Technologies

In DS-CDMA systems, due to multipath fading and shadowing as well as distance-dependent path loss, severe multiple access interference is often produced, which significantly reduces the link capacity. In the forward link, although the orthogonality among the same propagation channels is achieved by using orthogonal variable spreading factor channelization codes, the multipath interference, especially from high rate users, is severe. Also, demand for higher capacity in the forward link is high because it can lead to high-speed Internet access and broadcast services from information sites. Interference canceller or multiuser detection [39, 40] is an effective technique for reducing multiple access interference but is exclusively exploited for considering the application to the reverse link. The use of adaptive antenna arrays is a more promising and practical technique than MI or multiuser detection, because the coherent adaptive antenna array diversity receiver [41, 42] at the BS is effective in decreasing multiple access interference and thereby decreasing the transmitting power of MSs in the reverse link. Adaptive antenna array transmit diversity [43–46] is effective in decreasing severe multipath interference in the forward link without changing the air interface and adding complexity to the MS. Thus, in this section, we focus on coherent adaptive antenna array diversity receivers in the reverse link and adaptive antenna array transmit diversities in the forward link.

1.6.1 Coherent Adaptive Antenna Array Diversity Receiver/Adaptive Antenna Array Transmit Diversity

Figure 1.28 shows an overall block diagram of the four-antenna adaptive antenna array diversity transceiver. The receiver antenna weights in the

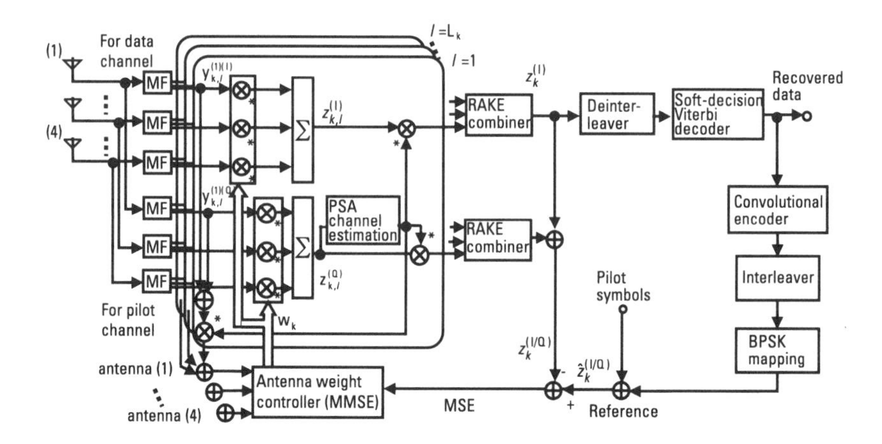

Figure 1.28 Block diagram of coherent adaptive antenna array diversity receiver.

reverse link are generated in the coherent adaptive antenna array diversity receiver block. In the scheme, transmitter antenna weights are generated by modifying the receiver antenna weights (i.e., performing calibrations that will be explained later) [41–44]. Our concern is *frequency-division duplex* (FDD) in WCDMA systems, thus, the instantaneous phase and amplitude variations due to fading in the reverse link have no correlation to those in the forward link. Noting that the distance-dependent macroscopic propagation factors determining the average signal power (i.e., path loss and shadowing) are not frequency dependent, the coherent adaptive antenna array diversity receiver was designed based on the following concept. The adaptive antenna array forms an antenna beam that tracks only slow changes in the directions of arrival and average powers of the desired and interfering users, and the RAKE combiner tracks the instantaneous variations in the channel conditions to maximize the instantaneous *signal-to-interference plus background noise power ratio* (SINR).

We perform two calibrations on the receiver antenna weights when generating transmit antenna weights in the forward link. We use the common array antennas with the antenna separation of half the carrier wavelength in the reverse link. Thus, when the direction of arrival of the incoming signal is not 0°, the phase difference between antennas is different between different carrier frequencies. Then, if the generated receiver antenna weights are used directly as transmitter antenna weights, the direction of the main lobe in the transmitter beam pattern is shifted from the original direction of arrival of the desired user, and the directions of the beam nulls are shifted from those of the interfering users. Therefore, in our approach, we shift the direction of the main lobe in the transmitter beam pattern so that it coincides with the main lobe in the receiver beam pattern. We call this modification in the transmitter beam pattern carrier frequency calibration.

Furthermore, because adaptive antenna array processing is done in the baseband, the generated reverse link weights are reflected on the phase/amplitude deviations in the transfer functions of the *radio-frequency* (RF) receiver circuitry associated with different antennas. In the transmitter, the RF transmitter circuitry of the different antenna branches is adversely influenced by the different transfer functions (see Figure 1.29). Therefore, the transmit antenna weights are generated by modifying the receiver antenna weights, taking into account the phase/amplitude variations in parallel RF receiver/transmitter circuitries. This calibration is called RF circuitry calibration. Let $w_{\text{ideal}}^{(i)}$, $w_{\text{R}}^{(i)}$, and $w_{\text{RX}}^{(i)}$ be the complex-valued antenna weight for the ideal case (no phase/amplitude errors exist in the RF receive circuitry transfer functions), the complex-valued receiver antenna weight generated in coherent adaptive antenna array diversity, and complex-valued transfer function of the RF receive circuitry of the ith antenna branch, respectively. For the ideal case (i.e., $x_{\text{RX}}^{(i)}$ = const. for all antenna branches), received signal $r_{\text{RX}}^{(i)}$ should be weighted by $w_{\text{ideal}}^{(i)}$ to be combined for beam forming. However, in a real receiver, $r_{\text{RX}}^{(i)}$ goes through the RF receiver circuitry and then, weighted by $w_{\text{R}}^{(i)}$, produces $w_{\text{R}}^{(i)} \cdot x_{\text{RX}}^{(i)} \cdot r_{\text{RX}}^{(i)}$, before combining. The resultant signal must be the same as that of the ideal case. Therefore, we obtain

$$w_{\text{ideal}}^{(i)} = w_{\text{R}}^{(i)} x_{\text{RX}}^{(i)} \tag{1.28}$$

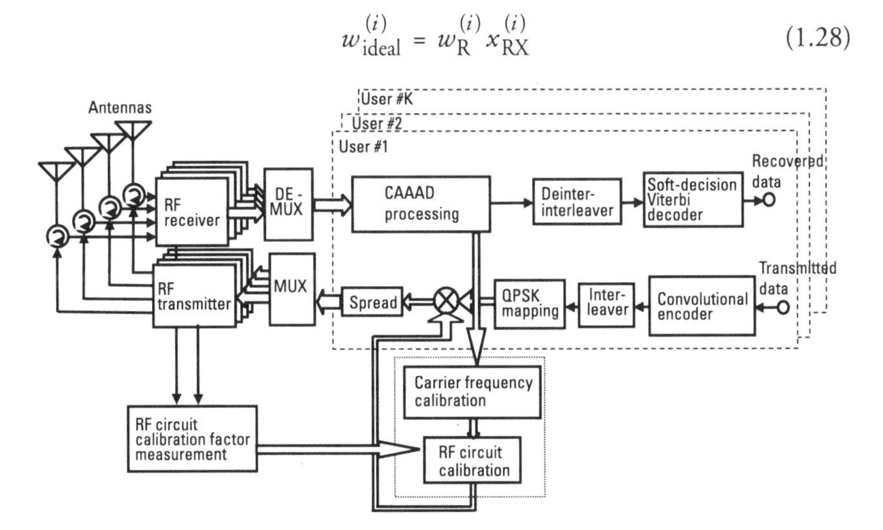

Figure 1.29 Overall block diagram of the coherent adaptive antenna array diversity (CAAAD) transceiver.

The weights provided by (1.28) cannot be directly used to form the transmitter beam because the signal weighted at the baseband stage suffers phase/amplitude shift due to the RF transmit circuitry before transmission from an antenna. Let $x_{TX}^{(i)}$ be the transfer function of the RF transmit circuitry for the ith antenna branch. The signal to be transmitted from the ith antenna is first weighted using transmit antenna weight $w_T^{(i)}$ and then suffers phase/amplitude shift (equivalent by multiplication of $x_{TX}^{(i)}$). The equivalent antenna weight becomes, therefore, $w_T^{(i)} \cdot x_{TX}^{(i)}$, and this must be equal to $w_{ideal}^{(i)}$. As a consequence, we obtain

$$w_T^{(i)} = w_{ideal}^{(i)} / x_{TX}^{(i)} = w_R^{(i)} \left(x_{RX}^{(i)} / x_{TX}^{(i)} \right) \tag{1.29}$$

Using (1.29), the transmit antenna weights (at the baseband beam-forming stage) can be obtained from adaptively generated receive antenna weights. Both $x_{RX}^{(i)}$ and $x_{TX}^{(i)}$ can be measured even during the operation mode. By performing the above two calibrations on the generated receiver antenna weights in the coherent adaptive antenna array diversity receiver, the maximum gain is obtained toward the desired signal direction and the nulls are nearly directed toward the interfering sources.

Figure 1.28 showed a block diagram of the digital beam former and RAKE combiner in the coherent adaptive antenna array diversity receiver [47, 48]. This block is made up of matched filters, a beam former, pilot symbol-assisted-coherent RAKE combiner, and weight controller. The MF output signal samples of all antennas for each resolved path are weighted by the receiver antenna weight, which is common to all paths for RAKE combining, and then combined. Because it was reported that the angle spread among paths is within 10°, in an urban area with a high-elevation antenna configuration such as in a cellular system [49, 50], we used the common antenna weights, that is, common receiver beam pattern, for all paths. The resulting composite channel gain at the beam former output is estimated by using the pilot symbols multiplexed into the Q channel of several successive slots for coherent RAKE combining of different resolved paths. The receiver antenna weights in the coherent adaptive antenna array diversity receiver are updated so that the mean squared error between the RAKE-combined signal and the reference signal is minimized. As a reference signal, we used the decision-feedback data symbols after forward error correction decoding in addition to pilot symbols for generating refined mean squared error and improving channel estimation accuracy.

1.6.2 Experiments

We first present the laboratory experimental results for a coherent adaptive antenna array diversity receiver in the reverse link. In the experiments, fading simulators were used to simulate multipath signals that follow independent two-path Rayleigh fading with the maximum Doppler frequency f_D in hertz. The measured average BER performance of a coherent adaptive antenna array diversity receiver with four antennas in a six-user environment as a function of the average received SIR at each antenna, is shown in Figure 1.30 with the average received E_b/N_0 as a parameter with angle spread between two paths $\alpha = 0$. The directions of arrival of users 1, 2, 3, 4, 5, and 6 were set to −50°, −30°, −10°, 0°, +20°, and +40°, respectively (user 1 was a desired user). The BER performance of a four-branch space diversity

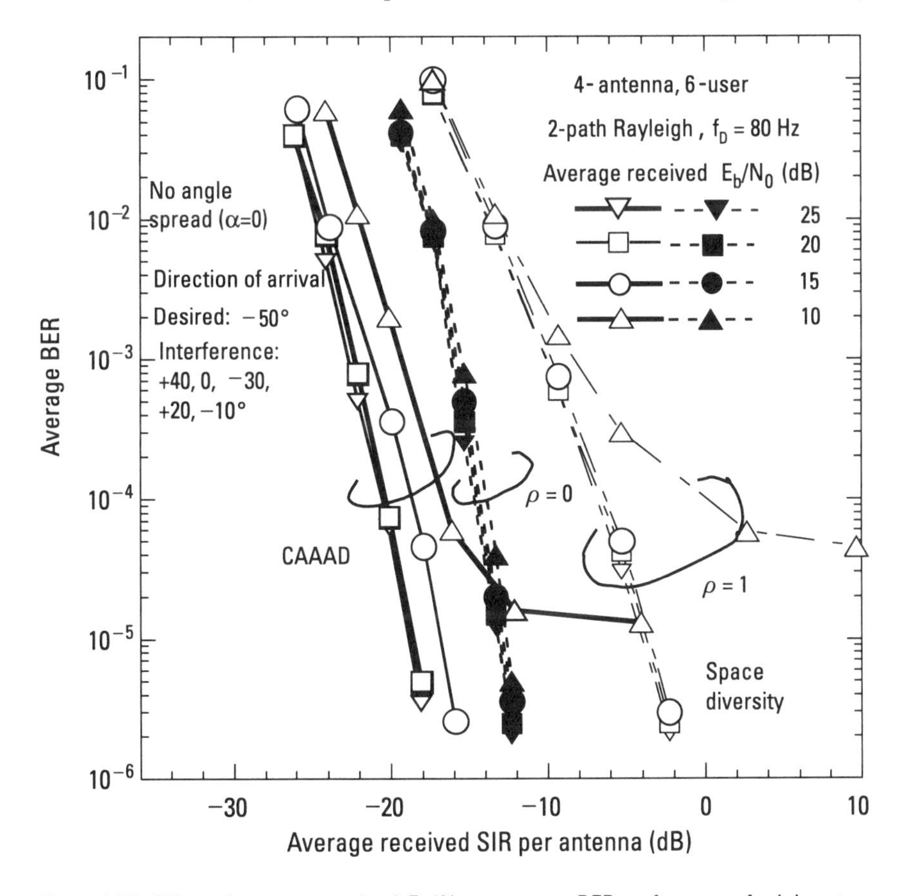

Figure 1.30 Effect of average received E_b/N_0 on average BER performance for laboratory experiment.

receiver using MRC with the fading correlation among antennas ρ = 0 and 1 was also plotted for comparison. From Figure 1.30, even when the number of interfering users is larger than that of the beam nulls, it is clear that the required average received SIR of the coherent adaptive antenna array diversity (CAAAD) receiver for achieving the average BER of 10^{-3} can be reduced by approximately 6 dB compared to the MRC space diversity reception when the average received E_b/N_0 = 20 dB.

Next, we investigated the effect of receiver antenna weight updating using decision-feedback data symbols after forward error correction decoding in addition to pilot symbols and performance differences for channel coding. The measured average BER performance with the coherent adaptive antenna array diversity receiver with fast TPC using convolutional coding and turbo coding is shown in Figure 1.31 as a function of the average transmit E_b/N_0 in the reverse link. The directions of arrival of the desired and interfering users were set to 0, and +40°, respectively, in a two-user environment. The ratio of the target E_b/I_0 for fast TPC of the desired user to interfering user was $\Delta E_b/I_0$ = −12 dB. The performances with receiver weight updating using decision-feedback data symbols as well as pilot symbols, and pilot symbols only, are plotted as solid and dotted lines, respectively. It was assumed that f_D was 5 Hz and T_{CHL} was 20 ms. Figure 1.31 shows that the required transmit E_b/N_0 for satisfying the average BER of 10^{-3} and 10^{-6} with the coherent adaptive antenna array diversity receiver using decision-feedback antenna weight updating and channel estimation is decreased by approximately 0.8 dB compared to the case using only pilot symbols both for convolutional coding and turbo coding. These results also confirm that for T_{CHL} = 20 ms, the required transmit E_b/N_0 at the average BER of 10^{-6} of the coherent adaptive antenna array diversity receiver using decision-feedback antenna weight updating and channel estimation with turbo coding is decreased by approximately 0.5 dB compared to the case with convolutional coding.

Moreover, based on laboratory experiments, we evaluated the forward link performance using a four-antenna adaptive antenna array transmit diversity with SIR-based fast TPC. The average BER performance of an MS using adaptive antenna array transmit diversity with carrier frequency and RF calibrations in the forward link is plotted in Figure 1.32 as a function of the average transmit E_b/N_0. The experimental configuration of adaptive antenna array transmit diversity has been described [45]. The number of users was four (three interfering users) and the directions of arrival of the desired and interfering users were set to θ_D = −50° and θ_U = −20°, +15°, +45°. The BER performance with one antenna transmitter is plotted for

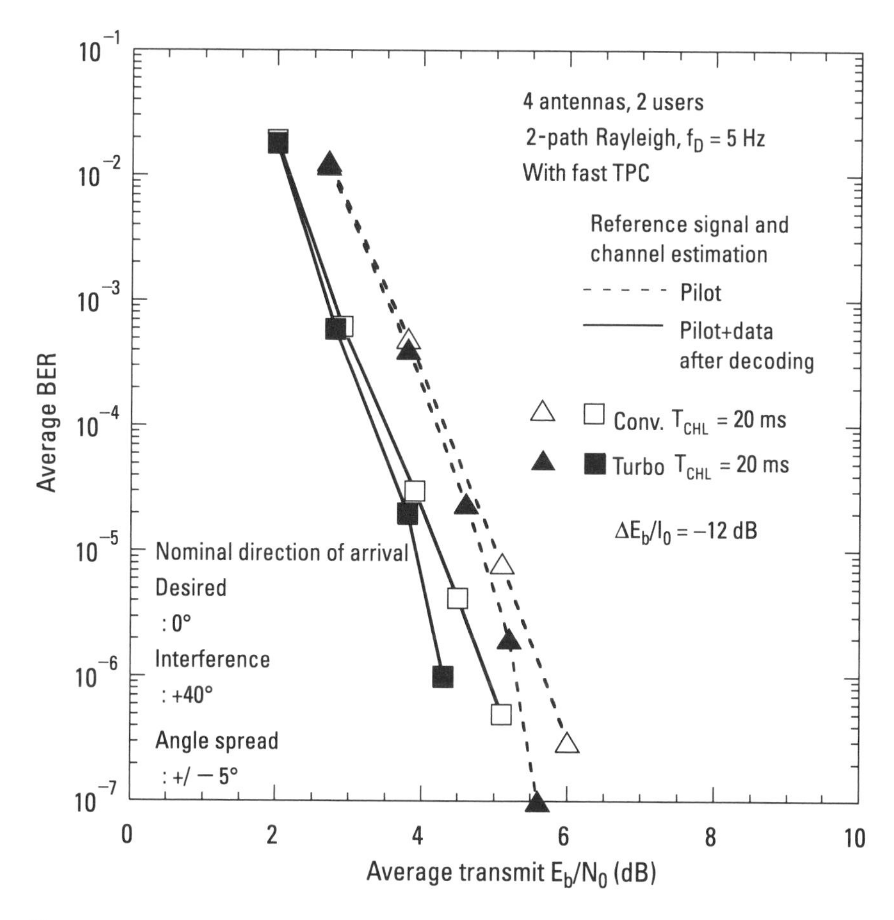

Figure 1.31 Average BER performance as a function of average transmit E_b/N_0 (with fast TPC and two users).

comparison. The ratio of the target E_b/I_0 of the desired user to the interfering users, $\Delta E_b/I_0$, is −3, −5, and −12 dB and the single-user case is shown. Figure 1.32 clearly shows that when using the one-antenna transmitter, as the interfering power is increased, the BER performance is significantly degraded; that is, an error floor is observed, due to the severe MPI from the interfering users. Meanwhile, by applying adaptive antenna array transmit diversity, almost identical BER performance to that in the single-user case was achieved even for the case of $\Delta E_b/I_0 = -12$ dB owing to the significant interference suppression effect. The results demonstrate that the RF circuitry calibration and carrier frequency calibration are working satisfactorily and that consequently adaptive antenna array transmit diversity is very effective in decreasing strong MPI especially from high-rate physical channels.

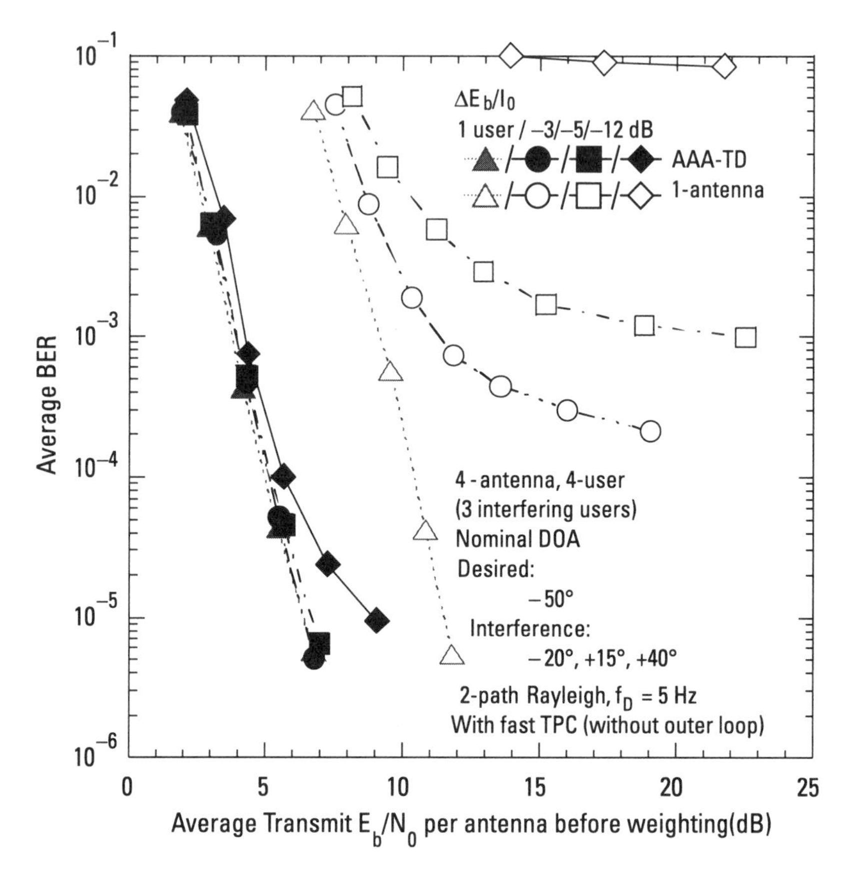

Figure 1.32 Average BER performance of adaptive antenna array transmit diversity as a function of average transmit E_b/N_0 (with fast TPC). DOA: direction of arrival.

Next, the results of field experiments conducted near Tokyo are evaluated. In the coherent adaptive array antenna diversity receiver experiments, MS 1 (desired user) moved along the measurement course, which was located approximately 600m to 850m away from the BS. The height of the BS antenna was 50m from the ground, and 120° sectored antennas were used. When MS 1 moved along the measurement course, the direction of arrival toward the BS changed from −10° to +10°. In the measurement course, one-to-two-path fading appeared in the first half of the course, followed by two-to-three-path fading with an average power difference of approximately 3 dB. In contrast, MS 2 (interfering user) was located at a fixed point 600m away from the BS and almost in the line-of-sight path (i.e., single path). The direction of arrival of MS 2 was approximately +40°. The generated beam pattern associated with one of two resolved paths of MS 1 is plotted in Figure 1.33 for the average received E_b/N_0 of MS 1 of 25 dB. The figure shows that the direction of the main lobe shifts along with the movement of MS 1. Although MS 2 is located at the fixed point at +40°, the directions of the beam nulls are also shifted according to the movement of MS 1. This is because the receiver antenna weights are updated to enhance the gain toward the direction of arrival of MS 1 in order to maximize the received SIR after weight combing.

The measured BER performance is plotted in Figure 1.34 as a function of the average received signal power with the average received SIR of MS 1 as a parameter. For comparison, the results of space diversity using MRC are also plotted (antenna separation was 10λ, where λ is the carrier wave in the reverse link). Figure 1.34 shows that the required average received signal power for obtaining the average BER of 10^{-3} is decreased by approximately 8 to 10 dB using the coherent adaptive antenna array diversity receiver compared to the case using space diversity.

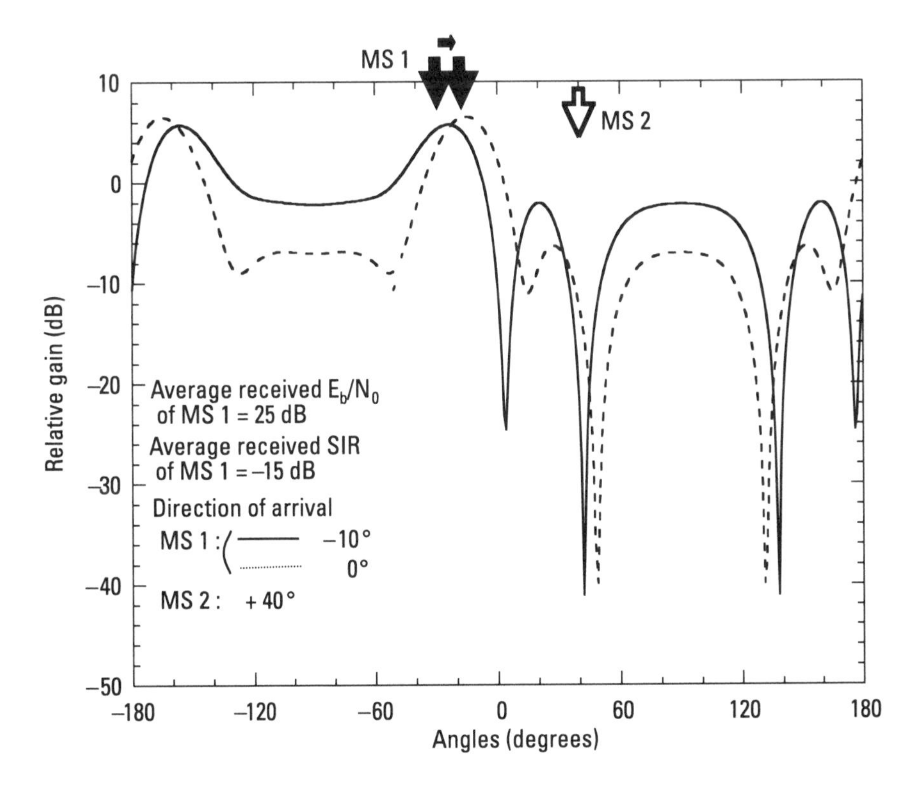

Figure 1.33 Generated beam patterns for field experiments.

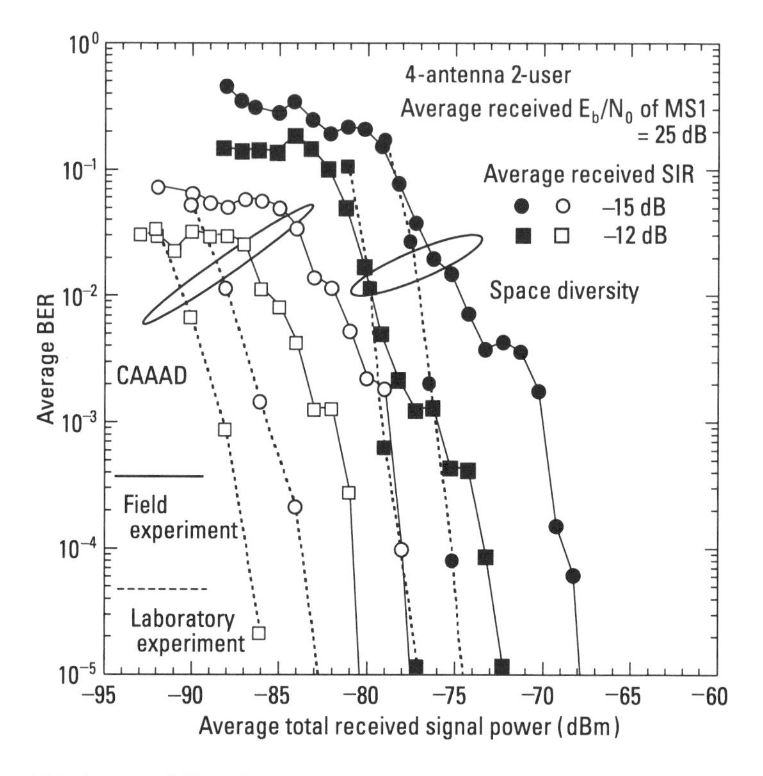

Figure 1.34 Average BER performance of the coherent adaptive antenna array diversity (CAAAD) receiver as a function of average received signal power for a field experiment.

The average BER performance measured at an MS using adaptive antenna array transmit diversity (AAA-TD) in the forward link is plotted in Figure 1.35 as a function of the average received signal power when the average received SIR of MS 1 is 0 dB. The transmitted SIRs of MS 1 before weighting are −5, −10, and −12 dB. The BER performance of a one-antenna transmitter is also shown in the figure. The figure clearly shows that although the BER performance with the one-antenna transmitter is severely degraded as the transmitted SIR is decreased, the performance is significantly improved by using adaptive antenna array transmit diversity due to beam and null steering. When adaptive antenna array transmit diversity is used, the increase in the required transmit power at the average BER of 10^{-3} from the case without an interfering user is within 5 dB when the transmitted SIR is −12 dB. These results in the reverse and forward links verify the effects of reducing the interference from high-rate users since the receiver and transmitter antenna weights can precisely track the changes in the directions of arrival of the desired signal.

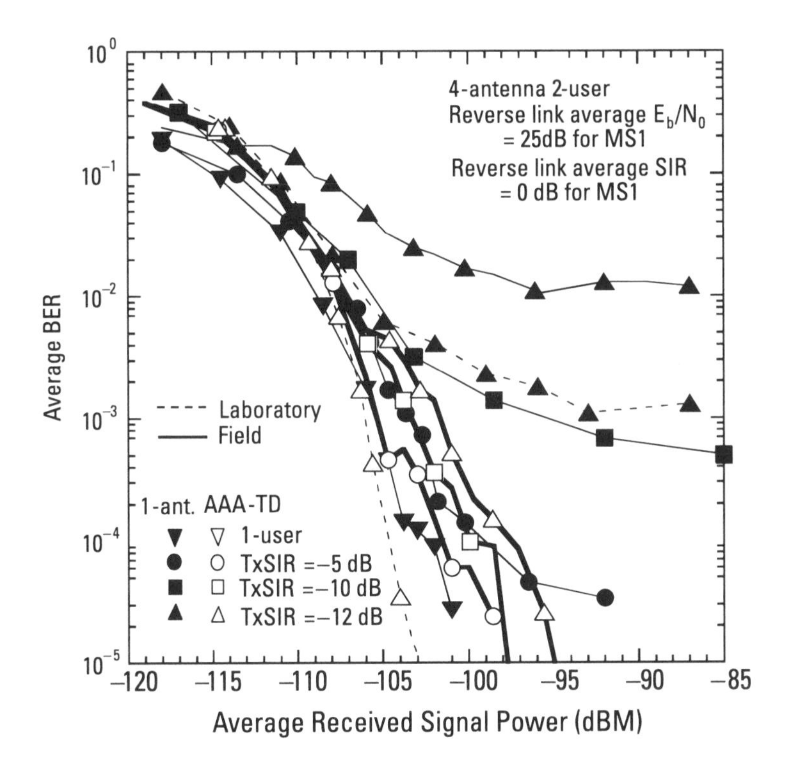

Figure 1.35 Average BER performance as a function of average received signal power for a field experiment (reverse link SIR = 0 dB).

References

[1] Adachi, F., M. Sawahashi, and H. Suda, "Wideband DS-CDMA for Next Generation Mobile Communication System," *IEEE Communications Magazine,* Vol. 36, Sept. 1998, pp. 56–69.

[2] Dahlman, E., et al., "UMTS/IMT-2000 Based on Wideband CDMA," *IEEE Communications Magazine,* Vol. 36, Sept. 1998, pp. 70–80.

[3] 3GPP RAN, 3G TS 25.301 V3.5.0, June 2000.

[4] 3GPP RAN, 3G TS 25.211 V3.4.0, Sept. 2000.

[5] 3GPP RAN, 3G TS 25.213 V3.3.0, Sept. 2000.

[6] Okawa, K., and F. Adachi, "Orthogonal Forward Link Using Orthogonal Multi-Spreading Factor Codes for Coherent DS-CDMA Mobile Radio," *IEICE Trans. on Communication,* Vol. E81-B, April 1998, pp. 777–784.

[7] IMT-2000 Study Committee, AIF/SWG2-25-4, July 1998.

[8] 3GPP RAN, 3G TS 25.302 V3.5.0, June 2000.

[9] 3GPP RAN, 3G TS 25.212 V3.4.0, Sept. 2000.

[10] Berrou, C., A. Glavieux, and P. Thitimajshima, "Near Shannon Limit Error-Correcting Coding and Decoding: Turbo-Codes," *Proc. IEEE ICC'93,* Geneva, Switzerland, May 1993, pp. 1064–1070.

[11] 3GPP RAN, 3G TS 25.402 V3.2.0, June 2000.

[12] Higuchi, K., M. Sawahashi, and F. Adachi, "Fast Cell Search Algorithm in DS-CDMA Mobile Radio Using Long Spreading Codes," *IEICE Trans. on Communication,* Vol. E81-B, July 1998, pp. 1527–1534.

[13] Higuchi, K., et al., "Experimental Evaluation of 3-Step Cell Search Method in WCDMA Mobile Radio," *Proc. VTC'2000,* Tokyo, Japan, May 2000, pp. 1527–1534.

[14] Higuchi, K., et al., *Experiments on 3-Step Fast Cell Search Method Using TSTD in WCDMA Mobile Ratio,* IEICE Technical Report RCS99-155, Nov. 1999, pp. 99–105 (in Japanese).

[15] Hanada, Y., et al., "Fast Cell Search Algorithm in Idle Mode for Intercell Asynchronous WCDMA Mobile Radio," *IEICE Trans. on Communication,* Vol. E83-B, Aug. 2000, pp. 1610–1618.

[16] 3GPP RAN, 3G TS 25.214 V3.4.0, Sept. 2000.

[17] Ariyavisitakul, S., "Signal and Interference Statistics of a CDMA System with Feedback Power Control—Part II," *IEEE Trans. on Communication,* Vol. 42, Feb./April 1994, pp. 597–605.

[18] Seo, S., T. Dohi, and F. Adachi, "SIR-Based Transmit Power Control of Reverse Link for Coherent DS-CDMA Mobile Radio," *IEICE Trans. on Communication,* Vol. E81-B, No. 7, July 1998, pp. 1508–1516.

[19] Ling, F., "Coherent Detection with Reference-Symbol Based Estimation for Direct Sequence CDMA Uplink Communications," *Proc. VTC'93,* New Jersey, May 1993, pp. 400–403.

[20] Andoh, H., M. Sawahashi, and F. Adachi, "Channel Estimation Filter Using Time-Multiplexed Pilot Channel for Coherent Rake Combining in DS-CDMA Mobile Radio," *IEICE Trans. on Communication,* Vol. E81-B, No. 7, July 1998, pp. 1517–1526.

[21] Abeta, S., M. Sawahashi, and F. Adachi, "Adaptive Channel Estimation for Coherent DS-CDMA Mobile Radio Using Time-Multiplexed Pilot and Parallel Pilot Structure," *IEICE Trans. on Communication,* Vol. E82-B, No. 9, Sept. 1999, pp. 1505–1513.

[22] Higuchi, K., et al., "Experimental Evaluation of Combined Effect of Coherent RAKE Combining and SIR-Based Fast Transmit Power Control for Reverse Link of DS-CDMA Mobile Radio," *IEEE J. on Selected Areas of Communication,* Vol. 18, Aug. 2000, pp. 1526–1535.

[23] Higuchi, K., et al., "Experimental Evaluations of High Rate Data Transmission Using Turbo/Convolutional Coding in WCDMA Mobile Radio," *Proc. Wireless 2000,* Calgary, Canada, July 2000, pp. 687–693.

[24] Shibutani, A., H. Suda, and F. Adachi, "Multi-Stage Interleaver for Turbo Codes in DS-CDMA Mobile Radio," *Proc. IEEE APCC/ICC'98,* Nov. 1998, pp. 391–395.

[25] Shibutani, A., H. Suda, and Y. Yamao, "Performance of WCDMA Mobile Radio with Turbo Codes Using Prime Interleaver," *Proc. VTC'2000,* Tokyo, Japan, May 2000, pp. 1570–1574.

[26] Gilhousen, K. S., et al., "On the Capacity of a Cellular CDMA System," *IEEE Trans. on Vehicular Technology,* Vol. VT-40, No. 5, May 1991, pp. 303–312.

[27] Wong, D., and T. J. Lim, "Soft Handoffs in CDMA Mobile Systems," *IEEE Personal Communications Magazine,* Vol. 36, Dec. 1997, pp. 6–17.

[28] "Mobile Station-Base Station Compatibility Standard for Dual-Mode Wideband Spread Spectrum Cellular System," TIA/EIA/IS-95, Telecommunication Industry Association, July 1993.

[29] Fukumoto, S., et al., "Combined Effect of Site Diversity and Fast Transmit Power Control in WCDMA Mobile Radio," *Proc. VTC'2000,* Tokyo, Japan, May 2000, pp. 1527–1534.

[30] Morimoto, A., et al., "Experiments on Intercell Site Diversity Using Two-Step Selection Combining in WCDMA Reverse Link," *IEICE Trans. on Communication,* Vol. E84-B, No. 3, March 2001, pp. 435–445.

[31] "Adjustment Loop in Down Link Power Control During Handover," TSG-RAN Working 1 Meeting, NEC, Oct. 1999.

[32] "Down Link Power Control During Soft Handover," 3GPP TSG RAN WG1 Meeting, Nortel, July 1999.

[33] Furukawa, H., K. Hamabe, and A. Ushirokawa, "SSDT—Site Selection Diversity Transmission Power Control for CDMA Forward Link," *IEEE J. Selected Areas of Communication,* Vol. 18, Aug. 2000, pp. 1546–1554.

[34] Hottinen, A., and R. Wichman, "Transmit Diversity by Antenna Selection in CDMA Downlink," *Proc. IEEE ISSSTA'98,* Sept. 1998, pp. 767–770.

[35] Fukumoto, S., M. Sawahashi, and F. Adachi, "Performance Comparison of Forward Link Transmit Diversity Techniques for WCDMA Mobile Radio," *Proc. PIMRC'99,* Sept. 1998, pp. 1139–1143.

[36] Alamouti, S. M., "A Simple Transmit Diversity Technique for Wireless Communications," *IEEE J. Selected Areas of Communication,* Vol. 16, No. 8, Oct. 1998, pp. 1451–1458.

[37] Fukumoto, S., et al., "Experiments on Space Time Block Coding Transmit Diversity (STTD) in WCDMA Forward Link," *IEICE Trans. Fundamentals,* Vol. E84-A, Dec. 2001, pp. 3045–3057.

[38] Fukumoto, S., et al., "Field Experiments on Closed Loop Mode Transmit Diversity in WCDMA Forward Link," *Proc. IEEE ISSSTA'2000,* Sept. 2000, pp. 433–438.

[39] Duel-Hallen, A., J. Holtzman, and Z. Zvonar, "Multiuser Detection for CDMA Systems," *IEEE Personal Communications Magazine,* April 1995, pp. 46–58.

[40] Moshavi, S., "Multi-User Detection for DS-CDMA Communications," *IEEE Communications Magazine,* Oct. 1996, pp. 124–136.

[41] Tsoulos, G. V., M. A. Beach, and J. McGeehan, "Wireless Personal Communications for the 21st Century: European Technological Advances in Adaptive Antennas," *IEEE Communications Magazine,* Sept. 1997, pp. 102–109.

[42] Compton, R. T., Jr., "An Adaptive Antenna in a Spread-Spectrum Communication System," *Proc. IEEE,* Vol. 66, No. 3, March 1978, pp. 289–295.

[43] Harada, A., et al., "Performance of Adaptive Antenna Array Diversity Transmitter for WCDMA Forward Link," *Proc. IEEE PIMRC'99,* Osaka, Japan, Sept. 12–15, 1999, pp. 1134–1138.

[44] Tanaka, S., et al., "Transmit Diversity Based on Adaptive Antenna Array for WCDMA Forward Link," *Proc. CDMA Int. Conf., CIC'99,* Korea, Sept. 1999, pp. 282–286.

[45] Harada, A., et al., "Experiments on Adaptive Antenna Array Transmit Diversity in WCDMA Forward Link," *Proc. CDMA Int. Conf., CIC2000,* Korea, Nov. 2000, pp. 47–51.

[46] Taoka, H., et al., "Experiments on Adaptive Antenna Array Transmit Diversity with Carrier Frequency Calibration in Transmit Power-Controlled Forward Link for WCDMA Mobile Radio," *Proc. 3G Wireless and Beyond,* San Francisco, CA, May 2001.

[47] Tanaka, S., M. Sawahashi, and F. Adachi, "Pilot Symbol-Assisted Decision-Directed Coherent Adaptive Array Diversity for DS-CDMA Mobile Radio Reverse Link," *IEICE Trans. Fundamentals,* Vol. E80-A, Dec. 1997, pp. 2445–2454.

[48] Tanaka, S., et al., "Experiments on Coherent Adaptive Antenna Array Diversity for Wideband DS-CDMA Mobile Radio," *IEEE J. Selected Areas of Communication,* Vol. 18, Aug. 2000, pp. 1495–1504.

[49] Kozono, S., and S. Sakagami, "Correlation Coefficient on Base Station Diversity for Land Mobile Communication System," *IEICE Trans. on Communication,* Vol. J70-B, April 1987, pp. 476–482 (in Japanese).

[50] Martin, U., "Spatio-Temporal Radio Channel Characteristics in Urban Macrocells," *IEE Proc. Radar, Sonar Navigation,* Vol. 145, Feb. 1998.

2

Downlink Performance Enhancements in CDMA Cellular Systems

Hiroshi Furukawa

As data communication continues to become a more prevalent addition to our society, 3G cellular systems will be called upon to handle data communication more effectively. When we consider the present Internet access in offices and homes, we will see a need for much more downlink traffic than uplink traffic. The 3G system thus must offer high-capacity downlinks that can accommodate multimedia traffic.

The initial studies of CDMA cellular systems focused on the uplink, especially on the power control needed to solve the near-far problem. The power levels of the uplink signals received at a BS vary a great deal because of the path-loss differences resulting from the different positions of MSs within a cell [1]. Power control can ensure that the power levels of all received uplink signals are nearly the same, thereby helping to keep these signals from interfering with each other. Under such a framework, the study of the downlink was historically subjugated to that of the uplink. In this chapter we focus our attention on a detailed assessment of downlink characteristics and on the enhancement of downlink performance. In Section 2.1 we first present the results of computer simulations investigating downlink capacity and then address the barriers faced by those who try to design effective downlinks: the lack of orthogonal codes and the increased interference due to SHO. These intensive evaluations are followed by two proposals to enhance downlink performance: in Section 2.2, an optimum allocation of downlink orthogonal code, and in Section 2.3, site selection diversity transmission power control.

2.1 Evaluation of WCDMA Downlinks

In this section we describe the results of computer simulations evaluating the downlink capacity of WCDMA systems [2]. To explore the detailed behavior of complex CDMA systems, we developed a dynamic multicell simulator that realizes time-domain simulation with slot-time resolution. TPC and SHO are exactly emulated in a slot-by-slot fashion, and the clarification of downlink capacity by such an elaborate simulator clearly identifies the essential issues.

2.1.1 Assumptions in System Evaluation

The evaluations were done in line with the guidelines defined in ITU-R Recommendation M.1225 [3]. This recommendation specifies in detail the evaluation assumptions made regarding service classes, transmission power, data rates, and cell deployment models so that different radio transmission technologies for IMT-2000 can be compared fairly under a variety of simulation conditions.

Service Classes

The service classes evaluated were circuit-switched voice with an activity of 0.5, circuit-switched data with an activity of 1, and packet-switched data. A complex traffic model for packet-switched data was assumed using a Pareto distribution model. Data rates were uniquely set for respective combinations of service class and environment model. The actual settings of data rates are described later in conjunction with other important simulation parameters.

Environmental Models

Three simulation environmental models were provided for the evaluations: an outdoor vehicular model, an outdoor pedestrian model, and an indoor pedestrian model. The outdoor vehicular model comprised classic macrocells with hexagonal shapes as shown in Figure 2.1. We assume a composition of cells concentrically located with two cell rings for a total of 19 cells and 57 sectors, that is, each BS deploys three sectored antennas. The cell radius

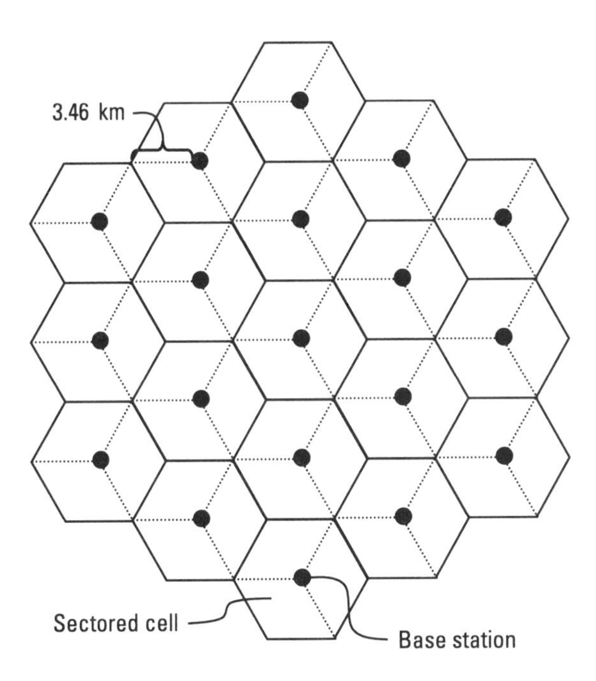

Figure 2.1 Cell layout of the outdoor vehicular model.

was assumed to be 3.46 km. In the calculation of interference, we used the wraparound cell layout and sampled data from all sectors. The MSs moved at 120 km/hr, and the mobility was modeled as a pseudorandom model with semidirected trajectories. The direction of the MSs is updated with a probability of 0.2 for every 20-m movement, and we use a uniform distribution of angle for the direction update having a range of –45° to +45°. All of the MSs on the map move with the same velocity, which is determined from a given Doppler frequency.

The outdoor pedestrian model was a Manhattan-like environment with a block size of 200m and low-speed users (3 km/hr). The geographical layout of the model is shown in Figure 2.2. Each of its 72 BSs had an omnidirectional antenna deployed 10m above the ground, and the pedestrians moved in the middle of streets 30m wide. To exclude the buffer cells, we evaluated only data sampled from the inner six cells.

The indoor pedestrian model represented a three-floor office building in which the mobile terminals moved slowly (3 km/hr) back and forth between offices and corridors. The floor layout of the model is shown in Figure 2.3. Evaluation data were sampled from the second floor.

Different downlink transmission powers were assumed in each of these environments. The actual settings are specified later in this section.

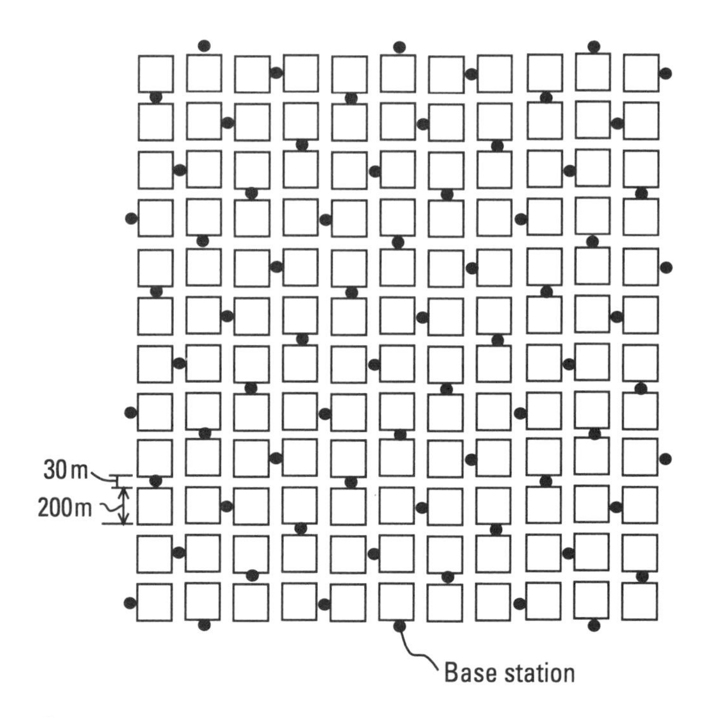

Figure 2.2 Geographical layout of the outdoor pedestrian model.

Radio Propagation Models

We did not use a deterministic approach like ray-tracing but instead assumed the path loss to be proportional to d^{α}, where α is the path-loss exponent and d is the distance between the transmitter and receiver. The path-loss exponent in the vehicular model was assumed to be 3.75, and the path-loss exponents in the outdoor and indoor pedestrian models were changed depending on terminal position. In the outdoor pedestrian model, radio waves propagating on rooftops as well as along roads were taken into account as radio propagation along the road [4]. In all cases, the path-loss calculation included random shadow fading with a log-normal distribution.

SHO

Intercell/intracell SHO was assumed for the circuit-switched services but not the packet-switched services. The intercell/intracell SHO algorithm simply connected the strongest BSs within the SHO window (denoted by the so-called T_add threshold [5]). The maximum active set size was two and the handover window threshold was set to 3 dB. The algorithm was executed every 0.5 second for the pedestrian models and 0.01 second for the vehicular

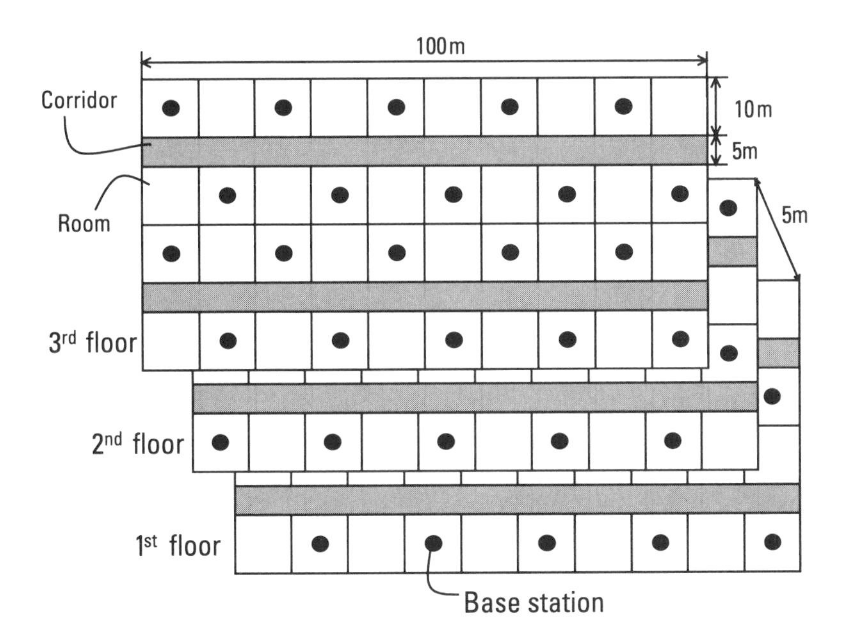

Figure 2.3 Floor layout of the indoor pedestrian model.

model. The short update period was used in the vehicular model because MSs could not capture the strongest BSs when a longer period was used. A longer period could, however, be used if the handover window threshold or the maximum active set size was increased. No measurement errors in the active set detection were included, and no SHO was used in the packet simulations: The user was simply connected to the best BS (i.e., the one for which the path loss was smallest).

Orthogonal Code Assumption

The most attractive feature of the CDMA downlink is that its use of orthogonal codes results in low intracell interference [6]. Orthogonal transmission can be used only in the downlink because the downlink signals transmitted by a single BS travel to each MS through the same channel synchronously. Each downlink source signal is modulated by a different spreading code, and that code is orthogonal to the codes assigned to the other downlink source signals. Then all the products within the cell are gathered and further modulated by a scrambling code unique to that cell.

The downlink capacity is limited because only a finite number of orthogonal codes are available in actual systems, and we therefore evaluated downlink performance with and without code limitations. When codes were limited and no new orthogonal codes were available, a new downlink signal was spread using a new scrambling code. Multiple scrambling codes could thus be allocated in a single cell. Advances of such a code allocation scheme will be intensively studied in Section 2.2.

RAKE Receiver

The SINR of the downlink signals received at MSs was calculated assuming multiple-finger RAKE reception. Because radio channels are subject to multipath distortion, the orthogonality of downlink signals from a single BS cannot be assured. This is because a signal is not orthogonal to all the other signals propagating via different paths, not even to its own copies. The calculation of the intercell interference power due to multipath propagation was based on a predetermined path intensity ratio. The path intensity ratios to be described later in this section are specified for each environmental model.

Power Control

Fast SINR-based power control was assumed. That is, the power of the transmitter was balanced so as to maintain a target SINR. The power control can track fast fluctuation of the reception power well due to fast fading for the 3-km/hr case (pedestrian models). However, because power control cannot track the fast fading in the outdoor vehicular model (120 km/hr), the power control for this case was assumed to cope only with shadow fading. The power control range was 20 dB, and the maximum transmission powers to be used varied between service classes and data rates.

Packet Transmission

Packets were queued when no resources were available, and it was assumed that a retransmission packet could be transmitted in a time slot next to one sent for the original packet. That is, it was assumed that the transmission of *acknowledgment* (ACK)/*negative acknowledgment* (NAK) packets was error-free and that there was virtually no delay in delivery.

Transmission Parameters

Important transmission parameters including data rate and maximum transmission power per traffic channel are summarized in Table 2.1. The maximum transmission power and data rates were set as specified in [3], and the broadcast channel powers were set so as to ensure that the broadcast signal reached the area outside the cell coverage. The packet frame structures for the various data rates were taken into account when determining what the packet block lengths should be [4].

Table 2.1
Summary of Transmission Parameters

	Environment **Path Intensity Ratio**	**Outdoor Vehicular** **0.60:0.40**	**Outdoor Pedestrian** **0.94:0.06**	**Indoor Pedestrian** **0.90:0.10**
Circuit-switched voice	Data rate and information rate (Kbps)	8	8	8
	BS max power per traffic channel (dBm)	30	20	10
	Broadcast CH power (dBm)	30	23	10
Circuit-switched data	Data rate and information rate (Kbps)	144	384	2,048
	BS max power per traffic channel (dBm)	30	20	10
	Broadcast CH power (dBm)	30	23	–4
Packet-switched data	Data rate (Kbps)	144	384	2,048
	BS max power per traffic channel (dBm)	30	20	10
	Broadcast CH power (dBm)	30	28	7
	Packet block length (ms)	5	1.25	0.625

Capacity Measure

The capacity measures for circuit-switched services differed from those for packet-switched services [4]. The capacity for the circuit-switched services was defined as the number of users offered in a cell while 98% of the users (C_1) were not blocked when accessing the system; (C_2) had communication of sufficient quality more than 95% of the session time (the quality threshold was defined as a BER of 10^{-3} for voice and 10^{-6} for data); and (C_3) were not dropped. Voice users were dropped if the BER was greater than 10^{-3} for 5 seconds, and data users of services at 2,048, 384, and 144 Kbps, respectively, were dropped if the BER was greater than 10^{-6} for 5, 26, and 69 seconds.

Assuming one CDMA carrier with a 5-MHz bandwidth, the number of users for which these conditions were satisfied was found by computer simulation.

The capacity for the packet-switched services, however, was defined as the total throughput—that is, the sum of throughputs due to all users in a cell—while 98% of the users (P_1) were not blocked; and (P_2) maintained a throughput greater than 10% of maximum transmission rate.

The time spent waiting for ACK/NACK packets was not included in the calculation of the throughput. The use of different capacity measures for circuit-switched and packet-switched services makes it hard to compare capacities of the circuit-switched services to those of the packet-switched services. We therefore used a unified measure in which capacity is specified in units of kilobits per megahertz per cell (Kb/MHz/cell). For circuit-switched services, this measure of capacity is obtained by multiplying the data rate by the number of allowable users per cell and then dividing by the CDMA carrier bandwidth (5 MHz). For a packet-switched service, however, this measure of capacity is obtained by simply dividing the total throughput by 5 MHz.

2.1.2 Downlink Capacities of WCDMA

The capacity results are listed in Table 2.2, where uplink capacities are indicated by values in brackets. We used the required E_b/N_0 and uplink capacities shown in [4]. The items marked by a plus sign denote the case of orthogonal code limitation. The number of orthogonal codes is 128 for the circuit-switched voice at 8 Kbps, 4 for the circuit-switched data at 384 Kbps, and 1 for the packet-switched data at 2,048 Kbps.

The values listed in Table 2.2 show that, for the same type of service, the capacity in the indoor pedestrian environment is lower than that in the outdoor pedestrian environment, and the capacity is lowest in the outdoor vehicular environment. The relatively high capacity obtained in the outdoor pedestrian environment is due to both intercell and intracell interference being low there. The low intercell interference is a result of the presence of corner roads in propagation paths. The low intracell interference is a result of the low delay spread characteristics assumed for this environment model.

In many cases the downlink capacities are greater than the uplink ones. The high capacities of the downlink are due to the orthogonal transmission of the downlink signals from a single BS. In fact, according to Table 2.2, when the code limitation is given, the uplink capacity is greater than the downlink capacity. The capacity restraint due to the code limitation is more serious as the processing gain is smaller. We can understand this trend from the fact that deviation of traffic load distribution relative to its average becomes higher because the processing gain becomes smaller.

As shown so far, the orthogonal code limitation affects CDMA performance: The fewer the orthogonal codes, the smaller the capacity. This code limitation problem is also discussed extensively in [7], which takes a theoretical approach.

Two general approaches are used to cope with the code limitation: the call-blocking approach and the multiple-scrambling-code approach. In call blocking, the arrival of a new call is blocked if the number of MSs connected

Table 2.2
Downlink Capacities of WCDMA

		Required E_b/N_0 (dB)	Capacity in Users/Cell	Capacity in Kbps/MHz/Cell
Circuit-switched voice	Outdoor vehicular (8 Kbps)	8.8 (6.8)[a]	73.8 (81.0) 68.2[b]	59.1 (65.0) 54.6[b]
	Outdoor pedestrian (8 Kbps)	6.8 (4.8)	173 (109) 128[b]	138 (87.0) 103[b]
	Indoor pedestrian (8 Kbps)	6.7 (4.8)	152 (117) 114[b]	122 (93.0) 91.0[b]
Circuit-switched data	Outdoor vehicular (144 Kbps)	2.5 (3.1)	6.9 (7.3)	198 (208)
	Outdoor pedestrian (384 Kbps)	1.1 (1.3)	4.8 (3.5) 1.55[b]	372 (269) 119[b]
	Indoor pedestrian (2,048 Kbps)	1.6 (1.8)	0.89 (0.43) 0.115[b]	365 (176) 47.1[b]
Packet-switched data	Outdoor vehicular (144 Kbps)	2.9 (3.0)	N/A	314 (202)
	Outdoor pedestrian (384 Kbps)	0.1 (0.4)	N/A	897 (449)
	Indoor pedestrian (2,048 Kbps)	0.1 (0.6)	N/A	358 (273)

[a]Items in parentheses denote the uplink cases.
[b]The orthogonal code limitation cases.

exceeds the number of orthogonal codes available. This approach sacrifices some of the soft capacity feature of CDMA because a call will be rejected in case of orthogonal code shortage even though the interference has not reached the specified limit. However, the multiple-scrambling-code approach that was used in the aforementioned capacity results for the code limitation cases fully utilizes the soft capacity feature of CDMA; if the orthogonal code is short in a cell, then a new orthogonal code set is made appending a new scrambling code. However, we can no longer expect to avoid intracell interference between signals assigned different scrambling codes, and therefore the assignment of the orthogonal codes to the respective signals is an important study item if we continue to consider the multiple-scrambling-code approach. An optimum orthogonal code allocation in case of the multiple-scrambling-code approach is discussed in Section 2.2 [8].

2.1.3 Impact of SHO Window Threshold on Downlink Capacity

To investigate the impact of the SHO window threshold on downlink capacity, we used a more detailed simulation scenario: We took into account the 250-ms handover processing delay (it takes 250 ms to complete the link connection process). In this case, we assumed that the MSs moved at 60 km/hr, that the active set size was 3, and that the path intensity ratio was 0.55:0.45. Three kinds of handover situations were investigated: (1) real hard handover, (2) ideal hard handover, and (3) real SHO. We take into account the handover processing delays in the real hard handovers and the real SHOs, but not in the ideal hard handover. Normalized capacity is plotted against the SHO window threshold in Figure 2.4, where each capacity curve is normalized by the one for the real hard handover.

The capacity is highest with ideal hard handover, lowest with real hard handover, and intermediate with real SHO. Clearly, the low capacity obtained with a real hard handover is due to the handover processing delay [9]: In keeping its connection to the nonminimum path-loss BSs, the MS raises its power because the new minimum path-loss BS is behind in being updated due to the processing delay of handover. SHO can mask the processing delay because candidate BSs, which are to be minimum path-loss BSs in the future, can be included in the active set beforehand.

From Figure 2.4, local peak capacity is found in the case of real SHO. This is due to the tradeoff of two conflicting factors with respect to the

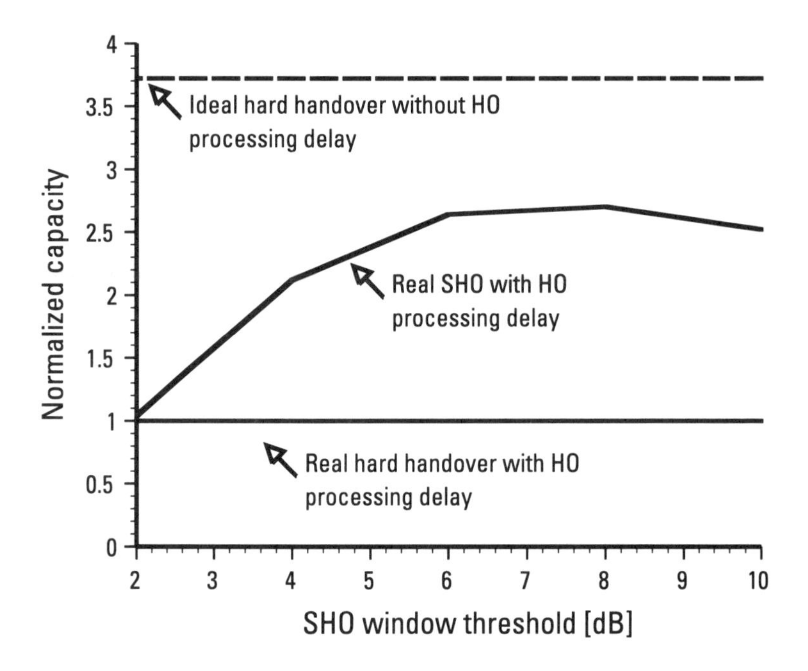

Figure 2.4 Capacity versus SHO window threshold.

SHO window threshold. The first factor is that, in the case of insufficient SHO window threshold, the transmission power of MSs is forced to be high because the MSs cannot connect to minimum path-loss BSs due to the handover processing delay. The second factor is that, in the case of excess SHO window threshold (this means that many BSs are contained in the active set), even though a minimum path-loss BS can likely be included in the active set, the number of downlink branches is increased, whereby downlink interference increases. The real SHO cannot provide as much capacity as ideal hard handover can because it causes more interference.

The results disclosed by Figure 2.4 remind us of one important thing: Hard handover in downlinks is beneficial even though it provides no site diversity gain because, unlike SHO, it never increases downlink interference. It is worth considering a new paradigm in which we combine the advantage of hard handover, which never increases downlink interference, and the advantage of SHO, which is its ability to mask the processing delay. This issue is further discussed in Section 2.3, which describes a new form of transmission power control in the SHO mode [10].

2.2 Optimum Orthogonal Code Allocation in CDMA Downlinks

In cellular CDMA systems, downlink signals are spread by orthogonal spreading codes in order to minimize the interference between the signals [6]. As explained in Section 2.1.2, however, because the number of orthogonal codes is limited, the downlink capacity is also limited once the number of MSs connected exceeds the number of orthogonal codes available. The impact of code limitation depends on such factors as the data transmission rate and the FEC coding rate and is especially significant when SHO is used. This is because SHO accelerates the consumption of the orthogonal codes in accordance with the number of base stations connected to an MS.

The number of downlink channels can be increased by enabling multiple scrambling codes to be allocated to a single BS. But a downlink signal is then subject to strong interference from the other signals assigned different scrambling codes. In this section we discuss, with the help of a mathematical tool based on the variation method, an optimum code allocation maximizing the average SIR measured at MSs within a cell.

Average SIR

The term *code allocation* throughout this section means an allocation of codes given as products of the multiplication of a basic orthogonal code set and multiple scrambling codes. We denote as S the number of basic orthogonal codes and denote as N the number of orthogonal code sets provided for a cell. The transmission power for downlink signals is fixed, and the orthogonal code occupancy ratio k_i is the ratio of the number of MSs assigned the ith orthogonal code set to the number n of all MSs within a cell, where the sum of $k_1, k_2, \ldots,$ and k_N is 1. Defining the code occupancy ratio vector of $\mathbf{k} = (k_1, k_2, \ldots, k_N)$, we can write the following equation for the average SIR measured at MSs within a cell:

$$\overline{\Gamma}(\mathbf{k}) = \sum_{i=1}^{N} k_i \frac{P_g}{\epsilon k_i n + (1 - k_i)n} = \frac{P_g}{n} \sum_{i=1}^{N} \frac{1}{\epsilon + 1/k_i - 1} \tag{2.1}$$

where P_g and ϵ denote the processing gain and the interference figure. The value of ϵ ranges from 0 to 1, and $\epsilon = 0$ denotes no multipath distortion in the downlink channel. In (2.1), $\epsilon = 1$ is assumed between downlink signals assigned different scrambling codes, hence the signals with different scrambling codes interfere completely with each other. We also assume in (2.1) a single-cell environment, that is, we exclude the intercell interference in measurement of the average SIR.

Extreme Value of Average SIR

When $\mathbf{k}^* = (k_1^*, k_2^*, \ldots, k_N^*)$ is the code occupancy ratio vector giving an extreme value of $\overline{\Gamma}(\mathbf{k})$ and when $\mathbf{d} = (d_1, d_2, \ldots, d_N)$ is the variation vector, the equation

$$\left. \frac{d\overline{\Gamma}(\mathbf{k}^* + \alpha \mathbf{d})}{d\alpha} \right|_{\alpha = 0} = 0 \tag{2.2}$$

is satisfied for an arbitrary variation vector $\mathbf{d}$ because $\overline{\Gamma}(\mathbf{k})$ takes an extreme value at $\mathbf{k} = \mathbf{k}^*$. From (2.1) and (2.2), we can derive

$$\frac{P_g}{n} \sum_{i=2}^{N} d_i \left[-1/\left\{ (\epsilon - 1)\left(1 - \sum_{j=2}^{N} k_j^*\right) + 1 \right\}^2 + 1/\{(\epsilon - 1)k_i^* + 1\}^2 \right] = 0 \tag{2.3}$$

which is an identical equation with respect to d_i because the equation must be satisfied for an arbitrary variation vector $\mathbf{d}$, and hence all coefficients of d_i must be 0. Eventually, simultaneous equations with respect to k_i are given as follows:

$$-\frac{1}{\left\{(\epsilon - 1)\left(1 - \sum_{j=2}^{N} k_j^*\right) + 1\right\}^2} + \frac{1}{\{(\epsilon - 1) k_i^* + 1\}^2} = 0 \qquad (2.4)$$

Solving simultaneous equations led by (2.4), we have

$$k_i^* = 1 - \sum_{j=2}^{N} k_j^* = k_1^* \qquad \because i = 2 \sim N \qquad (2.5)$$

which leads to a unique solution to the simultaneous equations of interest, $\mathbf{k}^* = (1/N, 1/N, \ldots, 1/N)$. To determine which $\mathbf{k}^* = (1/N, 1/N, \ldots, 1/N)$ gives a minimum or maximum SIR, we use Figure 2.5, which is a three-dimensional plot of the average SIR against k_1 and k_2 in the case for which $N = 3$ is assumed and ϵ is arbitrarily given. The assumed case gives $\mathbf{k}^* = (k_1, k_2, k_3) = (1/3, 1/3, 1/3)$. From Figure 2.5, we can see that the point at $(k_2, k_3) = (1/3, 1/3)$ shows the lowest SIR. The sequence developed so far tells us that the average SIR is smallest when each orthogonal code set is equally used.

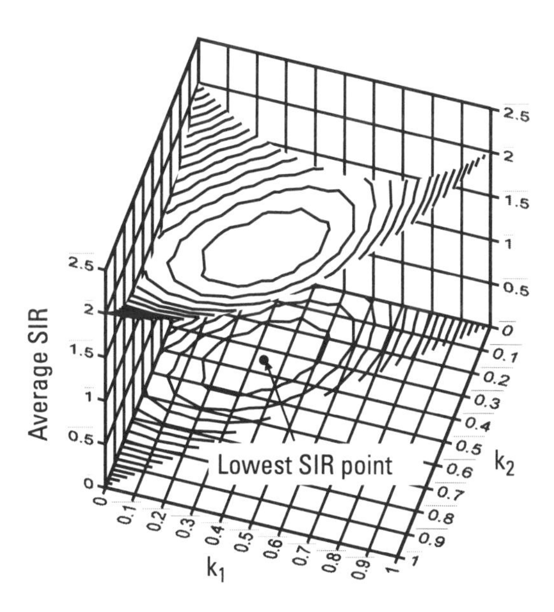

Figure 2.5 The average SIR for $N = 3$.

Optimum Code Allocation

Figure 2.5 also shows that the average SIR $\overline{\Gamma}(\mathbf{k}^* + \alpha\mathbf{d})$ becomes larger as the absolute value of α increases. There is, however, a limit to the value of α because all elements of the vector $\mathbf{k}^* + \alpha\mathbf{d}$ have to be between 0 and S/n ($n \geq S$). Defining the limit of α as α_l, we can write the code occupancy vector $\mathbf{k}^* + \alpha_l\mathbf{d}$ at α_l as

$$\mathbf{k}^* + \alpha_l\mathbf{d} = \left(\frac{S}{n}, \frac{S}{n}, \ldots, \frac{S}{n}, r_1, r_2, \ldots, r_R, 0, 0, \ldots, 0\right) = \mathbf{k_r} \tag{2.6}$$

where $0 < r_1 < r_2 < \ldots < r_R < S/n$. We call $\mathbf{k_r}$ a code occupancy bound vector, and the code occupancy vector maximizing $\overline{\Gamma}$ is one of these code occupancy bound vectors.

We generate a new code occupancy vector $\mathbf{k'_r}$ by replacing the element of r_R in (2.6) with S/n and compensating for the shortage of the code occupancy ratio due to this replacement by arbitrarily decreasing respective $r_1, r_2, \ldots$ and r_{R-1} so as to keep the sum of all the elements of (2.6) at 1, and then consider $\overline{\Gamma}(\mathbf{k'_r}) - \overline{\Gamma}(\mathbf{k_r})$. This difference can be written as

$$\begin{aligned}
&\Gamma(\mathbf{k'_r}) - \Gamma(\mathbf{k_r}) \\
&= \frac{P_g}{n}\left\{\frac{M+1}{\epsilon + \frac{n}{S} - 1} + \sum_{i=1}^{R-1}\frac{1}{\epsilon + \frac{1}{r'_i} - 1}\right\} \\
&\quad - \frac{P_g}{n}\left\{\frac{M}{\epsilon + \frac{n}{S} - 1} + \sum_{i=1}^{R}\frac{1}{\epsilon + \frac{1}{r_i} - 1}\right\} \\
&= \frac{P_g}{n}\left\{\frac{1}{\epsilon + \frac{n}{S} - 1} + \sum_{i=1}^{R-1}\frac{1}{\epsilon + \frac{1}{r'_i} - 1}\right\} \\
&\quad - \frac{P_g}{n}\left\{\frac{1}{\epsilon + \frac{1}{r_R} - 1} + \sum_{i=1}^{R-1}\frac{1}{\epsilon + \frac{1}{r_i} - 1}\right\}
\end{aligned} \tag{2.7}$$

$$= \frac{P_g}{n} \frac{\frac{S}{n} - r_R}{\left\{(\epsilon - 1)\frac{S}{n} + 1\right\}\{(\epsilon - 1)r_R + 1\}}$$

$$+ \frac{P_g}{n} \sum_{i=1}^{R-1} \frac{r_i' - r_i}{\{(\epsilon - 1)r_i' + 1\}\{(\epsilon - 1)r_i + 1\}}$$

$$= \frac{P_g}{n} \frac{1}{\left\{(\epsilon - 1)\frac{S}{n} + 1\right\}\{(\epsilon - 1)r_R + 1\}}$$

$$\left\{\left(\frac{S}{n} - r_R\right) + \sum_{i=1}^{R-1} \frac{\left\{(\epsilon - 1)\frac{S}{n} + 1\right\}\{(\epsilon - 1)r_R + 1\}}{\{(\epsilon - 1)r_i' + 1\}\{(\epsilon - 1)r_i + 1\}}(r_i' - r_i)\right\}$$

$$= \frac{P_g}{n} \frac{1}{Q_1 Q_2}\left\{\left(\frac{S}{n} - r_R\right) + \sum_{i=1}^{R-1} \frac{Q_1 Q_2}{Q_3 Q_4}(r_i' - r_i)\right\}$$

where M denotes the number of S/n elements in (2.6) and where the terms Q_1, Q_2, Q_3, and Q_4, respectively, are defined as follows:

$$Q_1 = (\epsilon - 1)\frac{S}{n} + 1 \tag{2.8}$$

$$Q_2 = (\epsilon - 1)r_R + 1 \tag{2.9}$$

$$Q_3 = (\epsilon - 1)r_i' + 1 \tag{2.10}$$

$$Q_4 = (\epsilon - 1)r_i + 1 \tag{2.11}$$

Because $r_R > r_i$ for any r_i with $i = 1, 2, \ldots, R - 1$ is assumed,

$$1 > Q_4 > Q_2 > Q_1 > 0 \quad \text{and} \quad 1 > Q_3 > Q_1 > 0 \tag{2.12}$$

Thus,

$$0 < \frac{Q_1 Q_2}{Q_3 Q_4} < 1 \tag{2.13}$$

Using the relation

$$\sum_{i=1}^{R-1} (r_i' - r_i) = r_R - \frac{S}{n} = -A \tag{2.14}$$

where we introduce A (≥ 0) for the sake of expression simplicity, we can further modify (2.7) to

$$\Gamma(\mathbf{k}_\mathbf{r}') - \Gamma(\mathbf{k}_\mathbf{r}) = \frac{P_g}{n} \frac{1}{Q_1 Q_2} \{A - BA\} > 0 \tag{2.15}$$

This equation also introduces B, whose value is between 0 and 1. As seen in (2.15), $\overline{\Gamma}(\mathbf{k}_\mathbf{r}')$ is always greater than $\overline{\Gamma}(\mathbf{k}_\mathbf{r})$. This tells us that as the number of fully occupied orthogonal code sets increases, the average SIR can be larger.

In contrast, we can generate a new code usage vector, $\mathbf{k}_\mathbf{r}''$ by replacing the element of r_1 in (2.6) with 0 and compensating for the excess of code occupancy ratio due to this replacement by arbitrarily increasing respective $r_2, r_3, \ldots, r_R$, and then investigate $\overline{\Gamma}(\mathbf{k}_\mathbf{r}'') \cdot \overline{\Gamma}(\mathbf{k}_\mathbf{r})$ in the same way described above in the development of (2.7) through (2.15). Eventually, we find that $\overline{\Gamma}(\mathbf{k}_\mathbf{r}'')$ is always larger than $\overline{\Gamma}(\mathbf{k}_\mathbf{r})$. This tells us that as the number of empty orthogonal code sets increases, and hence as the use of scrambling codes is reduced, the average SIR can be larger.

Finally, an optimum code allocation can be described as follows: Optimum code allocation is an allocation scheme that maximizes the number of code sets fully occupied and minimizes the use of scrambling codes.

Evaluation of Optimum Code Allocation

The average SIRs for the optimum code allocation were compared with those for the worst code allocation in which every orthogonal code set enabled by multiple scrambling codes is equally occupied. When $N = 5$, $S = 128$, $n = 270$, $P_g = 512$, and $\epsilon = 0.5$, the average SIR was found to be 3.24 dB for the worst allocation and 3.90 dB for the optimum allocation. That is, the optimum code allocation improved the average SIR by about 0.7 dB. And when $N = 2$, $S = 128$, $n = 150$, $P_g = 512$, and $\epsilon = 0.5$, the average SIRs given by the optimum and worst code allocations, respectively, were 6.58 and 7.50 dB; the average SIR was about 1 dB better with the optimum allocation.

2.3 Site Selection Diversity TPC

TPC is key to a high-capacity and reliable CDMA cellular system [1, 11, 12]. Uplink TPC is required to combat the near-far problem because the BSs receive uplink signals that are asynchronous and nonorthogonal. Downlink TPC, however, is mainly used to reduce the transmission power used by BSs. In both uplinks and downlinks, TPC reduces intercell interference and thus increases system capacity [1, 13, 14]. TPC enables the BS in a cell to maintain its output power at an adequate level such that interference with other cells is minimized. This feature of TPC greatly enhances system capacity in conjunction with the minimum path-loss BS connection achieved by SHO.

We can find many studies about the downlink TPC [11, 15, 16] and the downlink performance of systems using TPC [14], but the effect of SHO has not been considered in this context. The feedback in the downlink TPC used in the first commercial cellular CDMA system capable of working in SHO mode [5] is slower than the feedback in the uplink TPC. Compensation of shadowing has thus been considered, but fast fading has not. Such an implementation, however, cannot optimize the downlink as well as one that compensates fast fading, and in some cases the result will be that the downlink capacity is less than the uplink capacity [17, 18]. WCDMA provides an efficient downlink by using a fast closed-loop TPC that modifies the BS output power in a cycle that is comparable to the cycle for the uplink even though the additional signaling required for the TPC reduces the uplink capacity [19]. In this section, we focus on such downlink TPCs, that is, on those that can adjust the transmission power rapidly.

In the straightforward implementation of downlink TPC with fast power adjustment in the SHO mode, each active BS in the same active set modifies its output power equally in accordance with TPC commands that are sent by the MS [19]. The TPC commands transmitted by an MS request a decrease in BS output power if the quality of reception at the MS is better than a target quality and request an increase if the quality is worse than the target quality. Each active BS then modifies its output power in steps of fixed size, ΔP, in decibels. Although such a TPC is going to assign the same transmission power to all the active BSs while maintaining the forward link in the target quality, the actual output powers of the active BSs will vary greatly because of the initial settings for transmission power and because of errors in TPC command reception. Throughout this chapter we refer to this scheme as conventional TPC.

SHO can increase a system's uplink capacity by a factor greater than 2 [20], but three major problems are involved in implementing conventional TPC in the downlink during SHO:

1. The multiple-site transmission required for SHO increases the interference affecting other radio links and thus limits the downlink capacity [21–23]. This fact has also been addressed in Section 2.1.3.
2. SHO reduces the path capturing efficiency of a RAKE receiver. The number of RAKE fingers is limited in the implementation of MSs; this limitation prevents the RAKE receiver from capturing all of the resolved paths because SHO increases the number of paths to be necessarily resolved by an MS. The missing paths interfere with the paths captured at the receiver. Unlike the SHO of a downlink, however, the SHO of an uplink never increases the number of paths to be resolved.
3. The transmission powers of active BSs will become imbalanced because of errors in receiving TPC commands. Such errors cannot be avoided and result in the transmission powers of active BSs becoming imbalanced even though their initial settings were the same. The biggest problem here is that these errors may cause the transmission powers of some of the active BSs to become excessively high and thus increase the interference.

A new closed-loop downlink TPC scheme for the SHO mode has been developed in order to overcome the first problem [24, 25]. In this scheme, an MS sends to the active BSs commands that are used to select the BS that provides the minimum path loss for the MS, and the other BSs in the same active set reduce their transmission power to a minimum level. The BS selection signaling is transmitted to the active BSs periodically in a cycle identical to one used for signaling that causes the MSs to reduce or increase transmission power. The signals for BS selection and BS output power control are combined into a composite control signal that requires no overhead to deliver the primary BS information. The selected BS is power controlled so that the signal maintains a constant quality at the MS. Multiple-site transmission can thus be avoided and the total transmission power for all active BSs minimized.

The capacity of this downlink TPC method was compared with that of conventional TPC, and the results indicated some promise in terms of improved capacity. Intensive link level analysis also confirmed the benefits of such a BS selection [23]. But although the composite control signal is beneficial in terms of signaling overhead, the complexity of the messages increases the likelihood of failure in decoding them and this reduces the benefits of the BS selection. Earlier evaluations of the improved TPC scheme [24, 25] took into account neither errors in the decoding of the BS selection message nor the scheme's effects on problems 2 and 3 listed above.

This section describes SSDT power control, an advanced form of downlink TPC that can solve the problems incurred when conventional TPC is used. The principle is identical to that of the TPC proposed in [24, 25]: The best BS is chosen adaptively, and the output power of the other BSs is reduced. The distinctive difference between SSDT power control and the TPC proposed in [24, 25] is that it uses a signaling method that improves accuracy of selecting a minimum path-loss BS while achieving a good degree of adaptation to the WCDMA standard [19, 26].

2.3.1 Overview

SSDT power control provides site selection transmission diversity during the SHO mode instead of the full-site transmission diversity used in conventional TPC. The principles of SSDT power control and conventional TPC are compared in Figure 2.6. In SSDT power control an MS periodically chooses as the transmitting site one of the active BSs having minimum path loss[1] to the MS and then sends the identification number of this BS to all the active BSs so that the output power of the other active BSs can be reduced. In the rest of this chapter, the BS with the minimum path loss is referred to as the *primary BS*.[2]

The advantages of SSDT power control can be summarized as follows:

1. Interference is reduced because one of the BSs within an active set provides a downlink signal with adequate power to the connected MS and the output power of the other BSs within the same active set is reduced.
2. In conventional TPC, multiple BSs transmit the same downlink signal with adequate power, so an MS receiver needs to have enough RAKE fingers to collect all the paths due to the active BSs. In SSDT power control, however, only one active BS becomes a serving site,

1. "Minimum path loss" here refers to the minimum downlink path loss. The term *minimum path loss* is used throughout this chapter for a path loss including fast fading. Because the fading fluctuation differs between the downlink and uplink paths, the minimum path-loss BSs for the two links are not necessarily the same.
2. Note that the term *primary BS* here is short for *primary downlink BS*.

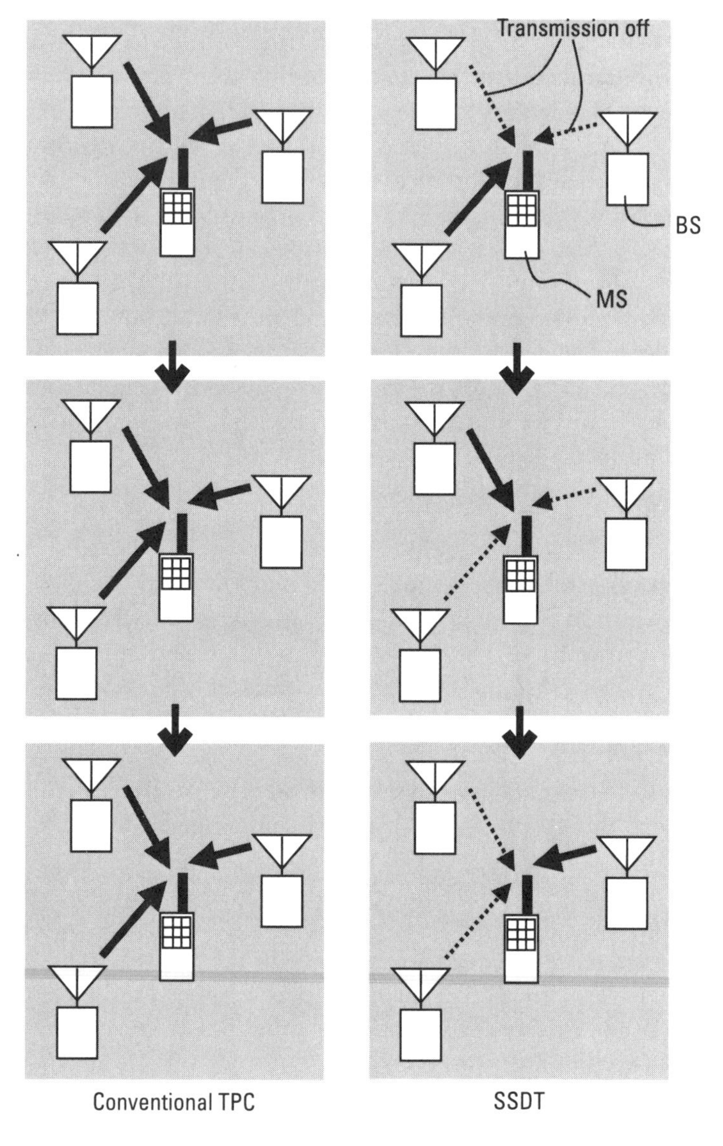

Figure 2.6 Principles of SSDT power control and conventional TPC. (*Source:* [10], © 2000, IEEE. Reprinted with permission.)

so the number of paths that should be collected by the RAKE receiver is smaller than that required for conventional TPC. This feature makes the path-capturing efficiency of SSDT power control better, for a given number of RAKE fingers, than that of conventional TPC.

3. The power imbalance problem due to TPC command reception error never happens because only one BS transmits the downlink signal during SHO.

Fast Site Selection and Minimum Path Loss

SSDT power control was developed in order to cope with the fast changes of the primary BS due to fast fading. This feature of SSDT power control is discriminated from that of the hard handover. An MS must therefore send site selection messages to the active BSs as frequently as possible. The required cycle of site selection depends mainly on a fading pitch, namely, the speed of the MS and the carrier frequency.

One way to signal the site selection is by inserting the corresponding command in the higher layer signaling messages, for example, the Layer 2 or Layer 3 signaling messages WCDMA uses to control the system operations. Although the higher layer signaling messages can be accurately delivered, this signaling method cannot be used for SSDT power control because the length of these messages (nearly 100 ms) is too great for them to be used for frequent site selection. In addition, the messages are delayed until reaching the active BSs because the message is in practice designed to reach the higher control station first and then transfer the message to the active BSs. So the use of low-layer signaling such as TPC signaling is promising for SSDT power control.

The quality with which low-layer signaling messages are received, however, is not good enough for reliable site selection—especially when they are received at some of the active BSs that do not have the minimum path loss on the uplink. This is due to a feature of uplink TPC during SHO: The BSs that do not have the minimum path loss in the uplink are forced to accept poor reception because the uplink TPC works for the uplink minimum path-loss BS [4]. This problem can be solved by adding some redundancy to the site selection messages, but this redundancy as well as the frequent signaling of it increases the overhead for the uplink control signals. This means that the site selection cycle must be determined by balancing the redundancy against the overhead.

2.3.2 Detailed Operation

In the work described here, we assumed the same channel structures defined in the WCDMA system [26]. Although a number of channels are defined in WCDMA, we are primarily concerned here with the two major channels: the common control channel and the dedicated channel. The common control channel is a downlink channel by which the BS broadcasts system information and pilot signals. The dedicated channel is a channel dedicated to each radio link that connects an MS with a BS. Figure 2.7 shows the frame and slot structures of an uplink dedicated channel. The slot length and the number of slots in one frame are 0.625 ms and 16, and hence the length of the frame is 10 ms. The uplink dedicated channel carries data and control signals, respectively, transmitted as the in-phase and quadrature components of this channel. The control signal contains a pilot, TPC, FBI, and TFCI. The TPC, FBI, and TFCI consist of 1 or 2 bits [26]. One TPC command produced by an MS is sent as one TPC symbol in a slot of the uplink dedicated channel, and thus downlink transmission powers are controlled with a cycle of slot length.

In SSDT power control, each BS within an active set is assigned a temporary *identification number* (ID) and the BS with the minimum path loss to the connected MS is selected as a primary BS. An MS measures the pilot reception level of the common control channel to detect the primary BS and then sends the primary BS ID to all the active BSs within the same active set. All the signal processing required for the common control channel measurement share ones done in the typical path search for the traffic channel demodulation. An MS using SSDT power control does not need to assign RAKE fingers to the signals arriving by paths from nonprimary BSs, but these paths also have to be searched ordinarily in order to prepare for the quick change of primary BS. This is because the path searcher cannot quickly resolve the paths.

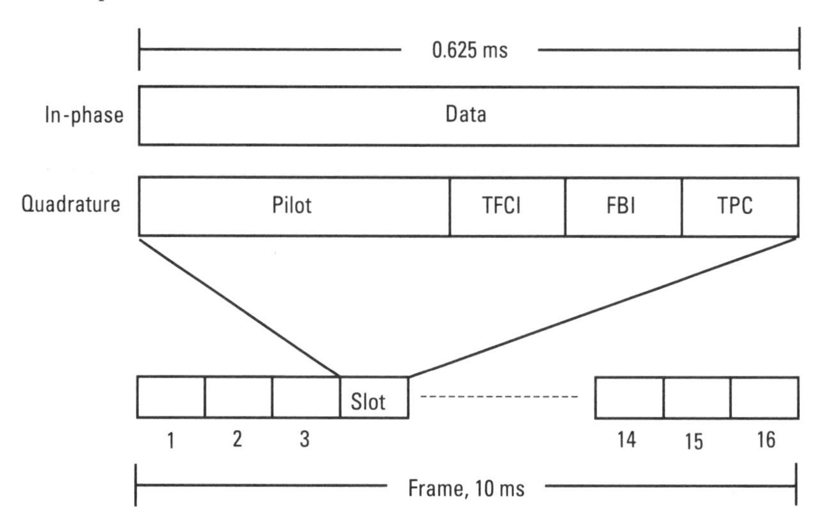

Figure 2.7 Uplink frame and slot structures. (*Source:* [10], © 2000, IEEE. Reprinted with permission.)

To realize an effective RAKE receiver, the number of paths to be searched for a single active BS is greater than the number of RAKE fingers. Later in this chapter, we assume that the path searcher can capture all the paths given by the profile models to be used, and all these paths are gathered in the measurement of pilot reception power. The BS selected as the primary station transmits its signal in the downlink dedicated channel with adequate power and the other BSs' output power is minimized.

A BS manages two transmission power levels: a hidden power P_1 and a real power P_2. A BS maintaining minimum power can know how much power is adequate by referring to P_1 if the BS is chosen as the primary BS. Any BSs equally update P_1 whether each of them is primary or not. The downlink dedicated channels are actually transmitted with power P_2, and powers P_1 and P_2 are updated in accordance with a downlink TPC command sent by the MS (Table 2.3). The downlink TPC command is produced in such a way that if the reception quality of the downlink at an MS is below a predetermined threshold, the MS produces a "power-up" signal to make the active BSs increase their transmission power levels. And if the quality is above the threshold, the MS produces a "power-down" signal to make the BSs decrease their transmission power levels.

In Table 2.3, ΔP and P_{min}, respectively, denote the power control step size and the minimum transmission power level. Note that P_{min} can be set to $-\infty$, and thus the transmission of nonprimary BSs can be switched off. P_1 and P_2 are updated for every TPC event. A BS initially deals with P_1 and then P_2, limiting both to ensure that they neither exceed the maximum level nor fall below the minimum level. Also note that if we set P_2 equal to P_1 for nonprimary BSs, the power update map becomes identical to that used for conventional TPC.

To provide fast and accurate site selection while taking into account the issues described in the previous subsection, we can use a codeword with a long bit length for expressing the ID (i.e., ID of a BS in an active set)

Table 2.3
Updating Method of P_1 and P_2

State of BS	Contents of the Forward Link TPC Signal	P_1 (dBm)	P_2 (dBm)
Nonprimary	Down	$P_1 - \Delta P$	P_{min}
	Up	$P_1 + \Delta P$	P_{min}
Primary	Down	$P_1 - \Delta P$	$= P_1$
	Up	$P_1 + \Delta P$	$= P_1$

and FBI field in the uplink dedicated channel for delivering codewords. If the number of codeword bits is greater than the number of FBI bits per slot, the codeword is divided into a number of parts and each part is distributed to each FBI field of continuous slots. As a result, the primary ID can be delivered every several slots and the number of slots affected by a single primary ID depends on the length of codeword and the number of FBI bits per slot. This cycle of site selection is shorter than that obtained using the higher layer signaling method. Table 2.4 shows an example of a set of codewords for which a codeword length of 16 bits and an active set size of 3 are assumed. This set gives a minimum Hamming distance of 8.

In WCDMA the FBI field is also used to deliver the antenna weight coefficient for the downlink antenna transmission diversities [19, 27]. If neither SSDT nor the transmission diversities are used, the FBI field need not be set. Because the FBI field is made by the replacement of the pilot bits, its use degrades the uplink performance. It has been reported, however, that the degradation estimated from simulations is only 0.1–0.2 dB [28], which is small enough because SSDT can increase the downlink capacity by more than 0.2 dB in a pedestrian environment. The capacity improvements provided by SSDT and by the combination of SSDT and transmission diversities are demonstrated in Section 2.3.5.

Before SSDT power control is started, or if the active set changes during operation, the ID of each active BS is assigned by the network and then the assignment result is sent to an MS and the associated active BSs. The ID assignment process is carried out with the help of the network protocols defined in the interface between a radio network controller and a BS and in the interface between a radio network controller and an MS. An MS detects a primary BS by monitoring the pilot reception levels of the common control channels transmitted by all active BSs. The BS with the highest level is chosen as the primary BS and then its ID codeword is periodically transmitted to the active BSs by the MS. A BS recognizes its state as nonprimary if the following two conditions are satisfied:

Table 2.4
Examples of ID Codewords

BS Label	Codeword
BS-A	0000000000000000
BS-B	1111111111111111
BS-C	0000000011111111

1. The reception quality of downlink exceeds a threshold level Q_{th}.
2. A received ID does not match its own ID.

Condition 1 is used to ensure a reliable site selection so that all the active BSs do not reduce their output power simultaneously because of ID reception error.

2.3.3 Simulation Conditions for Evaluation of Performance

SSDT power control and conventional TPC were compared according to the evaluation guidelines for IMT-2000 systems by using a system-level simulator [2]. Although several environment models are defined in this reference, in the work described here we used the same vehicular model described in Section 2.1.1 (19 cells in two concentric rings; 57 sectors). The shadowing was modeled as a log-normal random variable with a standard deviation of 10 dB, and the shadowing's autocorrelation function with respect to the distance was given as an exponential function with the decorrelation distance of 20m [29].

The simulator we developed for this evaluation could dynamically emulate the system operation with slot-time resolution, allowing closed-loop TPCs to be evaluated in detail—for example, including delays in delivery and errors in the reception of control bits. To limit the number of iterations required for such a high time resolution, trials with a very short period of observation were repeated and the data from these trials were processed statistically. In each trial a given number of MSs were distributed uniformly and call termination and call dropping did not occur (i.e., all MSs were connected continuously during the trial). The active set management—that is, the processes for adding and deleting BSs—was based on the reception level of the pilot signal broadcast by each BS via the common control channel. A new BS was added to the active set when the measured pilot signal strength as normalized by the maximum one was greater than or equal to −5 dB. If the normalized pilot signal strength for a connected BS fell below −7 dB, the connection was terminated. The difference between −5 and −7 dB is the so-called hysteresis margin that is used to overcome the ping-pong effect.

We assumed that the maximum number of active BSs within an active set is 3. It is obvious that SSDT power control will be activated whenever there is more than one BS in the active set. We took into account the BS processing delay for addition and deletion, which is thought to have a major impact on system capacity [9]. The time needed to complete the link connection or deletion process was 250 ms.

Other major parameters and their settings are listed in Table 2.5. The reception error probability of TPC and FBI bits in an uplink dedicated channel was assumed to be 5% at the uplink minimum path-loss BS. An error probability greater than 5% is assumed at the other BSs in the same

Table 2.5
Simulation Parameters and Settings

Parameter	**Setting**
Service type	Circuit switching data, 100% activity
Slot duration	0.625 ms
Frame duration	10 ms
Number of slots per frame	16
Information bit rate	384 Kbps
Spread bandwidth	5 MHz
Carrier frequency	2 GHz
Processing gain	10.7
Cell radius	1.73 km
Downlink reception diversity (MS reception diversity)	Yes/no
Target SINR (signal power to interference + noise power ratio)	0 dB (with downlink reception diversity), 3 dB (without downlink reception diversity)
ΔP, transmission power control step size	1 dB
P_{min}, transmission power of nonprimary BS	0 ($-\infty$ dBm)
Control range of transmission power	10–30 dBm
Transmission power of common control channel	20 dBm
FBI and TPC bit reception error probability at BS	5% in uplink primary BS, probability greater than 5% is set in uplink nonprimary BS according to path loss difference between the primary and nonprimary BSs
Q_{th}, threshold level for detecting nonprimary state	10 dB below target SINR in uplink channel
Path and fading model	Independent two- or six-path Rayleigh fading model; respective path power ratios [23] are 0.60:0.40 and 0.506:0.252:0.126:0.063:0.028:0.025
RAKE finger assignment	Highest *n* paths are assigned to RAKE fingers; *n* denotes the number of RAKE fingers

active set according to path-loss difference from that of the uplink minimum path loss BS. This assumption would be feasible considering the behavior of uplink TPC. The transmission power of nonprimary BSs, P_{min}, was set to 0. The step size ΔP was assumed to be 1 dB for both conventional TPC and SSDT power control using TPC. We assumed that in the SSDT mode the frame length delay from the primary ID reception to the primary state update was 10 ms. Unless otherwise noted, the parameters listed in Table 2.5 are used in the following description.

The capacity is defined as the system load at an interference probability of 5% [4]. System load is calculated as follows:

System load (Kbps/MHz/sector = Number of offered MSs per sector
× Information bit rate (Kbps)/Spread bandwidth (MHz)

Interference probability was defined as the ratio of the number of "interfered MSs" to the total number of MSs offered in the system. An interfered MS was defined as one that could not maintain the target SINR. To determine whether the target SINR could be maintained at an MS, a distribution of SINR is taken for each offered MS and then an MS is detected as an interfered user if the probability of SINR being less than a tolerable SINR is greater than a predetermined probability threshold. The tolerable SINR was set at 0.5–1.5 dB below the target SINR, depending on MS speed. The probability threshold, however, was set to 0.4 regardless of MS speed. The tolerable SINR was determined for the respective speed of the MS assuming a single-cell environment. According to this criterion, both the mean and the spread of the SINR are taken into account in the evaluation of interference. This allows performances to be compared more precisely than when only the mean value of the SINR is used.

2.3.4 Codeword Set

The codeword sets of codewords used in the simulations are listed in Table 2.6. Because the maximum active set size was assumed to be 3, we consider the minimum length of a codeword to be 2 bits. For increased codeword redundancy we also consider codewords with lengths up to 16 bits. It is clear that the shorter the codeword, the shorter the site selection cycle. But short codewords are more likely to result in site selection error than long ones. The performance of the codeword sets with respect to Doppler frequency is evaluated in detail later. To minimize the deterioration of the uplink performance, we assumed 1 bit of FBI per slot. This means that 16-, 8-,

Table 2.6
ID Codewords Used in the Evaluation

BS Label	16-Bit Codeword	8-Bit Codeword	4-Bit Codeword	2-Bit Codeword
BS-A	0000000000000000	00000000	0000	00
BS-B	1111111111111111	11111111	1111	11
BS-C	0000000011111111	00001111	0011	01

4-, and 2-bit codewords, respectively, yield 1, 2, 4, and 8 site selections per frame. As shown in Table 2.6, the minimum Hamming distances of codewords were 8 for 16 bits, 4 for 8 bits, 2 for 4 bits, and 1 for 2 bits.

Figure 2.8 shows ID reception error probability versus the reverse link SINR normalized by that at the uplink minimum path-loss BS. The data plotted in Figure 2.8 were obtained assuming stationary MSs and a one-path profile model. Note that as the uplink TPC maintains the target SINR at the uplink minimum path-loss BS, the reception error probabilities of

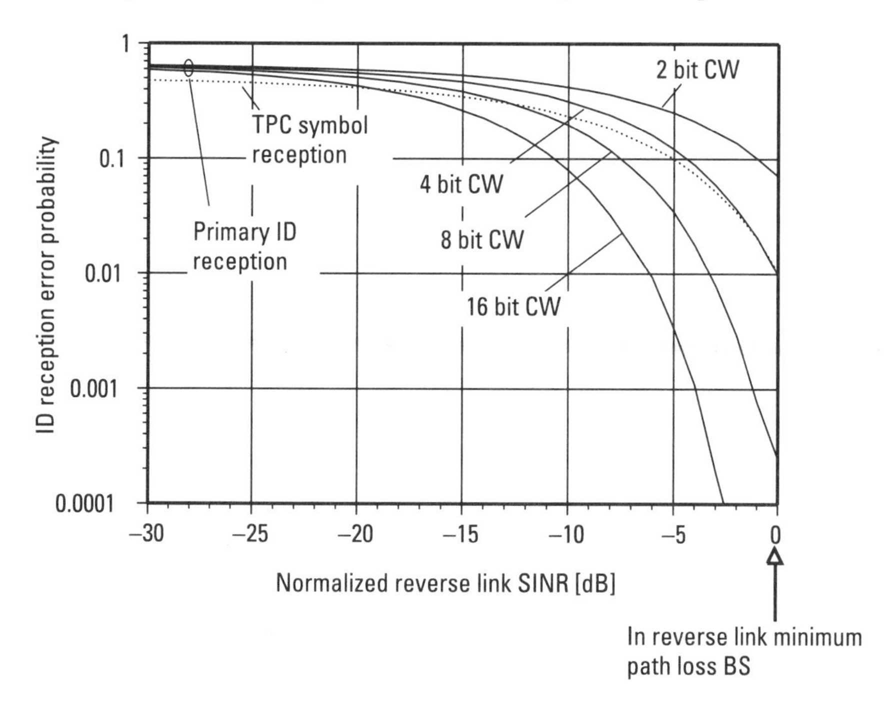

Figure 2.8 ID reception error probability versus normalized uplink SINR. CW: codeword. (*Source:* [10], © 2000, IEEE. Reprinted with permission.)

the uplink dedicated signals at this BS are also maintained at a fixed level. The reception error of the TPC symbol delivered to the uplink is also shown in this figure. The TPC symbol consists of 2 bits, "11" for a power-up message and "00" for a power-down message. As seen in Figure 2.8, the ID reception error probability decreases as the length of the codeword increases.

Figure 2.9 shows the distribution of the normalized uplink SINR under the SHO parameters listed in Table 2.5. The data in Figure 2.9 were also obtained assuming stationary MSs and a one-path profile model. Irrespective of the uplink state, the SINR for about 98% of the radio links was greater than (target SINR: 10 dB). The reception quality threshold Q_{th} for detecting the nonprimary state of an active BS was therefore set 10 dB below the target SINR. For this Q_{th}, the CW sets listed in Table 2.6, and the ID reception error probability shown in Figure 2.8, the probability that all the active BSs reduce their output power simultaneously was less than 10^{-6} in the case of a 16-bit CW. This probability is thought to be small enough because 10^{-6} is much smaller than the tolerable interference probability (5%).

2.3.5 Performance

Figure 2.10 shows an example of the power transition in two BSs in the same active set. We assume here that the initial transmission powers of active

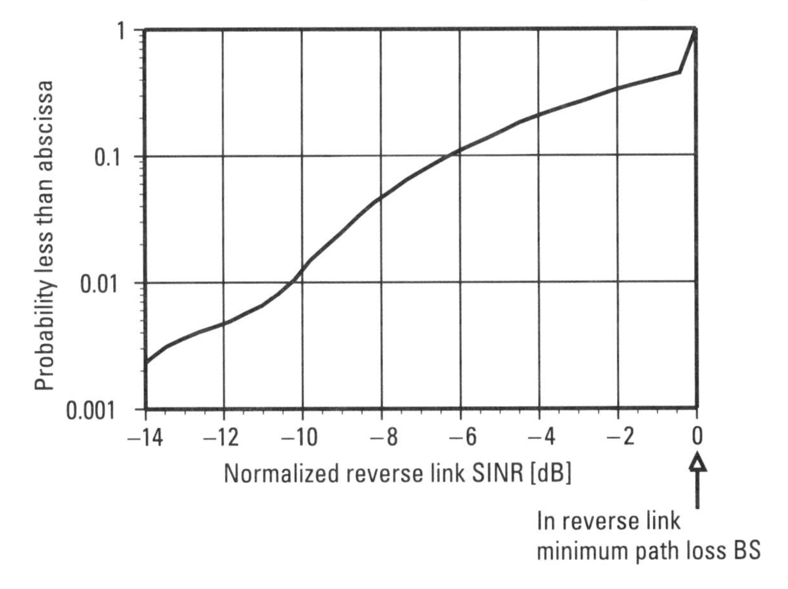

Figure 2.9 Distribution of normalized uplink SINR. (*Source:* [10], © 2000, IEEE. Reprinted with permission.)

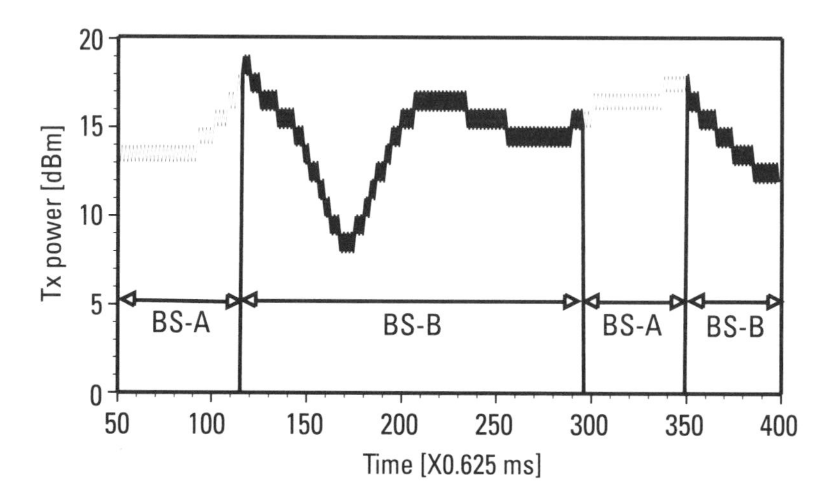

Figure 2.10 Power transition for two BSs in the same SHO situation during SSDT power control. (*Source:* [10], © 2000, IEEE. Reprinted with permission.)

BSs have been set to the same value and that there is no TPC command reception error. Sixteen-bit CWs and a two-path profile model are also assumed. As the figure shows, one of the active BSs transmits a downlink signal with adequate power, and the output power of the other BS is reduced. In addition, the transmission power of the preceding primary BS is smoothly handed over to the subsequent primary one. This is enabled by the management of the hidden power P_1.

The relation between maximum Doppler frequency and the capacity gains of SSDT power control and conventional TPC was compared assuming MS reception diversity, an infinite number of RAKE fingers, and a two-path profile model. The results for cases with 2-bit, 4-bit, and 8-bit codewords are shown in Figure 2.11, which also includes the curves for cases without ID reception errors. Because we assumed a 2-GHz carrier frequency, the maximum Doppler frequencies of 10 and 100 Hz, respectively, correspond to MS speeds of 5.4 and 54 km/hr. The capacity gain provided by SSDT power control is due to the reduction of multiple-site transmission, specifically, to the interference reduction provided by the site selection diversity.

It is readily understood that ID reception error greatly reduces the capacity gain of SSDT power control when the CW has only 2 bits because then the insufficient redundancy of ID messages makes the site selection error large. If the ID can be delivered without error, however, the best performance is obtained with the 2-bit codeword because it offers the shortest site selection cycle. With 4-bit and 8-bit codewords, on the other hand,

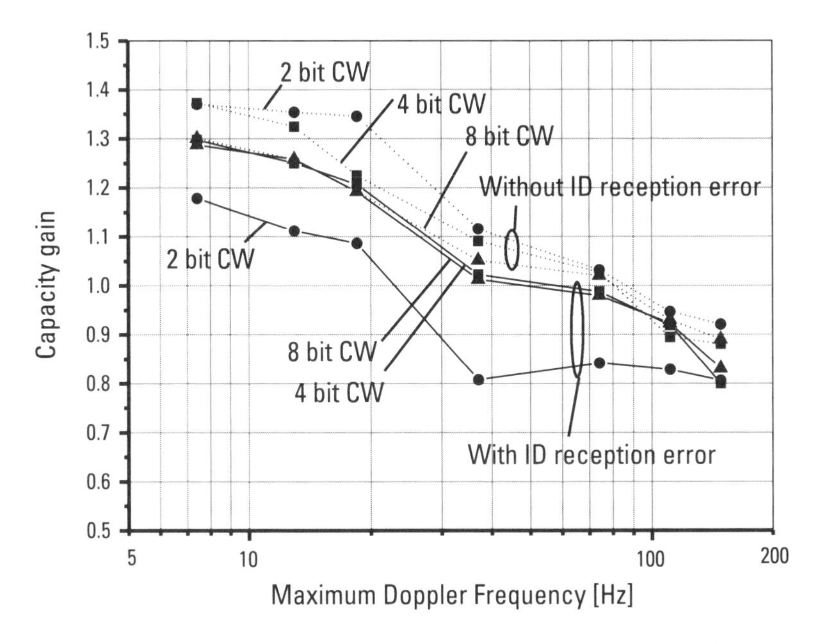

Figure 2.11 Capacity gains for SSDT power control and conventional TPC. CW: codeword. (*Source:* [10], © 2000, IEEE. Reprinted with permission.)

little capacity difference is seen between the cases with and without site selection error. This means that site selection error can be reduced effectively by using a codeword set with a minimum Hamming distance greater than or equal to 2 bits. The advantage of SSDT power control over conventional TPC decreases as the Doppler frequencies increase because the ability of the conventional TPC to track the changing primary BS improves with the decrease in Doppler frequency.

Regarding the performance of SSDT power control in terms of the site selection cycle in the case without ID reception errors, the capacity gain at low Doppler frequencies improves as the length of the codeword (i.e., the site selection cycle) decreases. But the benefit of the short site selection cycle was actually masked in the case of ID reception error because the tolerance to ID reception error is lower when the codeword is shorter. In Figure 2.11 the capacity gain due to SSDT power control can be seen to be greater than 1 for Doppler frequencies up to about 40 Hz, which corresponds to MS speeds of about 20 km/hr. A distinct capacity gain, however, is obtained around the pedestrian speed. These MS speed limits with ID reception errors do not differ much between 4-bit and 8-bit codewords.

Here we describe how we compared the path-capturing efficiency of SSDT power control with that of conventional TPC in which full site transmission diversity is carried out. We defined a *path collection efficiency ratio* (PCER) as follows:

$$\text{PCER} = \frac{(\text{Total power of paths captured by RAKE})}{(\text{Total power of all paths given by the active BSs})}$$

Figure 2.12 shows PCER versus the number of RAKE fingers for conventional TPC and SSDT power control assuming two-branch reception diversity at MS where the number of RAKE fingers is defined not in a unit of per antenna, but in a unit of per MS. It is obvious that the path collection efficiency of SSDT power control is higher than that of conventional TPC. This high efficiency enables the transmission power of the active BSs to be reduced, and this, as well as the reduction in the power output from non-primary BSs, can reduce interference and thereby increase the downlink capacity. Another attractive result offered by Figure 2.12 is that the same PCER can be achieved with a smaller number of RAKE fingers by using SSDT power control. For six-finger RAKE reception, conventional TPC loses 21% of the total power of the paths given by the active BSs, whereas SSDT power control loses only 12%.

Figure 2.13 shows the capacity versus the Doppler frequency limiting the number of RAKE fingers. The data plotted in this figure were obtained

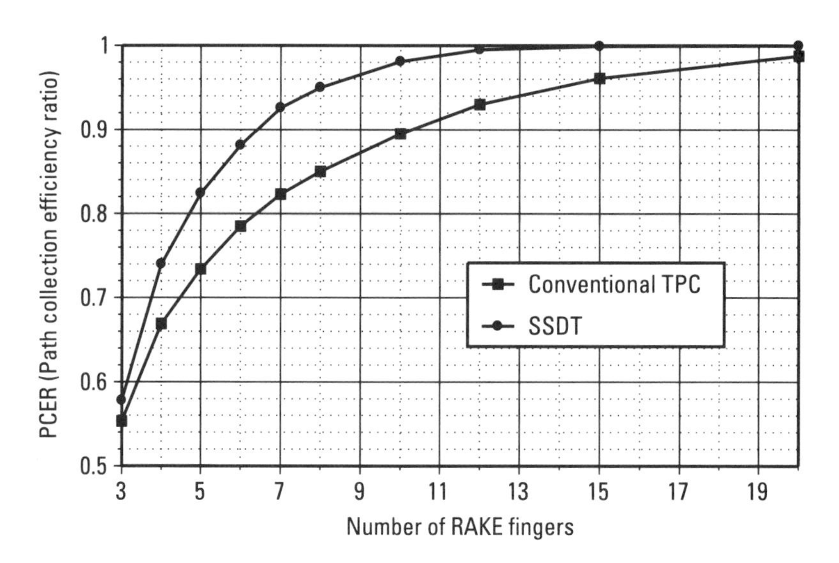

Figure 2.12 PCER versus the number of RAKE fingers. (*Source:* [10], © 2000, IEEE. Reprinted with permission.)

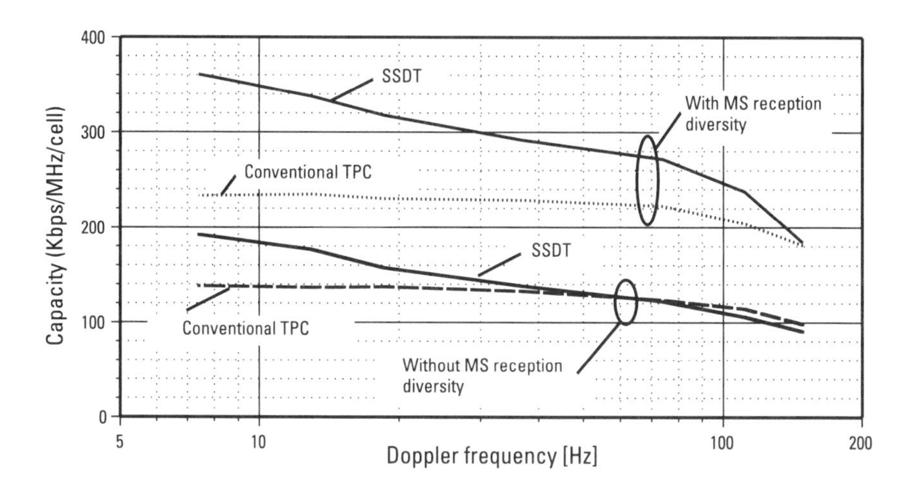

Figure 2.13 Capacity versus Doppler frequency in six-finger RAKE reception. (*Source:* [10], © 2000, IEEE. Reprinted with permission.)

by assuming the six-path profile model, the six-finger RAKE reception, and the 16-bit codeword set listed in Table 2.6. As the figure shows, SSDT power control outperforms conventional TPC at most of the Doppler frequencies. In this case, the increased capacity provided by SSDT power control is due not only to the elimination of multiple-site transmission but also to increased path capturing efficiency like that shown in Figure 2.12. For a Doppler frequency of about 7 Hz, which corresponds to a pedestrian MS, SSDT power control increases by about 43% without MS reception diversity and by about 55% with MS reception diversity. The lower the Doppler frequency, the larger the gain in capacity provided by SSDT power control. This is a tendency identical to that already shown in Figure 2.11. From Figure 2.13, we can understand that the ability of the power control to track changes in the primary BS in an active set is also increased by the use of MS reception diversity. This is because reception signal fluctuation due to fast fading can be reduced.

Finally, we investigated the relationship between capacity and the number of RAKE fingers when the MS speed is 4 km/hr (maximum Doppler frequency about 7 Hz) and MS reception diversity is used. The results obtained assuming a 16-bit codeword and the six-path profile model are listed in Table 2.7. The high RAKE reception efficiency obtained with SSDT power control makes the capacity gain over conventional TPC larger when the number of RAKE fingers is smaller. For an infinite number of RAKE fingers, for example, SSDT power control increases capacity by 36%, whereas

Table 2.7
Capacity (Kbps/MHz/Sector) of SSDT Power Control and Conventional TPC*

	Infinite RAKE Fingers	Six RAKE Fingers	Five RAKE Fingers
Conventional TPC	377	233	217
SSDT	479	361	341

*With respect to the number of RAKE fingers in case of pedestrian MSs.

for 6 and 5 RAKE fingers it increases capacity by only 55% and 57%. Even though the capacity of SSDT for the five-finger RAKE becomes slightly smaller than that of SSDT for the six-finger RAKE, the capacity gain achieved by SSDT for the five-finger RAKE is still 46% higher than that of conventional TPC for the six-finger RAKE.

2.3.6 Performance in Combination with Transmission Diversities

In this section we briefly view the performance of SSDT power control used in combination with transmission diversities. Transmission diversity can increase the downlink diversity gain without requiring the MSs to be equipped with additional antennas. This is desirable in light of the tight restrictions on the size and cost of mobile terminals.

Two modes of transmission diversity operation are used, closed-loop and open-loop modes. In the closed-loop mode—that is, the mode used with a *transmission antenna array* (TxAA)—an MS selects from candidate antenna weights the one that maximizes the SINR measured at the MS, and an index for that weight is sent to the active BSs [19]. Each active BS uses this index to set its antenna weight coefficient to an appropriate value. The antenna weight determination process carried out at MSs is the same procedure described in [19]. The open-loop mode, called STTD [19, 30], uses a novel technique for encoding the downlink signals transmitted by two antennas in such a way that the composite downlink signals received at the MS can be separated. In evaluating both modes, two transmission antennas at a BS and a single reception antenna at an MS were assumed. Channel estimation is ideally carried out at MSs. For the mode using the TxAA, four candidate antenna weight coefficients and a weight update cycle for every two slots were assumed. For the SSDT power control, the 8-bit ID codeword defined in Table 2.6 was assumed. The other simulation parameters were the same as those listed in Table 2.5.

Normalized downlink capacities for various combinations of SSDT power control and transmission diversities are listed in Table 2.8. The

Table 2.8
Normalized Downlink Capacities for Various Combinations of SSDT Power Control and Transmission Diversities*

Selection of Downlink Transmission Diversities		MS Speed		
Non-SHO mode	**SHO mode**	**4 km/hr**	**10 km/hr**	**60 km/hr**
Nontransmit diversity	Nontransmit diversity	0.72 (8)	0.82 (8)	1.07 (6)
Nontransmit diversity	SSDT	1 (4)	1 (5)	1 (8)
STTD	STTD	0.90 (6)	0.98 (6)	1.53 (2)
TxAA	Nontransmit diversity	0.76 (7)	0.87 (7)	1.11 (5)
TxAA	TxAA	0.95 (5)	1.02 (4)	1.12 (4)
TxAA	SSDT	1.08 (3)	1.08 (3)	1.03 (7)
STTD	SSDT + STTD	1.24 (2)	1.31 (2)	1.74 (1)
TxAA	SSDT + TxAA	1.30 (1)	1.37 (1)	1.14 (3)

*Values in parentheses show the order of capacity for the respective MS speed.

capacities for respective MS speeds are normalized by one achieved by SSDT power control. The parallel operation of SSDT power control and transmission diversities increases the capacity. When the MS speed is low (4 and 10 km/hr), the highest capacity is provided by the combination of SSDT power control and TxAA. When the MS speed is high (60 km/hr), on the other hand, the capacity can be maximized by the combination of SSDT and STTD. This is because feedback information regarding site selection for SSDT and antenna weight for TxAA is too delayed to correctly update the active BSs due to the fast changes occurring to the downlink channel state. Although SSDT itself shows no advantage in capacity in comparison with the other cases disclosed in Table 2.8 for an MS speed of 60 km/hr, the combination of SSDT and transmission diversities results in the capacity gain. This is because fast fluctuation of the downlink reception power is smoothed by the transmission diversities and thus the error in tracking the primary BS can be reduced.

2.4 Summary and Subjects for Future Study

This chapter, based on intensive analysis of CDMA downlinks, has clarified two major barriers to sophisticated CDMA: the limited number of orthogonal codes available and the increased downlink interference due to SHO. To cope with the orthogonal code limitation, we developed an optimum code allocation by using a mathematical approach. And to cope with the increased downlink interference, we developed a new form of transmission power control, SSDT power control.

With regard to a code allocation optimum with respect to maximizing the average SIR measured by MSs within a cell, we concluded that an optimum allocation scheme maximizes the number of orthogonal code sets fully occupied and minimizes the use of scrambling codes. However, we also showed that if the orthogonal code sets are equally used, the average SIR is minimized. The optimum code allocation increases the average SIR by 0.7–1.0 dB over the average SIR obtained with the worst code allocation.

The SSDT power control enables site selection transmission diversity to be used during the SHO mode. Computer simulations showed that the capacity gain provided by SSDT power control becomes greater as the Doppler frequency becomes smaller. The capacity increase obtained with SSDT power control is due to the elimination of multiple-site transmission and the increased path-capturing efficiency of the RAKE receiver. The capacity of a pedestrian MS with a six-finger RAKE can be increased about 55% by using SSDT power control in combination with MS reception diversity. The high path collection efficiency obtained with SSDT power control increases the capacity gain by 36% for an infinite number of RAKE fingers, but by 55% for a six-RAKE finger. The gained capacity was obtained even though the number of RAKE fingers for SSDT is less than that for conventional TPC. These results indicate that SSDT power control will be useful for reducing the number of RAKE fingers, and hence reducing the complexity of MS equipment.

The power imbalance problem that results from TPC command reception error is also avoided because only one BS transmits a downlink signal during SSDT power control. Our evaluation of the parallel operation of SSDT power control and various transmission diversities indicated that capacity can be further increased.

In the process of making the WCDMA specifications, the number of slots per frame was modified from 16 to 15 [26] because of the change in the chip rate of from 4.096 to 3.84 Mbps. The intention of this modification was to keep the frame length at 10 ms. In this chapter, we have used the previous slot number and chip rate (16 slots and 3.84 Mbps). Reduction of the processing gain induced by the lower chip rate will attenuate the capacity approximately in proportion to the reduction ratio. In addition, the slot number change reduces the update cycle for power control from

1.6 to 1.5 kHz. These unfavorable changes affect SSDT power control and conventional TPC equally, so the trend of the results shown in this chapter will not be changed.

References

[1] Gilhousen, K. S., et al. "On the Capacity of Cellular CDMA Systems," *IEEE Trans. on Vehicular Technology,* Vol. 41, No. 2, May 1992, pp. 303–312.

[2] Furukawa, H., Y. Miyamoto, and A. Ushirokawa, "Capacity Evaluation of the WCDMA Forward Link," *Proc. IEICE Nat. Conf.,* B-5-189, March 1999 (in Japanese).

[3] *ITU-R Recommendation M.1225,* Annex 2, International Telecommunication Union, Dec. 1997.

[4] ARIB IMT-2000 Study Committee, "Self Evaluation Report on Japan's Proposal for Candidate Radio Transmission Technology on IMT-2000: WCDMA, Part II Revised RTT Proposal," Version 1.1, Sept. 1998.

[5] "Mobile Station-Base Station Compatibility Standard for Dual-Mode Wideband Spread Spectrum Cellular System," TIA/EIA/IS-95 Interim Standard, July 1993.

[6] Adachi, F., "Effects of Orthogonal Spreading and Rake Combining on DS-CDMA Forward Link in Mobile Radio," *IEICE Trans. on Communication,* Vol. E80-B, No. 11, Nov. 1997, pp. 1703–1712.

[7] Furukawa, H., "Theoretical Capacity Evaluation of Power Controlled CDMA Downlink," *Proc. VTC 2000-Spring,* Vol. 2, May 2000, pp. 997–1001.

[8] Furukawa, H., "An Optimum Code Allocation Scheme for CDMA Downlink," *Proc. IEICE Nat. Conf.,* B-5-93, Oct. 2000 (in Japanese).

[9] Andersson, T., "Tuning the Macro Diversity Performance in a DS-CDMA System," *Proc. VTC94,* June 1994, pp. 41–45.

[10] Furukawa, H., K. Hamabe, and A. Ushirokawa, "SSDT—Site Selection Diversity Transmission Power Control for CDMA Forward Link," *IEEE. J. Selected Areas of Communication,* Vol. 18, No. 8, Aug. 2000, pp. 1546–1554.

[11] Lee, W. C. Y., "Overview of Cellular CDMA," *IEEE Trans. Vehicular Technology,* Vol. 40, May 1991, pp. 291–302.

[12] Viterbi, A. M., and A. J. Viterbi, "Erlang Capacity of a Power Controlled CDMA System," *IEEE. J. Selected Areas of Communication,* Vol. 11, No. 6, Aug. 1993, pp. 892–900.

[13] Zander, J., "Performance of Optimum Transmitter Power Control in Cellular Radio Systems," *IEEE Trans. on Vehicular Technology,* Vol. 41, No. 1, Feb. 1992, pp. 57–62.

[14] Stuber, G. L., and C. Kchao, "Analysis of a Multiple-Cell Direct-Sequence CDMA Cellular Mobile Radio System," *IEEE. J. Selected Areas of Communication,* Vol. 10, No. 4, May 1992, pp. 669–679.

[15] Gejji, R. R., "Forward-Link-Power Control in CDMA Cellular Systems," *IEEE Trans. on Vehicular Technology,* Vol. 41, No. 4, Nov. 1992, pp. 532–536.

[16] Zorzi, M., "Simplified Forward-Link Power Control Law in Cellular CDMA," *IEEE Trans. on Vehicular Technology,* Vol. 43, No. 4, Nov. 1994, pp. 1088–1093.

[17] Song, L., and J. M. Holtzman, "CDMA Dynamic Downlink Power Control," *Proc. IEEE VTC'98,* May 1998, pp. 1101–1105.

[18] Wallace, M., and R. Walton, "CDMA Radio Network Planning," *Proc. ICUPC'94,* Oct. 1994, pp. 62–67.

[19] "Physical Layer Procedures (FDD)," 3GPP TSG RAN WG1, TS25.214, Version 3.1.0, Dec. 1999.

[20] Viterbi, A. J., et al., "Soft Handoff Extends CDMA Cell Coverage and Increases Reverse Link Capacity," *IEEE. J. Selected Areas of Communication,* Vol. 12, No. 8, Oct. 1994, pp. 1281–1288.

[21] Soleimanipour, M., and G. H. Freeman, "A Realistic Approach to the Capacity of Cellular CDMA Systems," *Proc. VTC'96,* April 1996, pp. 1125–1129.

[22] Gorricho, J., and J. Paradells, "Evaluation of the Soft Handover Benefits on CDMA Systems," *Proc. ICUPC'96,* 1996, pp. 305–309.

[23] Daraiseh, A., and M. Landolsi, "Optimized CDMA Forward Link Power Allocation During Soft Handoff," *Proc. VTC'98,* May 1998, pp. 1548–1552.

[24] Furukawa, H., *Site Selection Transmission Power Control in DS-CDMA Downlinks,* IEICE Technical Report RCS97–218, Feb. 1998, pp. 39–46 (in Japanese).

[25] Furukawa, H., "Site Selection Transmission Power Control in DS-CDMA Cellular Downlinks," *Proc. ICUPC'98,* Oct. 1998, pp. 987–991.

[26] "Physical Channels and Mapping of Transport Channels onto Physical Channels (FDD)," 3GPP TSG RAN WG1, TS25.211, Version 3.0.1, Dec. 1999.

[27] Hottinen, A., and R. Wichman, "Transmit Diversity by Antenna Selection in CDMA Downlink," *Proc. ISSSTA'98,* Sept. 1998, pp. 767–770.

[28] CSELT, "Impact of the Number of Pilot Bits on the Uplink Performance," 3GPP TSG RAN WG1, TSGR1#3(99)170, March 1999.

[29] Gudmundson, M., "Correlation Model for Shadow Fading in Mobile Radio Systems," *Electron. Lett.*, Vol. 27, No. 23, Nov. 1991, pp. 2145–2146.

[30] Debak, A., S. Hosur, and R. Negi, "Space Time Block Coded Transmit Antenna Diversity Scheme for WCDMA," *Proc. IEEE Wireless Communications and Networking Conference (WCNC),* Vol. 3, 1999, pp. 1466–1469.

3

GSM/EDGE Radio Access Network: Evolution of GSM/EDGE Toward 3G Mobile Services

Benoist Sébire, Janne Parantainen, and Guillaume Sébire

The *Global System for Mobile Communications* (GSM)/*enhanced data rates for global evolution* (EDGE) standard has been taking steps to define a fully capable 3G mobile *radio access network* (RAN): The 3GPP standard, release 5, specifies a *GSM/EDGE RAN* (GERAN) that can connect to the same 3G core network and provide the same set of services as the *UMTS terrestrial radio access network* (UTRAN).

3.1 Introduction

GSM began in 1991. Since then it has successfully evolved to offer a wide variety of speech and data services. The development work has been divided into different releases; phase 1, phase 2, releases 1996, 1997, 1998, 1999, and finally releases 4 and 5. Major technological steps were taken in releases 1996 and 1997 when *high-speed circuit-switched data* (HSCSD) and *general packet radio service* (GPRS) were introduced. In release 1999, EDGE introduced the 8-PSK modulation, which provided a threefold increase in the peak data rates. Two new data features were created for the established GSM/GPRS architecture: *enhanced GPRS* (EGPRS) and enhanced circuit-switched data for packet-switched and circuit-switched domains, respectively. Until release 5, a GSM/EDGE *base station subsystem* (BSS) connected either via a Gb interface toward the 2G packet-switched core network or via an A interface toward the 2G circuit-switched core network. These interfaces will exist also in the future but the main evolution is taking place in another direction.

To deploy the GSM/EDGE technology in a full 3G environment, release 5 of 3GPP defines GERAN where the functional split between the RAN and the core network is harmonized with UTRAN. As part of the harmonization, the Iu interface of the UMTS system has been adopted, enabling GERAN to connect via the same interface to the same 3G core network and provide the same set of services as UTRAN.

This chapter provides an overview of GERAN with a focus on release 5 and harmonization with UTRAN. It covers widely different aspects of GERAN, including 3G services, the GERAN reference architecture, and protocol architectures. Emphasis is placed on the radio protocols that constitute the main change from GSM/EDGE to the Iu-capable GERAN. The chapter is organized as follows. First, the 3G traffic classes offered by GERAN are described in Section 3.2. Section 3.3 presents the new GERAN architecture resulting from the RAN/core network functional split alignment. Then, Section 3.4 describes the GERAN radio protocols. The physical layer is described in detail in Section 3.5 and, finally, security issues are addressed in Section 3.6.

3.2 3G Mobile Services

As a part of the RAN harmonization, GERAN supports the same four traffic classes as defined for UTRAN [1]: (1) conversational, (2) streaming, (3) interactive, and (4) background.

Conversational traffic carries speech or video between end users. The QoS requirements are set mainly by human perception in terms of delay and error rate. The limit for acceptable transfer delay is very strict. Failure to fulfill the requirements causes degradation of the user-perceived QoS.

Streaming traffic is one-way data transfer without any absolute delay requirements, but a process in which the time relation between information entities has to be preserved. During transmission, relatively large delay variations can be accepted.

Interactive traffic applies whenever a human or machine is on-line requesting data from a remote entity. Because responses are typically expected within a certain time, the round-trip delay is an important QoS parameter. Error-free delivery of the requested data also has to be guaranteed. As an example, Web browsing comes to mind.

Background traffic is the most delay-insensitive traffic class. It is assumed that the end user does not expect the data for some time. The data transfer can be made on background traffic when there is no other delay-critical data to be transferred. Typical traffic belonging to this category includes file transfer processes and e-mail.

As can be seen, the main difference between the traffic classes is their delay sensitivity. Conversational class is meant for traffic, which is very delay sensitive, whereas background class has the most relaxed delay requirements. Real-time traffic flows belong to the conversational or streaming traffic class. Traditional Internet applications like the *World Wide Web* (WWW), e-mail, Telnet, *File Transfer Protocol* (FTP), and news belong to the interactive or background traffic class. Looser delay requirements allow a very low error rate to be guaranteed by means of channel coding and retransmissions. In that way, the interactive and background traffic classes often provide lower error rates than the streaming and conversational traffic classes.

Internet Protocol (IP)-based multimedia is a new service provided by release 5 of GERAN. This covers both conversational speech and video. For circuit-switched domains, these services already existed in release 1999. GERAN release 5 makes it possible to provide all services (both real time and non-real time) in the packet-switched domain as well. Furthermore, GERAN supports *Session Initiation Protocol* (SIP), with which it is not only possible to provide basic call control functions that exist today, but also to provide a flexible and future-proof mechanism for applications combining voice, video, and multimedia.

3.3 Architecture

3.3.1 Introduction

A GSM/EDGE RAN consists of one or more BSSs with interfaces to communicate with its environment (MSs, core networks, other GERANs or UTRANs) [2]. BSSs and associated external interfaces constitute the GERAN reference architecture, which is depicted in Figure 3.1. A GERAN BSS consists of one or more *base transceiver stations* (BTSs) and a *base station controller* (BSC). Because the BSS implementation is not mandated by the specification, the BSS internal architecture is vendor specific. However,

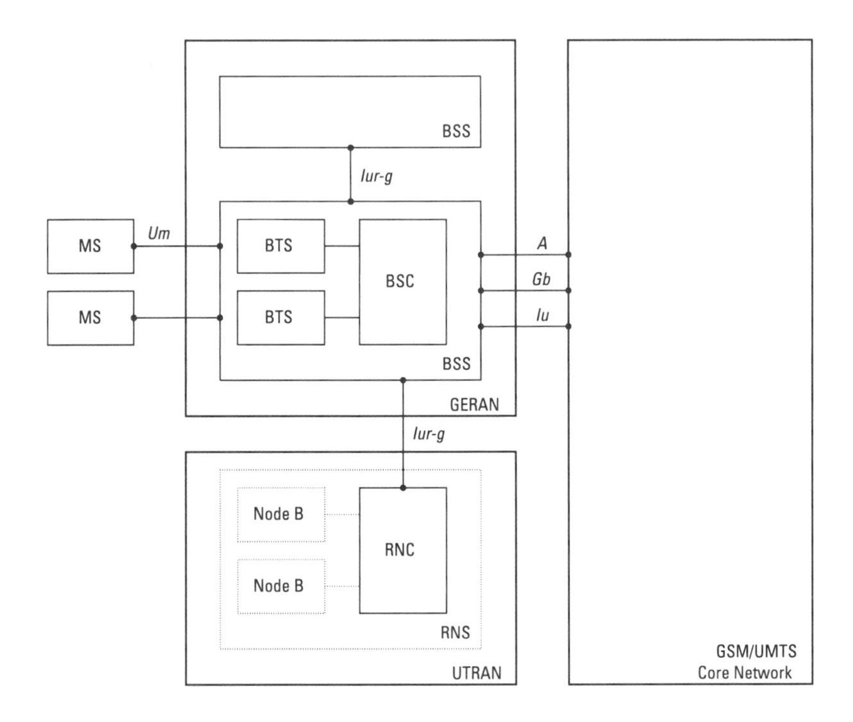

Figure 3.1 GERAN reference architecture.

the BSS external interfaces are open multivendor interfaces to allow for interworking between equipment from different vendors. On a given interface, the communications between two elements is ensured by means of layered protocols that follow a specific protocol architecture.

3.3.2 BSS External Interfaces

3.3.2.1 Radio Interface: Um

The radio interface (Um) is the connection point between an MS and the network. GERAN release 5 Um is based on the GSM/EDGE radio interface and is backward compatible with the existing services.

3.3.2.2 Core Network Interfaces: A, Gb, and Iu

The three different core network interfaces are A, Gb, and Iu (Figure 3.1). The A interface is the traditional GSM circuit-switched interface toward a 2G *mobile switching center* (MSC). The Gb interface connects to a 2G *serving GPRS support node* (SGSN) and provides general packet radio services. The Iu interface is new and is designed for connecting the RAN to a 3G SGSN and a 3G MSC, which includes both packet-switched (Iu-ps) and circuit-switched (Iu-cs) domains. The Iu interface is common to both UTRAN and GERAN.

3.3.2.3 Inter-BSS, BSS-Radio Network Subsystem Interface: Iur-g

Iur-g is an open interface aimed primarily at supporting the exchange of signaling information between two GERAN BSSs. It may exist as a logical interface if no direct physical connection is available. The existence of Iur-g is optional in the sense that it should not affect the behavior of a mobile station.

This interface, inherited from UTRAN as a result of the harmonization between GERAN and UTRAN, consists of a subset of the procedures available on the Iur interface of UTRAN, the main difference being that in the case of GERAN only the control plane is supported. The intention is for the Iur-g interface to also exist between GERAN and UTRAN, that is, between a BSS and a radio network subsystem. The status of this interface was open as of this writing.

3.3.3 Modes of Operation

Depending on the core network interfaces, the behavior of an MS changes and can be described by either of the two following operating modes:

1. *A/Gb mode:* This mode is applicable to MSs (including legacy ones) that are connected to the core network via an A and/or Gb interface.
2. *Iu mode:* This mode is applicable to mobile stations only when connected to the core network via the Iu interface. It is not applicable to legacy MSs.

Depending on the capabilities of the GERAN and of the MS, an MS may be simultaneously connected to A and Gb interfaces[1] or to Iu-ps and Iu-cs interfaces. Other combinations, for example, Iu-cs and Gb, are not allowed.

3.3.4 Protocol Architectures

3.3.4.1 MS-Core Network Protocol Architectures

This section provides an overview of the protocol architectures of the user and control planes when connected to the core network through the different interfaces.

---

1. This mode of operation is referred to as the dual transfer mode in release 1999.

User Plane

Figures 3.2 to 3.5, respectively, show the user plane of GERAN when connected to the core network through the Gb, Iu-ps, A, and Iu-cs interfaces. Note that the figures include only the existing GSM and UMTS release 99 transport layer options for A and Iu-cs.

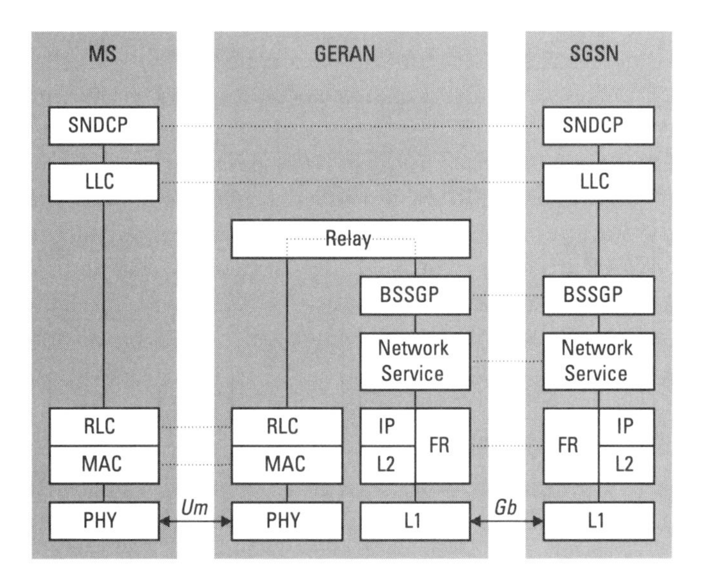

Figure 3.2 User Plane Protocols through the Gb interface.

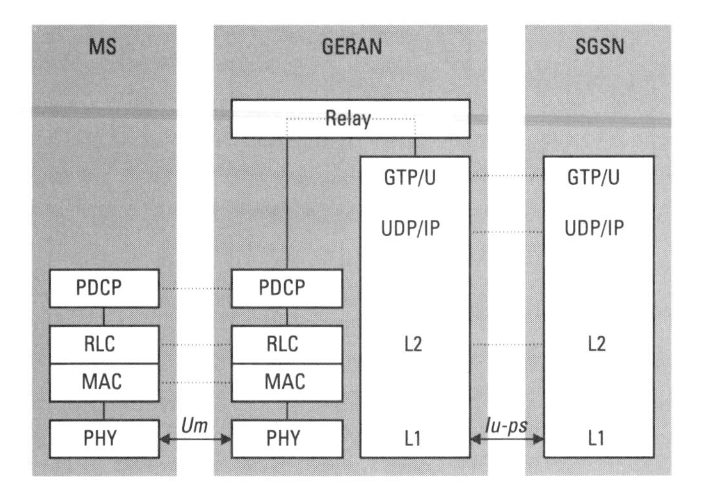

Figure 3.3 User Plane Protocols in the PS domain through the Iu-ps interface.

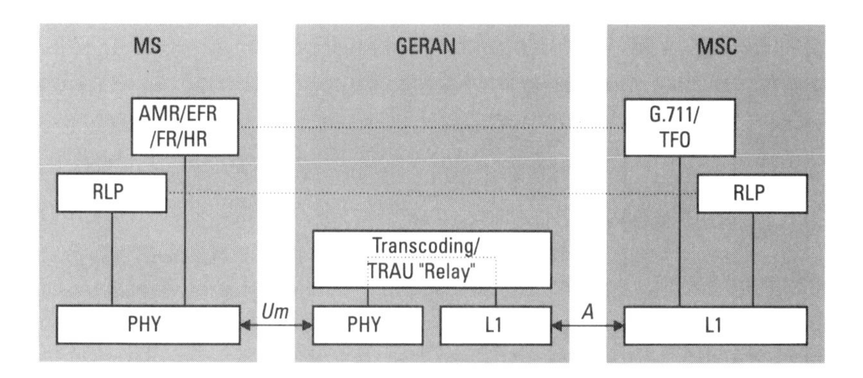

Figure 3.4 User Plane Protocols in the circuit-switched domain through the A interface.

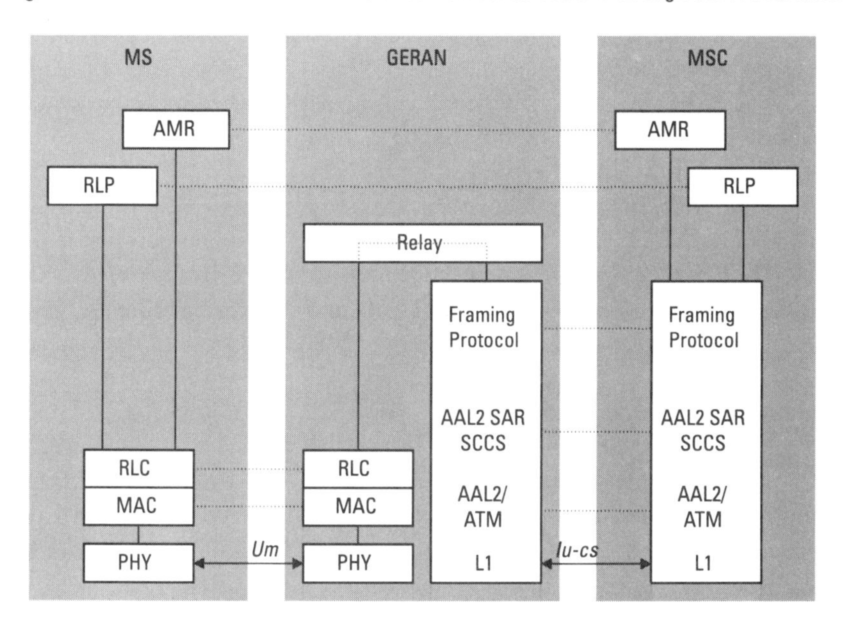

Figure 3.5 User Plane Protocols in the circuit-switched domain through the Iu-cs interface.

The *Subnetwork Dependent Convergence Protocol* (SNDCP) adapts and multiplexes several *Network Layer Protocols—Packet Data Protocols* (PDP) for transmission over the radio interface in a channel efficient manner, and therefore provides a compression/decompression technique for user data and protocol control information [3].

The *Logical Link Control* (LLC) Protocol provides a reliable ciphered logical link between the MS and the SGSN [4]. The relay function transfers LLC *Protocol data units* (PDUs) between the Um and the Gb interfaces.

The BSS *GPRS Protocol* (BSSGP) has been designed primarily to provide the information (radio-related, QoS, and routing) required to transmit user data between a BSS and an SGSN [5].

The *network service* (NS) provides network management functionality enabling the operation of the Gb interface. The NS is used to carry BSSGP PDUs between a BSS and an SGSN [6].

The *frame relay* (FR) is used for signaling and data transmission and constitutes the link layer of the Gb interface. It allows for multiplexing LLC PDUs from many users on virtual communication paths established between the SGSN and BSS.

The IP is the backbone network protocol used for routing user data and control signaling, and is based on IPv4 (IPv6) [7].

The *radio link control* (RLC) protocol provides a reliable radio-technology specific link [8].

The MAC Protocol provides access-signaling procedures for the radio channel. It defines a means for sharing a common transmission medium between several MSs [8].

The *GPRS tunneling protocol for the user plane* (GTP-U) tunnels user data between *GPRS support nodes* (GSNs) in the backbone network [9].

The *User Datagram Protocol* (UDP) transfers user data between GSMs [10]. UDP/IP is used for routing user data and control signaling.

The *Packet Data Convergence Protocol* (PDCP) adapts IP data streams for transmission over Radio Interface Protocols in a channel-efficient manner and therefore provides compression/decompression of protocol control information but not of user data [11].

The *Radio Link Protocol* (RLP) provides circuit-switched transmission within the GSM *public land mobile network* (PLMN). RLP covers the Layer 2 functionality of the ISO OSI reference model (IS 7498) and provides the OSI data link service (IS 8886) [12].

The *transcoding/transcoder and rate adapter unit* (TRAU) relay transcodes the speech signal into standard 64-Kbps pulse code modulation (PCM) samples for transport over the fixed part of the network (using 64-Kbps traffic links) [13, 14].

Control Plane

Figures 3.6 through 3.9, respectively, show the control plane of GERAN when connected to the core network through the Gb, Iu-ps, A, and Iu-cs interfaces.

GPRS mobility management and session management (GMM/SM) provides mobility management functions (e.g., attach and routing area update).

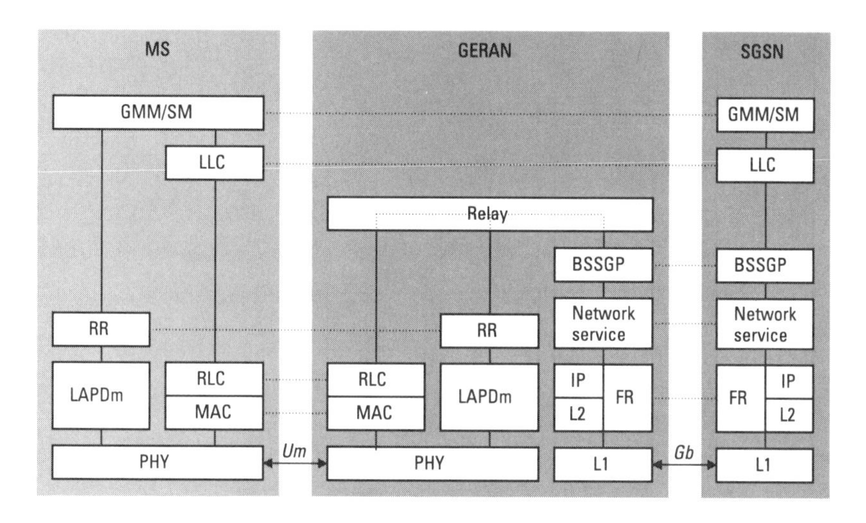

Figure 3.6 Control Plane Protocols in the PS domain through the Gb interface.

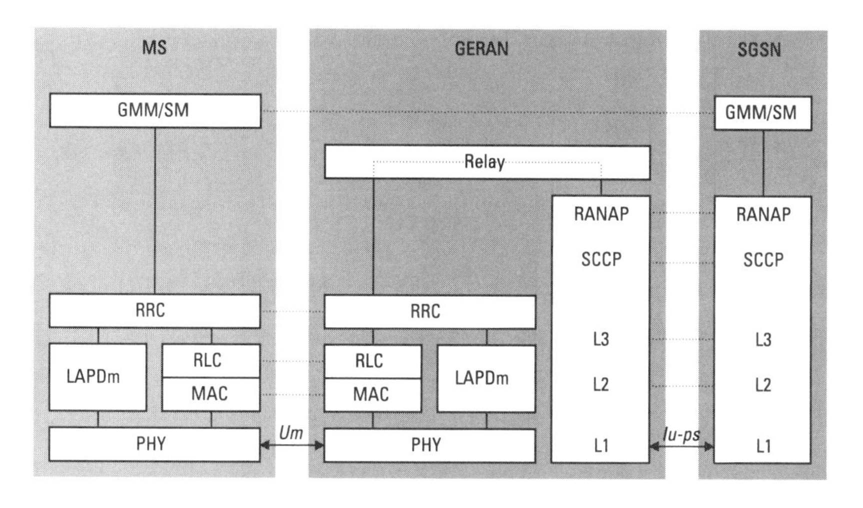

Figure 3.7 Control Plane Protocols in the PS domain through the Iu-ps interface.

The *RAN application part* (RANAP) handles higher layer signaling and signaling between SGSN and GERAN and also controls the GTP connections on the Iu-ps interface [15].

The *signaling connection control part* (SCCP) provides virtual connections and connection-less signaling.

The *mobility management/UMTS mobility management* (MM/UMM) allows the core network to track and find a given MS. The *call control* (CC)

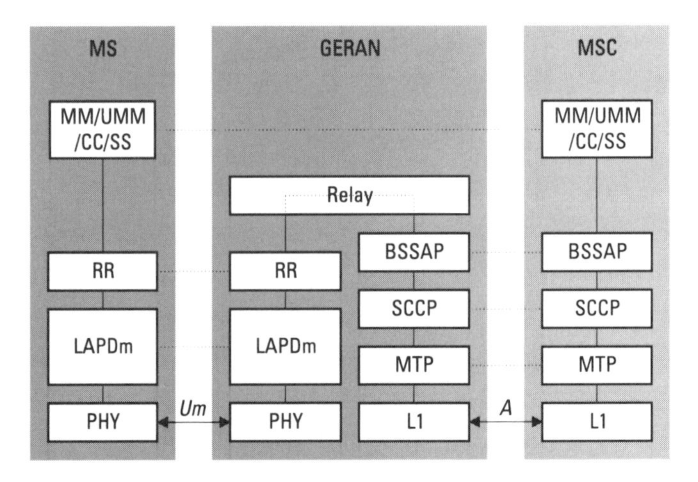

Figure 3.8 Control Plane Protocols in the circuit-switched domain through the A interface.

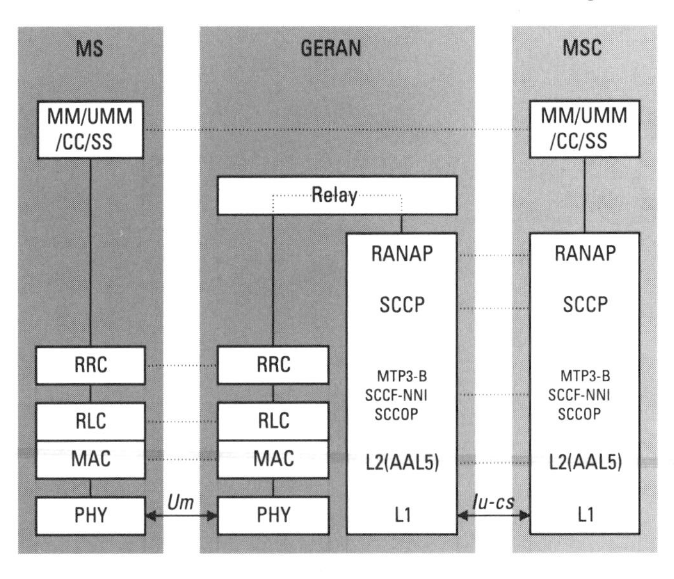

Figure 3.9 Control Plane Protocols in the circuit-switched domain through the Iu-cs interface.

feature provides the functions for circuit-switched call management, for example, call setup, incoming call notification, and end call. The *supplementary services* (SS) provide additional services such as location services and settings for call forwarding.

The *BSS application part* (BSSAP) provides signaling between MSCs and BSCs as well as signaling between MSCs and mobile stations. It is split into the *BSS management application part* (BSSMAP) and the *direct transfer application part* (DTAP) [16]. The BSSMAP supports the procedures between the MSC and the BSS that require interpretation and processing of information related to single calls and resource management. The DTAP is used to transfer call control and mobility management messages between the MSC and the MS.

The *message transfer part* (MTP) is responsible for transferring messages from one network element to another within the same network, and is one of the two initial parts of Common Channel Signaling System No. 7 (commonly referred to as SS7) from CCITT, the other part being the *telephone user part.*

3.3.4.2 Iur-g Protocol Architecture

In the Iur-g Protocol architecture, Radio Network Protocols are separated from the transport layer. The transport mechanism, for example, can be *asynchronous transfer mode* (ATM) or IP and has no impact on the radio network layer where the signaling information is provided. The protocol used for transferring signaling information across the interface is the radio *network subsystem application part* (RNSAP) [17]. In the case of UTRAN, RNSAP consists of four modules:

1. Basic mobility procedures;
2. RNSAP DCH procedures;
3. RNSAP common transport channel procedures;
4. RNSAP global procedures.

In GERAN, only basic mobility procedures and RNSAP global procedures are used. These modules contain the means to handle mobility within GERAN, and procedures that are not related to any particular MS, for example, error indication. Figure 3.10 shows the protocol architecture of the Iur-g interface. The presented Transport Network Layer Protocols have been taken from UTRAN release 1999. For GERAN release 5 these protocols may be different.

3.4 Radio Interface Protocols

This section describes the Radio Interface Protocol architecture (access stratum) of GERAN when connected to the core network through the Iu interface.

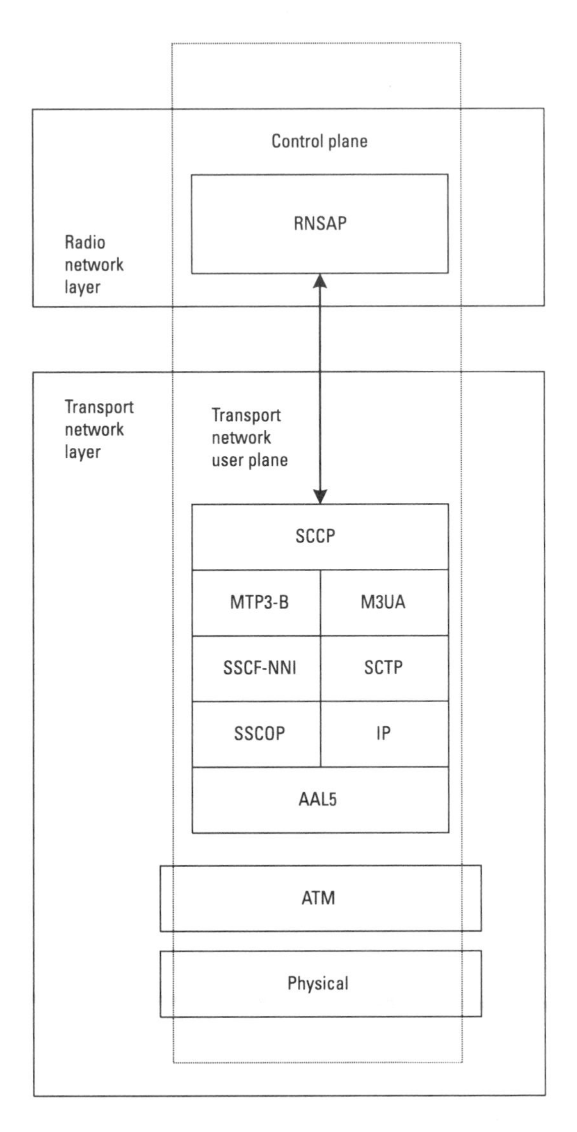

Figure 3.10 Iur-g Protocol architecture.

3.4.1 General

3.4.1.1 Protocol Architecture

The radio interface is layered as described in Table 3.1. Figure 3.11 shows the detailed protocol architecture. The ellipses represent the *service access points* (SAPs) at the interface between sublayers. The name of an SAP is

Table 3.1
Radio Interface Protocol Layers

Access Stratum	Control Plane			User Plane
Network layer	Duplication Avoidance*			—
	RRC			
Data link (DL) layer	RA	LAPDm	—	PDCP
			RLC	
			MAC	
Physical layer	PHY			

*Terminates in the core network.

determined from specifications in the GERAN standard. An SAP is left blank if no name is given in the specification.

The SAPs between the Layers 1 and 2 provide the logical channels, namely, *packet data traffic channel* (PDTCH), *packet associated control channel* (PACCH), *packet timing control channel* (PTCCH), *packet broadcast control channel* (PBCCH), *broadcast control channel* (BCCH), *random-access channel* (RACH), *traffic channel* (TCH), *fast associated control channel* (FACCH), *slow associated control channel* (SACCH), *standalone dedicated control channel* (SDCCH), and *cell broadcast channel* (CBCH). In Figure 3.11, CCCH stands for common control channel containing the *paging channel* (PCH) and *access grant channel* (AGCH); and PCCCH stands for packet common control channel containing the *packet random-access channel* (PRACH), *packet paging channel* (PPCH), and *packet access grant channel* (PAGCH).

The interface between duplication avoidance and higher Layer 3 sublayers in the core network (call control and mobility management) offers the following SAPs: *general control* (GC), *notification* (Nt), and *dedicated control* (DC) [18].

The other SAPs in the figure are not standardized, but are shown to illustrate the relationship between the various protocol instances that are configured to convey data flows. Further details are given in the next section. The interfaces and associated SAPs that allow the *radio resource control* (RRC) to control the configuration of the lower sublayers (PDCP, RLC, MAC, and PHY) are not shown.

3.4.1.2 Multiplexing

By means provided by the MAC and the physical layers, GERAN allows data from the same or different core network interfaces to be multiplexed on the same logical and *basic physical subchannels* (BPSCHs) as described in

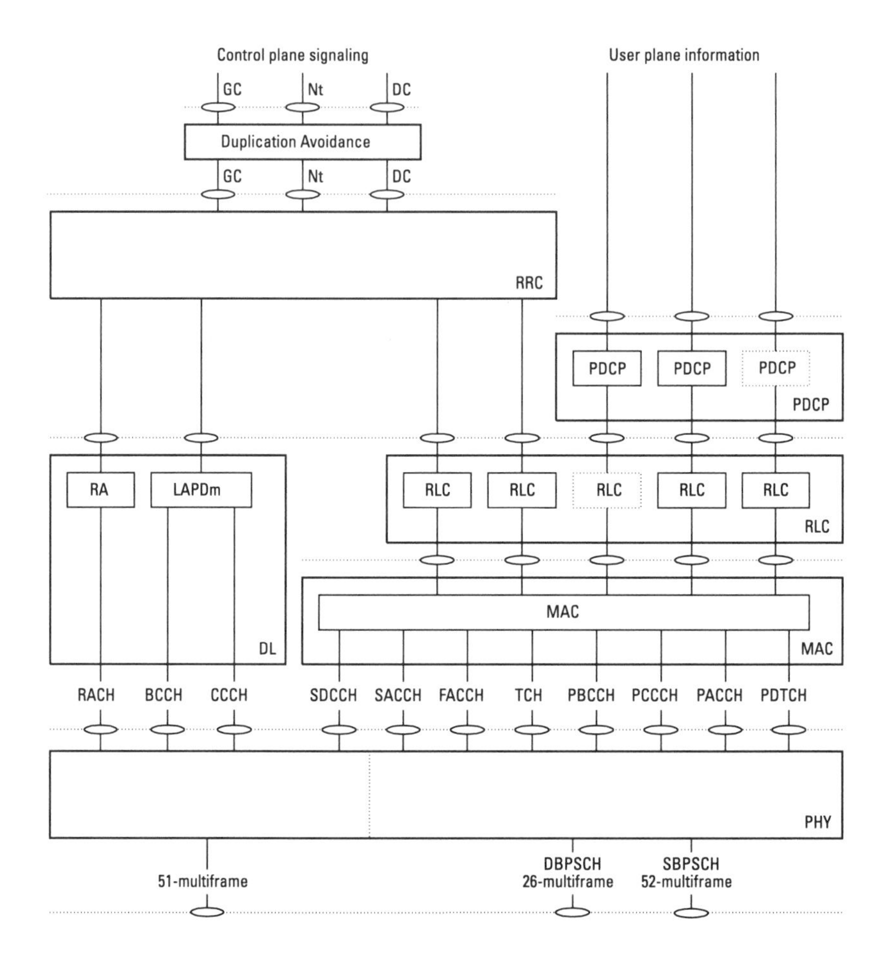

Figure 3.11 Radio Interface Protocol architecture when connecting through the Iu interface.

Figure 3.12. To optimize the use of radio resources, old and new MSs can be multiplexed on the same BPSCHs. As the physical layer is maintained, backward compatible, circuit-switched connections toward the A interface, (E)GPRS connections to the Gb interface, and GERAN Iu connections can be multiplexed on the same transceiver using physical layer multiplexing.

3.4.2 RRC Protocol

3.4.2.1 General

RRC procedures include functions related to the management and control of common transmission resources, for example, basic physical subchannels

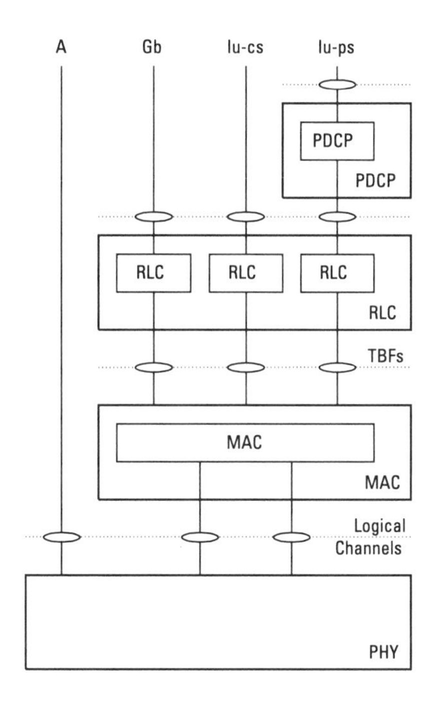

Figure 3.12 Multiplexing principles for data coming from different interfaces.

and data link connections on control channels. The general purpose of RRC procedures is to establish, maintain, and release RRC connections that allow data and signaling transfer between the network and the MS. RRC procedures also include the reception of BCCH and CCCH channels when an RRC connection does not exist.

The GERAN RRC layer is based on the GSM radio resource [19] and UTRAN RRC [20] Protocols. The basic GSM/(E)GPRS procedures apply in A/Gb mode, whereas in Iu mode, several new concepts, such as radio bearer, RRC connection, and RRC connection mobility management, have been adopted from UTRAN.

3.4.2.2 RR and RRC States

Due to the GERAN mobility concept and new functions introduced in the Iu mode that cannot be described by the GSM radio resource state machine alone, two state machines are needed to govern the RRC of GERAN: one for the A/Gb mode and one for the Iu mode. Figure 3.13 presents the basic radio resource state machine for a GPRS-capable MS that supports neither group transmit mode nor receive mode. Five different states can be distinguished:

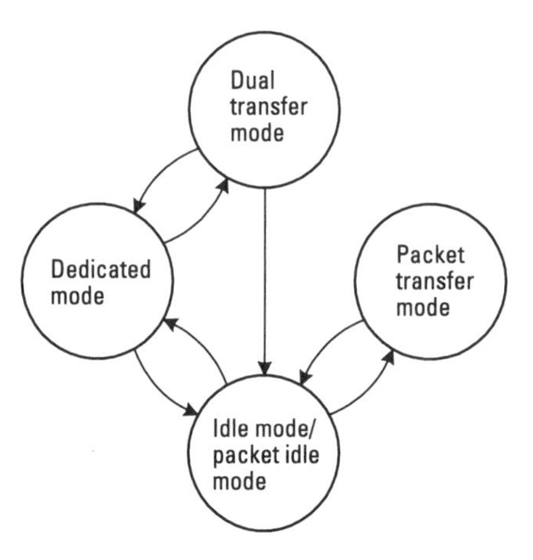

Figure 3.13 The A/Gb mode radio resource state diagram.

1. *Radio resource idle mode:* No radio resource connection exists. The MS does not have dedicated resources allocated and it camps on CCCH. Automatic cell selection and reselection procedures are used.
2. *Packet idle mode:* This is a state in which no temporary block flow exists. Applicable only for (E)GPRS-capable MSs.
3. *Dedicated mode:* The radio resource connection is a physical bidirectional point-to-point link.
4. *Packet transfer mode:* This is a state in which an MS is allocated a radio resource providing a temporary block flow on one or more PDCHs. Applicable only to (E)GPRS-capable MSs.
5. *Dual transfer mode:* In dual transfer mode the MS is simultaneously in dedicated mode and packet transfer mode. It is applicable only to dual transfer mode–capable MSs.

In A/Gb mode, the state transitions are rather limited. When a DBPSCH is assigned, the MS moves from the radio resource idle mode to the radio resource dedicated mode and returns to radio resource idle mode after the call has been terminated. GPRS-capable MSs may be in packet transfer or packet idle mode depending on whether a temporary block flow is established or not. Dual transfer mode–capable MSs may establish a temporary block flow for packet data transmission during a circuit-switched call. However, if the call terminates before the temporary block flow ends, the temporary block flow will be dropped (i.e., no transition from dual transfer mode to packet transfer mode is possible).

In the Iu mode an MS can be in RRC idle mode or in RRC connected mode. In RRC connected mode three different states are possible (Figure 3.14): RRC Cell_Dedicated state, RRC Cell_Shared state, and RRC GRA_PCH state.

- *RRC idle mode:* When in RRC idle mode the connection of the MS is closed on all layers of the access stratum and the MS is identified with nonaccess stratum identifiers: international mobile subscriber identity, temporary mobile subscriber identity, or packet TMSI. In this state GERAN does not have its own information about the individual MSs. It can only address, by means of broadcasting, all MSs in the cell or all MSs monitoring a paging occasion. MSs are in the RRC idle mode camp on the PCCCH if available in the cell, and are otherwise on the CCCH.
- *RRC Cell_Dedicated state:* When in the RRC Cell_Dedicated state the MS has at least one dedicated basic physical subchannel allocated. In addition, the MS may or may not have one or more shared basic physical subchannels. The position of the MS is known on a cell level.
- *RRC Cell_Shared state:* When in the RRC Cell_Shared state the MS does not have a dedicated basic physical subchannel but may or

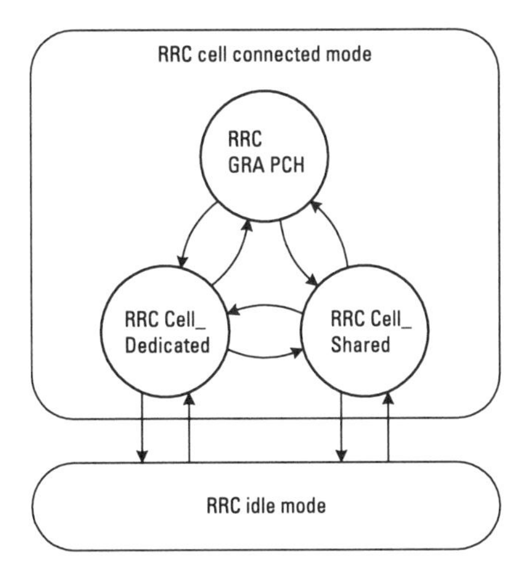

Figure 3.14 The Iu mode RRC state diagram.

may not have a shared channel. The MS position is known based on the cell where the last cell update was made.

- *RRC GRA_PCH state:* When in the RRC GRA_PCH state the MS has no basic physical subchannel allocated and no uplink activity is possible. The MS uses discontinuous reception for monitoring a common channel where it camps (on the CCCH or PCCCH). The MS position is known on the *GERAN registration area* (GRA) level.

In the Iu mode, one key difference from the A/Gb mode is the existence of a GRA concept. When in this state (RRC GRA_PCH), the MS location is not known on a cell level but on the GRA level, even though the RRC connection is maintained. Another difference is that the MS in Iu mode may have an RRC connection without any dedicated or shared channels allocated (RRC Cell_Shared state or RRC GRA_PCH state). Furthermore, the Iu mode provides more flexible state transitions and, hence, better service from the end user's perspective. For example, it is possible to move from the RRC Cell_Dedicated state to the RRC Cell_Shared state without dropping any packet data Layer 2 link (temporary block flow) on SBPSCH, whereas a similar transition in A/Gb mode between the RR dual transfer mode state and packet transfer mode state is not possible.

3.4.2.3 Services

The RRC offers three kinds of services to the upper layers: general control, notification, and dedicated control. The radio resource management sublayer provides a service to the mobility management entity. The radio resource services are used for establishing control channel connections, releasing control channel connections, and using control-data transfer.

3.4.2.4 RRC Functions and Procedures

The RRC performs the following functions:

- *Broadcast of nonaccess stratum information:* System information broadcasting to all MSs. This information may or may not be cell specific.
- *Broadcast of access stratum information:* System information broadcasting to all MSs. This information is typically cell specific.
- *Establishment, reestablishment, maintenance, and release of the RRC connection:* The establishment of an RRC connection is initiated by a request from higher protocol layers at the MS side to establish

the first signaling connection for the MS. The release of an RRC connection is initiated by a request from higher layers to release the last signaling connection for the MS or by the RRC itself in case of RRC connection failure. In case of a connection loss, the MS may request reestablishment, reconfiguration, and release of radio bearers; on request from higher layers, RRC can establish, reconfigure, or release radio bearers. When establishing or reconfiguring radio bearers, RRC performs admission control and, based on the information from higher layers, selects the parameters describing exhaustively the processing on lower Layers 1 and 2 for transmission over the radio interface. Two types of radio bearers are used depending on whether the control plane or the user plane is considered. *Signaling radio bearers* (SRBs) are used in the control plane for RRC signaling exchange between the network and one or more mobile stations. They are established at the RRC connection establishment and consist of SRB 0 used for all signaling messages sent on CCCH; SRB 1 used for all RRC messages sent in RLC unacknowledged mode; SRB 3 and optionally SRB 4 used instead of SRB 2 for all RRC messages carrying nonaccess stratum signaling in RLC ACK mode. Before an RRC connection is set up, RRC messages are sent with a predefined configuration (e.g., using RLC unacknowledged mode on BCCH). User plane radio bearers are used for user data exchange between the network and an MS. A user radio bearer is set up upon request from the core network to set up an RAB.

- *Assignment, reconfiguration, and release of radio resources for the RRC connection:* Depending on the RRC and MAC states, the RRC Protocol may handle radio resources needed for the RRC connection.
- *RRC connection mobility functions:* RRC performs the evaluation of serving and neighbor cell measurements and has the control over handover, cell reselection, and cell/GRA update procedures.
- *Paging/notification:* RRC may broadcast paging information from the network to selected MSs on CCCH. The procedure may be initiated by a higher layer request or during an established RRC connection by the RRC layer.
- *Listening to BCCH and CCCH:* The RRC layer on the MS side listens to the BCCH of the serving cell for decoding system information and the BCCH of neighbor cells for neighbor cell measurements (needed in case of, for example, handover). The RRC layer also monitors

paging occasions on CCCH according to the discontinuous reception cycle.

- *Routing higher layer PDUs:* This function performs routing of PDUs to correct higher layer entities and on the GERAN side to the correct RANAP entity.
- *Control of requested QoS:* The RRC guarantees that the requested QoS can be provided. This includes the allocation of sufficient radio resources to any given radio bearer.
- *MS measurement reporting and control of the reporting:* The RRC layer controls the measurements performed by the MS. On dedicated channels, the RRC layer performs measurement reporting using SACCH.
- *Power control:* The RRC layer controls the parameters of power control.
- *Control of ciphering:* The RRC layer provides the procedures for setting ciphering on/off.
- *Integrity protection:* This function performs integrity protection for RRC messages that are subject to integrity protection.
- *Support of location services:* The RRC layer provides signaling between MSs and the network for positioning the MS.
- *Timing advance control:* The RRC controls timing advance on dedicated channels.

3.4.3 PDCP

The PDCP is the layer where UTRAN and GERAN connect to the 3G core network. Although the two access networks have different Layer 2 and 1 designs, the PDCP is common to both. The sole exception is the provision of header removal in GERAN only. Above PDCP, the same Core Network Protocols and procedures are used.

PDCP exists in the user plane only and serves as a convergence layer between the Network Layer and the RLC Protocols. One of its main functions is to allow lower layers to be fully independent of the chosen Network Layer Protocol. Currently, at least two network layer protocols, IPv4 and IPv6, are supported, but new protocols must be able to be introduced without affecting GERAN or UTRAN.

The second important function of PDCP is to perform Network Layer Protocol header compression to improve spectral efficiency. PDCP can support several compression algorithms. The algorithms used have been standardized by the Internet Engineering Task Force and are not part of the PDCP itself.

3.4.3.1 Services

The PDCP layer provides service data unit delivery to upper layers, and expects the services from the RLC layer:

- Data transfer in acknowledged mode;,
- Data transfer in unacknowledged mode;
- Data transfer in transparent mode;
- Segmentation and reassembly;
- In-sequence delivery.

3.4.3.2 PDCP Functions

Two modes of operation are supported: transparent and nontransparent. In the transparent mode PDCP does not change the incoming service data unit, meaning that no header is added and no compression functions are performed. The nontransparent mode refers to the case where a PDCP header is added and header adaptation functions may be performed. The modes relate, respectively, to PDCP-no-header PDU and PDCP-data PDU.

In transparent mode, the PDCP has the following functions: transfer of user data, relocation of PDCP buffer, and PDCP service data unit buffering.

In nontransparent mode, PDCP has the following functions: header adaptation, transfer of user data, relocation of PDCP buffer, PDCP service data unit buffering, and adaptation of IP streams.

In GERAN header adaptation means either header compression or possibly header removal. In header compression, network level and higher layer protocols (mainly RTP/UDP/IP or TCP/IP) are compressed before sending the data over the radio. The compression is made such that the decompressed headers are semantically identical to the original headers. Header compression is suitable for generic applications that can be used either on fixed Internet or over the RAN.

Header removal is a function that removes the network and higher layer protocol headers before sending the data over the radio. In the uplink direction headers are generated in the network. This means that in the user plane the protocol end point is actually on the network side and part of the information might inevitably be lost. Header removal was introduced to maximize spectral efficiency of voice calls and to allow the use of existing channel coding schemes for IP telephony. Consequently, it is applicable only to GERAN and IP telephony with RTP/UDP/IP Protocols. For multimedia calls, header compression should be used to ensure the correct synchronization of media streams. With header removal this is not possible because the RTP Protocol terminates on the network side. As of this writing, it is not clear whether header removal will be part of GERAN release 5.

3.4.4 RLC

This section describes the RLC protocol in GERAN. As mentioned earlier, the RLC provides a reliable radio-technology specific link. It allows for data transfer in the transparent, acknowledged, or unacknowledged mode. The GERAN RLC is based on the (E)GPRS RLC but also provides a transparent mode. Some typical applications using the GERAN RLC modes are listed in Table 3.2.

3.4.4.1 Services

The RLC Protocol provides data transfer to higher sublayers (PDCP and RRC), in the transparent, acknowledged, or unacknowledged mode. When transparent, the RLC acts as a pipe and transmits higher layer PDUs *without* adding any protocol information. When nontransparent, the RLC may either operate in unacknowledged or acknowledged mode. In unacknowledged mode, the higher layer's PDUs are transmitted without any guarantee of delivery to the peer entity. In acknowledged mode, the RLC guarantees the delivery of the transmitted higher layer PDUs to the peer entity. Upper layers are notified of errors that cannot be resolved by the RLC.

3.4.4.2 Functions

The functions of the RLC protocol enable the provision of its services, and these functions vary depending on the RLC mode.

In transparent mode the incoming SDUs are transferred to the MAC layer without any removal of the upper layer's information or inclusion of

Table 3.2
Typical Applications Versus RLC Modes

Transparent	Unacknowledged	Acknowledged
Optimized speech, for example, GSM speech, Adaptive multirate (AMR) and wideband AMR speech Optimized video	Real-time video services Voice-over IP (VoIP)	On-line e-mail Web-enhanced short messages Real-time and non-real-time video services

RLC-specific overhead. This mode is used in case of, for example, optimized speech.

In unacknowledged and acknowledged modes, the RLC operates segmentation and concatenation of RLC SDUs into RLC data blocks, which are further transmitted as RLC PDUs over the air interface. In fact, RLC-specific overhead is added to enable the reassembly of RLC data blocks into RLC SDUs at the other RLC end.

In acknowledged mode, part of the overhead also controls the operation of the backward error correction procedures used for successful delivery of the RLC PDUs. These procedures enable the selective retransmission of RLC data blocks using either the (E)GPRS's selective type I *hybrid automatic repeat request* (HIARQ) or the EGPRS's selective type II hybrid ARQ (HII-ARQ, commonly referred to as incremental redundancy, or IR) depending on the capability of the network and the MS. In IR, upon reception failure, the retransmission of a data block is not identical with its initial transmission, but includes additional redundancy, which, when combined with the earlier (re)transmission(s), can be used to correct errors. In this way, IR does not need any information about the link quality to protect the data. In fact, protection of the data block is obtained incrementally, after transmitting additional redundancy and combining it with the prevailing (re)transmission(s). IR offers a drastic increase of performance over the HIARQ. For the same bit rate and channel conditions, it produces a higher and in most cases significantly higher data rate (throughput) than HIARQ, and therefore leads to a better spectral efficiency. To comply with the delay requirements of a radio bearer (e.g., for real-time streaming services), the transmitter RLC may discard transparently to the receiver the late RLC service data units not yet segmented into RLC PDUs, that is, those that are not yet sent to the peer RLC.

To further optimize the transmission over varying channel conditions, RLC provides a link adaptation mechanism that adapts the transmission of the user data to the link quality by means of channel quality measurement reports (controlled by MAC) and proper selection of channel coding schemes according to these measurements.

Finally, the RLC Protocol performs ciphering (see Section 3.6) of the RLC data blocks when in acknowledged or unacknowledged mode to ensure the security of the connection.

3.4.5 MAC

MAC provides access signaling procedures for the radio channel and defines means for sharing a common transmission medium among several data flows from one or more MSs. Hence, MAC allows a user to access various ongoing applications (e.g., WWW and VoIP) simultaneously and optimizes the use of the radio medium by sharing it dynamically among users according to their needs.

3.4.5.1 Services

The MAC provides data transfer (unacknowledged) of upper layer PDUs between peer MAC entities over the physical layer. It also controls access to and multiplexing onto basic physical subchannels. MAC can multiplex several data flows from one or more mobile stations in a fixed or dynamic manner onto one or more *shared basic physical subchannels* (SBPSCHs). However, on a dedicated basic physical subchannel MAC can multiplex data flows from a single MS only.

3.4.5.2 Functions

For each radio bearer service (for signaling or user plane), MAC defines which logical channels to use (e.g., PDTCH, PACCH, and PTCCH) that enable the system to control and operate the data transmission of this particular radio bearer over the radio interface. To fulfill this, MAC also configures the mapping of the logical channels onto the appropriate basic physical subchannels.

For multiplexing data flows on a common radio medium (DBPSCH or SBPSCH) MAC establishes Layer 2 logical links, temporary block flows, on which MAC PDUs are carried. TBF establishment for a given radio bearer may occur either explicitly (when on SBPSCH) or implicitly (when on DBPSCH), depending on whether or not Layer 2 signaling occurs between the mobile station and the network to establish the TBF. On SBPSCH, radio resources (assignment, release, and reconfiguration) used for a TBF are handled by MAC. On DBPSCH, RRC handles the radio resources, except when a TBF is reconfigured from a DBPSCH to an SBPSCH, in which case it is the responsibility of MAC. To make multiplexing possible, MAC generates its own in-band header and appends it to the MAC SDUs (RLC PDUs). This overhead consists of, for example, the temporary block flow identity and the uplink state flag, which are used, respectively, to address/identify a flow to and from an MS, and reserve uplink radio resources to a particular MS dynamically. In that way, MAC is able to identify and, hence, monitor different traffic flows to and from one or more MSs on the basic physical subchannels.

In addition, to meet the negotiated QoS of the data flows, MAC handles priorities between temporary block flows using, for example, the attributes of the corresponding radio bearer services.

MAC also handles specific control signaling to operate the data transmission and, for an optimal use of the radio resources, multiplexes it with the data with which it is associated (e.g., PDTCH and PACCH). This signaling consists of measurement reporting (link quality measurements and neighbor cell measurements) and broadcast information (e.g., on PBCCH). It also issues from packet timing advance procedures controlled by MAC on SBPSCHs.

When the RLC is in transparent mode, the MAC protocol performs ciphering of the incoming data to ensure the security of the connection. When the RLC is in the nontransparent mode, MAC also ciphers RLC/MAC control blocks performing radio resource management functions (e.g., assignment of resources and cell reselection).

3.4.5.3 MAC Operation

To better understand MAC operation and its interaction with RRC, the state-machine of the MAC protocol is described next, followed by the relation between MAC states and RRC states. The following states are defined to describe the behavior of an MS in Iu mode (Figure 3.15):

- *MAC idle state:* The MAC Protocol is said to be in the idle state when no radio resource (neither DBPSCH nor SBPSCH) is allocated to the MS.
- *MAC shared state:* The MAC Protocol is said to be in the shared state when radio resources on SBPSCH only are allocated to the MS.
- *MAC dedicated state:* The MAC Protocol is said to be in the dedicated state when radio resources on DBPSCH only are allocated to the MS.
- *MAC dual transfer mode state:* The MAC Protocol is said to be in the dual transfer mode state when radio resources on both DBPSCH and SBPSCH are allocated to the MS.

As of this writing, the state transition between the MAC shared state and MAC dedicated state is not defined.

Table 3.3 shows the relation between MAC states and RRC states and what protocol is responsible for new resource allocation. For instance, in

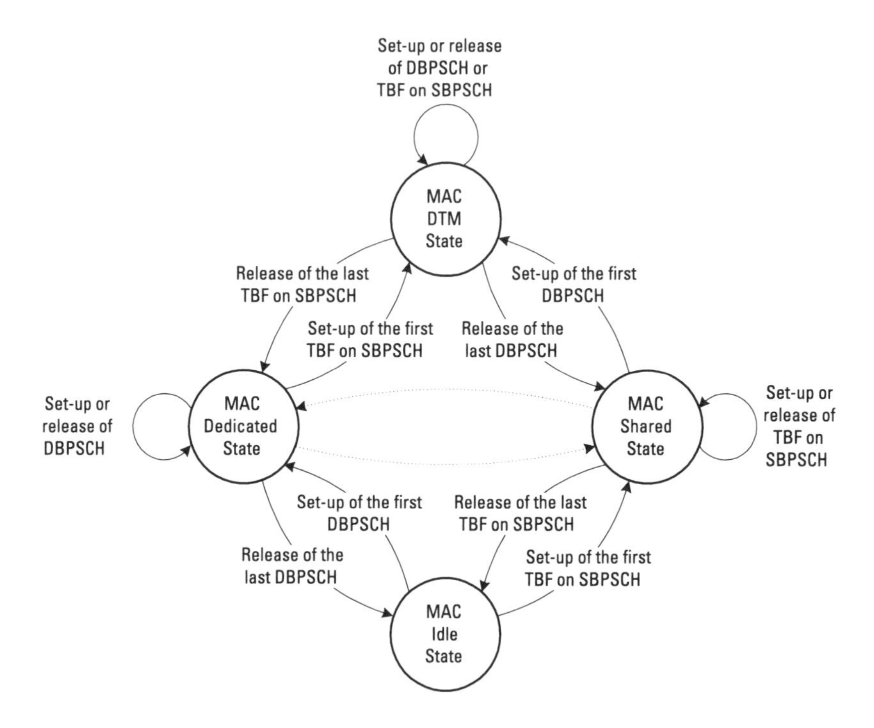

Figure 3.15 MAC state-machine.

Table 3.3
Relation Between MAC States and RRC States

Current Control Plane State		**Currently Allocated Radio Resources**		**Allocation of New Resources on**	
MAC state	RRC state	DBPSCH	SBPSCH	DBPSCH	SBPSCH
Dedicated	Cell, dedicated	x	—	RRC	RRC
dual transfer		x	x		MAC
mode	Cell, shared	—	x		
Shared		—	—	N/A	
Idle	GRA PCH	—	—		N/A
	Idle mode	—	—		

the MAC shared state (RRC cell shared state), for the MS having radio resources on SBPSCH only, the allocation of new resources on DBPSCH is RRC's responsibility, whereas it is MAC's responsibility on SBPSCH. Allocating new radio resources may result in RRC and MAC state transitions as shown previously (Figure 3.15).

3.5 GERAN Physical Layer

The physical layer or Layer 1 is the lowest layer of the protocol stack. Whereas all upper layers can rely on lower layers, the physical layer relies on the radio medium only to ensure data transmission [21]. Its main task is twofold: (1) convert logical channels delivered from Layer 2 into a radio signal, which can be sent over the air; and (2) translate the received radio signal into logical channels to be delivered to the Layer 2.

3.5.1 Services

The physical layer interfaces two upper layers: the MAC sublayer of Layer 2 and the RRC sublayer of Layer 3. It offers logical channels and the associated transmission services (including error protection and error detection functions) [22].

Furthermore, the physical layer ensures the measurement of the signal strength of neighboring cells. These measurements are transferred to the RRC. It also assesses the signal quality of the basic physical subchannel in use. On the MS side, measurement results are usually sent in the uplink for reporting to the BTS.

The physical layer, in close cooperation with Layer 3, also performs the cell/PLMN selection/reselection [23].

3.5.2 From GSM/EDGE to GERAN

For more than a decade, the physical layer of GSM/EDGE has been optimized for two different purposes: (1) the circuit-switched connection (through the A interface) and (2) the packet-switched connection (through the Gb interface).

For the circuit-switched connection, TCHs are used on the radio medium. They provide for optimized support of speech and data services. The speech channel coding includes, for instance, an unequal error protection scheme, which maximizes the user's perceived speech quality. For circuit-switched data, the channel coding comprises a long diagonal interleaving in order to benefit from the relaxed delay requirements inherent to circuit-switched data services.

For packet-switched connections, PDTCHs are used on the radio medium. Their channel coding is characterized by a short rectangular interleaving, which allows for efficient multiplexing.

The physical layer of GERAN is based on the physical layer of GSM/EDGE. It reuses TCH and PDTCH. But unlike GSM/EDGE, GERAN allows TCHs to be used for both circuit-switched and packet-switched connections. In GERAN, the optimized support of voice is thus available to both domains.

Furthermore, GERAN brings some major improvements to the physical layer:

- *8-PSK speech:* Although only data services were enhanced with the 8-PSK modulation in release 1999 of GSM/EDGE, release 5 of GERAN applies 8-PSK modulation to speech services as well. The narrowband AMR codec from GSM release 1998 is assumed to be a speech codec, and the new 8-PSK half-rate speech traffic channel is defined in order to increase the voice capacity. Note that an 8PSK quarter speech traffic channel is under consideration as well for even greater capacity gains [24].
- *AMR wideband (AMR-WB):* In release 1998 of GSM/EDGE, the AMR codec was introduced to improve the performance of narrowband speech (80–3,600 Hz) [25]. The introduction of wideband speech services (50–7,000 Hz) increases the speech quality even more, especially in terms of naturalness. The AMR-WB gives GERAN, and 3G core networks in general, a speech service for which the quality exceeds even that of wireline quality.
- *Fast power control for speech:* In release 1999 of GSM/EDGE, fast power control was standardized for the *enhanced circuit-switched data* [23]. In release 5 of GERAN, all speech services on DBPSCH now also benefit from fast power control (referred to as enhanced power control).

3.5.3 Physical Resource

The access to the physical resource is based on a TDMA scheme [26].

3.5.3.1 Time Slot

The physical resource is divided both in frequency and time (Figure 3.16). The bandwidth is partitioned into 200-kHz frequency bands: Each carrier is partitioned into 35/26 ms time slots. Thus, the basic radio resource unit is a time slot lasting ≈576.9 μs. A time slot is divided into 156.25 symbol periods. For GMSK modulation, a symbol is equivalent to 1 bit, whereas for 8-PSK modulation, one symbol corresponds to 3 bits.

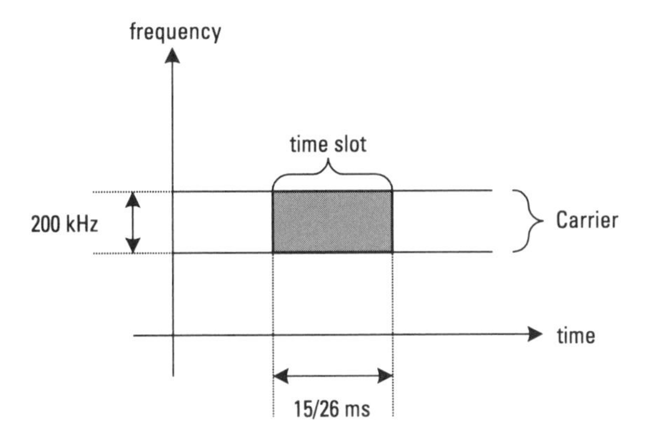

Figure 3.16 Physical resource.

3.5.3.2 Bursts

A burst carries the data that are transmitted within a time slot. Four different types of bursts exist (Figure 3.17):

1. *Normal burst (NB):* This burst is used to carry data. It contains 116 data symbols and includes a guard period of 8.25 symbols (30.46 μs). It can be either GMSK or 8-PSK modulated.

156.25 symbols—15/26 ms

Normal Burst

TS	Data Symbols	Training Sequence	Data Symbols	TS	GP
3	58	26	58	3	8.25

Frequency Correction Burst

TS	Fixed Pattern	TS	GP
3	58	3	8.25

Synchronization Burst

TS	Data Symbols	Training Sequence	Data Symbols	TS	GP
3	39	64	39	3	8.25

Access Burst

TS	Training Sequence	Data Symbols	TS	GP
8	41	36	3	68.25

GP : Guard Period / TS : Tail Symbols / numbers are shown in symbols

Figure 3.17 Bursts.

2. *Access burst (AB):* A long guard period of 68.25 symbols (252 μs) characterizes this burst. It can cope with mobiles, which do not know the timing advance at initial access. An access burst is always GMSK modulated.
3. *Frequency correction burst (FB):* This burst is equivalent to an unmodulated carrier, shifted in frequency. The mobile can use it to synchronize its frequency synthesizer after power on, or when accessing a new cell [27]. A frequency correction burst is always GMSK modulated.
4. *Synchronization burst (SB):* This burst contains a long training sequence and carries the actual TDMA frame number and base station identity code. It is broadcast together with the frequency correction burst on the same basic physical channel. The MS uses it for time synchronization and cell identification [23, 28]. A synchronization burst is always GMSK modulated.

3.5.3.3 TDMA Frame

Eight consecutive time slots of the same carrier form a TDMA frame (60/13 ms in duration). The time slots are numbered from 0 to 7 (see Figure 3.18).

3.5.3.4 Basic Physical Channels

Within a carrier, a series of time slots of consecutive TDMA frames form a basic physical channel (Figure 3.19). Because there are eight time slots per TDMA frame, eight basic physical channels exist per carrier.

Basic physical channels are organized into multiframes. Three types of multiframes are defined:

1. A 26-multiframe comprising 26 TDMA frames (120 ms). This multiframe is used to carry user data on DBPSCH only.

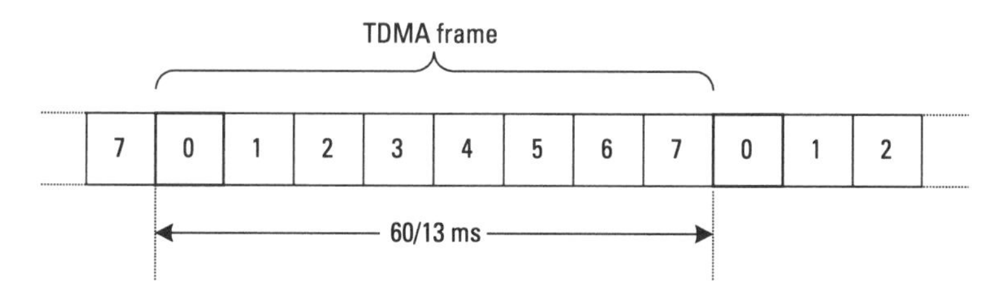

Figure 3.18 TDMA frame.

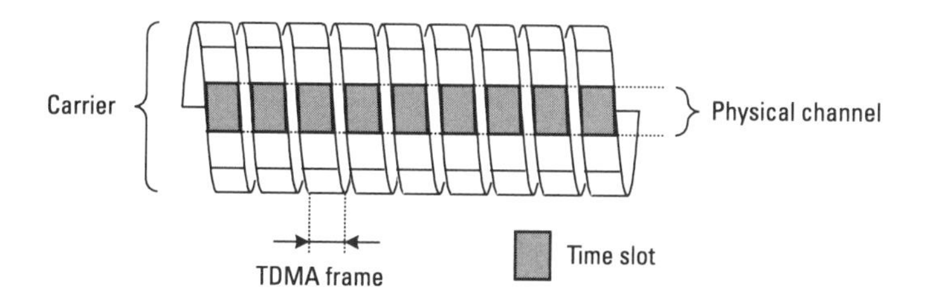

Figure 3.19 Basic physical channel.

2. A 52-multiframe comprising 52 TDMA frames (240 ms). This multiframe is used to carry packet data and control on SPSCH only.
3. A 51-multiframe comprising 51 TDMA frames (235.4 ms). This multiframe supports broadcast and common control. It always carries synchronization and frequency correction bursts. It also supports *short message service* (SMS).

3.5.3.5 Basic Physical Subchannels

A basic physical subchannel is defined as a basic physical channel or a part of a basic physical channel. Some BPSCHs are reserved by the network for common use (broadcast and common control), others are assigned to dedicated connections with MSs (DBPSCH) or are assigned to a shared usage between several MSs (SBPSCH[2]).

A DBPSCH always has a 26-multiframe structure (Figure 3.20). It can be either full rate (DBPSCH/F) or half rate (DBPSCH/H). Among the 26 time slots of a DBPSCH/F, one is always reserved for control (see SACCH in Section 3.5.4.2) and one is always idle. The idle slot allows the MS to read the frequency correction and synchronization bursts of neighboring cells for presynchronization [23, 27]. That leaves 24 time slots available for traffic every 120 ms. Each time slot is thus equivalent to 5 ms of traffic on DBPSCH/F. Similarly, each time slot is equivalent to 10 ms of traffic on DBPSCH/H.

A SPSCH always has a 52-multiframe structure (Figure 3.21). It can be either full rate (SBPSCH/F) or half rate (SBPSCH/H). An SBPSCH/H is available in the MAC dual transfer mode state only. Among the 52 time slots of a SBPSCH/F, two are always reserved for timing advance procedures (see PTCCH in Section 3.5.4.2) and two are always idle. That leaves 48

2. The SPSCH of GERAN is equivalent to the PDCH of GSM/EDGE.

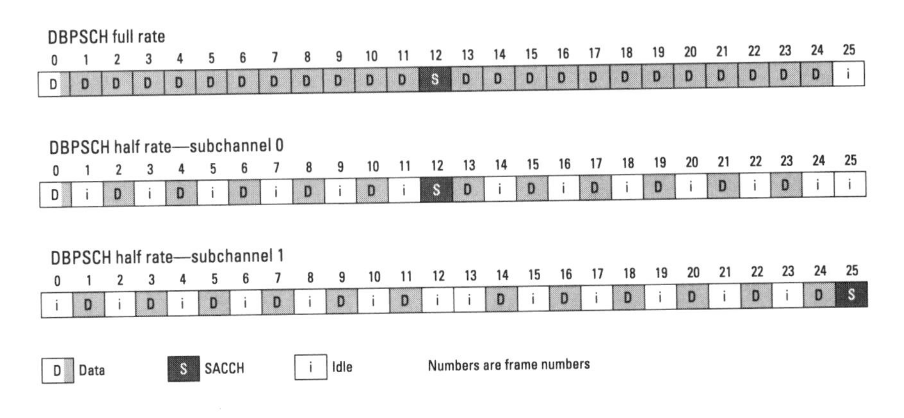

Figure 3.20 Dedicated basic physical subchannel.

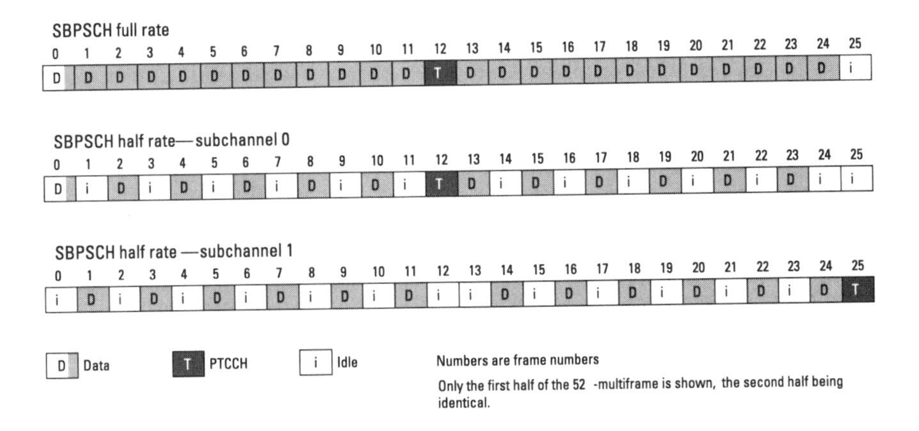

Figure 3.21 Shared basic physical subchannel.

time slots available for traffic every 240 ms. As on DBPSCH, each time slot is thus equivalent to 5 ms of traffic on SBPSCH/F. Similarly, each time slot is equivalent to 10 ms of traffic on SBPSCH/H. The types of logical channels that are allowed on a BPSCH depend on the mode of the BPSCH.

3.5.4 Logical Channels

The physical layer offers logical channels to upper layers. A logical channel is defined by the type of data it carries, and it is characterized by several parameters such as channel coding. It can be unidirectional or bidirectional. Logical channels are divided in two categories [26]: traffic channels and control channels.

Traffic channels are intended to carry either speech or user data, whereas control channels carry signaling or synchronization data. Four categories of control channels exist: broadcast, common, dedicated, and cell broadcast channels. All channels are bidirectional unless otherwise stated. All associated control channels have the same direction (bidirectional or unidirectional) as the channels with which they are associated.

The multiplexing of logical channels onto basic physical subchannels is dynamically performed by the MAC, or fixed according to the multiframe structure (e.g., SACCH on DBPSCH).

In the following, emphasis is placed on the traffic channels and their associated control channels. Note that when there is no need to distinguish between different subcategories of the same logical channel, only the generic name is used. For instance, TCH refers to all subcategories of TCH: TCH/F, TCH/H, O-TCH/F, O-TCH/H, and E-TCH, as defined next.

3.5.4.1 Traffic Channels

Traffic channels of type TCH are intended to carry either speech or user data on dedicated basic physical subchannels (for one user only). TCH can be either full rate (TCH/F) or half rate (TCH/H), and is GMSK modulated.

Traffic channels of type *octal traffic channel* (O-TCH) are intended to carry speech on dedicated basic physical subchannels (for one user only). O-TCH can be either full rate (O-TCH/F) or half rate (O-TCH/H) and is 8-PSK modulated.

Traffic channels of type enhanced traffic channel (E-TCH) are intended to carry user data on dedicated basic physical subchannels (for one user only). E-TCH is full rate and always 8-PSK modulated only.

Packet data traffic channels are intended to carry user data on either SBPSCH or DBPSCH. PDTCH is temporarily dedicated to one MS but one MS may use multiple PDTCHs in parallel. PDTCH allows several MSs to be multiplexed on the same SBPSCH. PDTCH also allows several traffic classes from the same MS to be multiplexed on the same BPSCH (shared or dedicated). PDTCHs can be either full rate (PDTCH/F) or half rate (PDTCH/H), and can be either GMSK or 8-PSK modulated.

3.5.4.2 Associated Control Channels

Fast Associated Control Channel

The FACCH[3] is used for various purposes such as handover, call establishment progress, subscriber authentication, notification, and paging [19]. The adjective fast refers to its ability to occur whenever required.

---

3. The FACCH can also be referred to as the main *dedicated control channel* (DCCH).

The FACCH is always associated with one TCH, and occurs on DBPSCH only. The FACCH is bidirectional. FACCH can be either full rate (FACCH/F) or half rate (FACCH/H). Its modulation depends on the TCH with which it is associated. FACCH is associated with TCH and is GMSK modulated, *octal fast associated control channel* (O-FACCH) is associated with O-TCH and is 8-PSK modulated, and enhanced fast associated control channel (E-FACCH) is associated with E-TCH and is GMSK modulated.

FACCH steals resources from the TCH with which it is associated. The FACCH is transmitted by preempting half or all of the information bits of the bursts. On full-rate speech traffic channels (TCH/F and O-TCH/F), FACCH/F is transmitted by stealing one speech frame. On half-rate speech traffic channels (TCH/H and O-TCH/H), FACCH/H is transmitted by stealing two consecutive speech frames. On GMSK circuit-switched data traffic channels (TCH/F and TCH/H), the FACCH is transmitted in the same way as that for the speech traffic channels. On 8-PSK full-rate circuit-switched data traffic channels (E-TCH), the FACCH is transmitted by preempting four consecutive bursts.

Slow Associated Control Channel

The SACCH is used for nonurgent procedures, mainly for the transmission of the radio measurement data needed for handover decisions [19]. Every SACCH block carries a Layer 1 header containing signaling for power control and timing advance procedures.[4] In the uplink, the MS sends measurement reports to the BTS, while in the downlink, the BTS sends commands to the MS. SACCH can also be used to convey SMS.

SACCH is associated with one TCH and/or PDTCH. It occurs on DBPSCH only and is bidirectional. Continuous transmission occurs for the SACCH in both the uplink and downlink. As a matter of fact, every 26 multiframes, one SACCH burst is always sent (Figure 3.20). One SACCH block requires four bursts to be transmitted (480 ms). Measurement reports, timing advance procedures, and power control signaling are thus exchanged on a 480-ms basis through the SACCH.

SACCH is always GMSK modulated and can be either full rate (SACCH/TF) or half rate (SACCH/TH) depending on the traffic channel with which it is associated.

---

4. Timing advance procedures are used to compensate for propagation delays.

Packet Associated Control Channel

The PACCH conveys signaling information related to packet data, including acknowledgments, power control information, measurements reports, resource assignment, and reassignment messages [8].

PACCH is associated with one PDTCH. It can occur on both DPSCH and SPSCH. The PACCH is bidirectional. PACCH/U is used for the uplink and PACCH/D for the downlink.

PACCH is always GMSK modulated and can be either full rate (PACCH/F) or half rate (PACCH/H) depending on the PDTCH with which it is associated.

Packet Timing Advance Control Channel

The PTCCH supports the timing advance procedures on SBPSCH [8]. The PTCCH is bidirectional. For description purposes PTCCH/U is used for the uplink and PTCCH/D for the downlink. In the downlink (PTCCH/D), timing advance updates for several MSs[5] are sent every 480 ms. In the uplink (PTCCH/U), access bursts are successively sent by MSs to allow estimation of the timing advance. Each MS sends one access burst on its PTCCH/U every 1,920 ms. One PTCCH/D is thus paired with several PTCCH/Us.

As for SACCH on DBPSCH, the PTCCH is part of the SBPSCH multiframe structure (Figure 3.21) and is always GMSK modulated.

3.5.5 Mapping of Logical Channels onto Basic Physical Subchannels

The multiplexing of logical channels onto physical subchannels is done according to the multiframe structure (e.g., SACCH on DBPSCH) or dynamically by the MAC. In the following paragraphs, only traffic channels and associated control channels are listed. Broadcast, common, and cell broadcast channels are not described here. A channel combination is defined as the combination of logical channels that is mapped onto a certain physical subchannel.

The following channel combinations are possible for DBPSCH full rate:

- TCH/F + FACCH/F + SACCH/TF (full-rate GMSK speech or circuit-switched data);
- PDTCH/F + PACCH/F + SACCH/TF (full-rate packet data);

5. A maximum of 16 MSs can be addressed at each occurrence of PTCCH.

- E-TCH/F + E-FACCH/F + SACCH/TF (enhanced circuit-switched data traffic);
- O-TCH/F + O-FACCH/F + SACCH/TF (full-rate 8-PSK speech).

The following channel combinations are possible for DPSCH half rate:

- TCH/H + FACCH/H + SACCH/TH (half-rate GMSK speech or circuit-switched data);
- PDTCH/H + PACCH/H + SACCH/TH (half-rate packet data);
- O-TCH/H + O-FACCH/H + SACCH/TH (half-rate 8-PSK speech).

The following channel combination is possible for SPSCH full rate:

- PDTCH/F + PACCH/F + PTCCH/F (full-rate packet data).

The following channel combination is possible for SPSCH half rate:

- PDTCH/H + PACCH/H + PTCCH/H (half-rate packet data).

3.5.6 Channel Coding

The physical layer of GERAN offers a wide range of traffic channels with different coding rates (from heavily coded to less protected) and optimized for different purposes. It ensures a flexible and efficient use of the radio medium under various channel conditions.

In this section, the main channel coding characteristics of the traffic channels are given. A complete description for all the channels can be found in [29].

3.5.6.1 Speech Traffic Channels

One speech frame is delivered to the channel encoder every 20 ms. The channel coding of the speech traffic channels is optimized to maximize the perceived speech quality. Output bits from the speech codec are divided into different classes based on their subjective importance. Bits are then unequally protected in channel coding according to their class. The coded speech frames are finally diagonally interleaved over 8 bursts on full-rate channels (Figure 3.22), and over 4 bursts on half-rate channels (Figure 3.23).

Earlier releases of the GSM standard already introduced a full-rate and half-rate speech on GMSK channels, an enhanced full-rate speech on GMSK

0	*1*	*2*	*3*	*4*	*5*	*...*	Frame number					
1	1	1	1	2	2	2	2	3	3	3	3	Even symbols
0	0	0	0	1	1	1	1	2	2	2	2	Odd symbols

Each column depicts the content of one burst.
Numbers in the table are speech frames.

Figure 3.22 Interleaving on full-rate speech traffic channels.

0	*2*	*4*	*...*	*Frame number*				
1	1	2	2	3	3	4	4	Even symbols
0	0	1	1	2	2	3	3	Odd symbols

Each column depicts the content of one burst.
Numbers in the table are speech frames.

Figure 3.23 Interleaving on half-rate speech traffic channels.

channels, and finally release 1998 introduced AMR codec on both full-rate and half-rate GMSK channels.

GERAN introduces the AMR-WB codec for high-quality speech services (Figure 3.24), and the 8-PSK narrowband AMR half-rate channel to increase the speech capacity (Table 3.4).

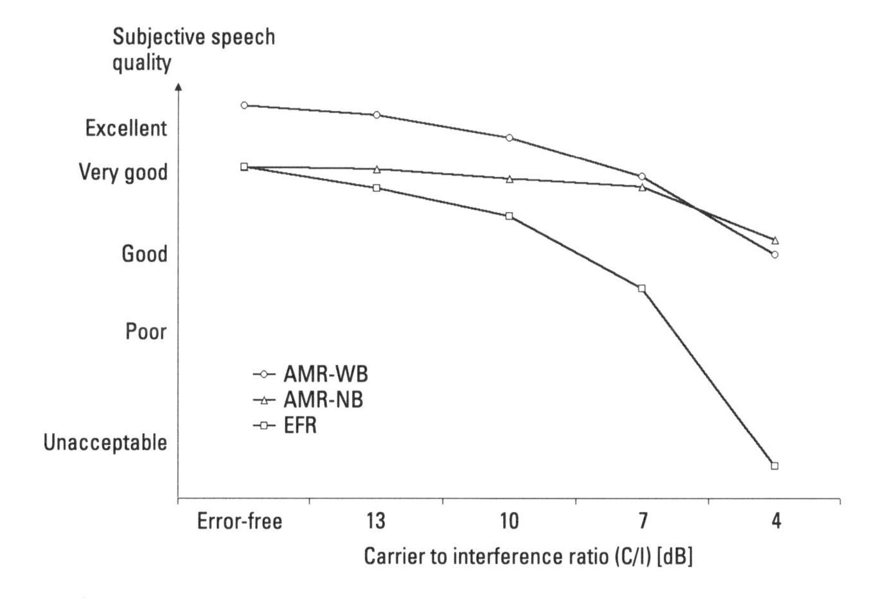

Figure 3.24 Subjective speech quality as a function of channel quality (AMR-NB versus AMR-WB versus enhanced full rate).

Table 3.4
Speech Traffic Channel Average Code Rate

Speech Codec		Bits per Speech Frame	Average Code Rate GMSK Channel Full Rate	GMSK Channel Half Rate	8-PSK Channel Full Rate	8-PSK Channel Half Rate
GSM full rate		260	0.57	X	X	X
GSM half rate		112	X	0.49	X	X
Enhanced full rate		244	0.54	X	X	X
AMR NB	12.2 Kbps	244	0.54	X	X	0.36
	10.2 Kbps	204	0.45	X	X	0.30
	7.95 Kbps	159	0.35	0.70	X	0.23
	7.4 Kbps	148	0.32	0.65	X	0.22
	6.7 Kbps	134	0.29	0.59	X	0.20
	5.9 Kbps	118	0.26	0.52	X	0.17
	5.15 Kbps	103	0.23	0.45	X	0.15
	4.75 Kbps	95	0.21	0.42	X	0.14
AMR WB	23.85 Kbps	477	X	X	0.35	0.70
	23.05 Kbps	461	X	X	0.34	0.67
	19.85 Kbps	397	0.87	X	0.29	0.58
	18.25 Kbps	365	0.80	X	0.27	0.53
	15.85 Kbps	317	0.70	X	0.23	0.46
	14.25 Kbps	285	0.63	X	0.21	0.42
	12.65 Kbps	253	0.55	X	0.18	0.37
	8.85 Kbps	177	0.39	X	0.13	0.26
	6.60 Kbps	132	0.29	X	0.10	0.19

The AMR and AMR-WB codecs provide a wide range of codec modes and two different channel modes (full rate/half rate). Each codec mode offers a different level of error protection; some are heavily encoded while some others are less protected. In good channel conditions, where little coding is necessary, the highest codec modes can be used, increasing the speech quality. Inversely, adverse channel conditions require the lowest codec modes to be used in order to maintain an acceptable level of speech quality.

Throughout a call, the channel conditions may vary considerably. When only one codec is used (e.g., GSM full rate), the frame erasure rate typically follows the quality of the channel (Figure 3.24): The worse the channel conditions are, the more frames are lost. In poor channel conditions, that may lead to unacceptable speech quality.

In AMR, by dynamically selecting the most appropriate codec mode and channel mode to the local channel conditions, the link adaptation minimizes the frame erasure rate, optimizing the speech quality and capacity.

We can see from Figure 3.24 that while the quality of the channel gets bad, the speech quality is maintained. Therefore, it provides substantial improved error robustness over the GSM full rate and enhanced full rate codecs. Table 3.4 summarizes the coding rate of the speech traffic channels.

The use of speech traffic channels allows GERAN to offer optimized support of voice services (conversational class) to both circuit-switched and packet-switched core networks.

3.5.6.2 Circuit-Switched Data Traffic Channels

Several circuit-switched data traffic channels had been standardized for GSM/EDGE, using both GMSK and 8-PSK modulation, and offering data rates from 9.6 to 43.2 Kbps in a single-slot configuration (Table 3.5). In a multislot configuration (high-speed circuit-switched data/enhanced circuit-switched data), the maximum bit rate increases to 345.6 Kbps. GERAN reuses circuit-switched data traffic channels, typically for services with relaxed delay requirements such as streaming.

Unlike speech traffic channels, the channel coding of circuit-switched data traffic channels equally protects the bits. Furthermore, thanks to more relaxed delay requirements inherent to data services, the channel coding is enhanced with longer interleaving (22 bursts instead of 8 in full-rate speech traffic channels). Table 3.5 lists the code rate and data rate of the circuit-switched data traffic channels. Here again, the wide range of coding scheme tolerates a wide range of channel conditions.

3.5.6.3 Packet Data Traffic Channels

The channel coding of PDTCH uses short rectangular interleaving in order to ensure efficient multiplexing. Every coded block is mapped onto four consecutive bursts. Each block can be allocated to a different user and can carry bits of a different traffic class. Figure 3.25 depicts a multiplexing example where three different users (A, B, and C) share the same SPSCH.

Table 3.5
Circuit-Switched Data Traffic Channels

TCH	Modulation	Code Rate	Data Rate
TCH/F9.6	GMSK	0.53	9.6 Kbps
TCH/F14.4		0.64	14.4 Kbps
E-TCH/F28.8	8-PSK	0.42	28.8 Kbps
E-TCH/F32.0		0.47	32.0 Kbps
E-TCH/F43.2		0.64	43.2 Kbps

	0 1 2 3	4 5 6 ..	*Frame number*		
...	User A streaming	User B background	User A streaming	User C background	...

Figure 3.25 Multiplexing example of SPSCH full rate.

In GPRS, four different coding schemes based on GMSK modulation were standardized (CS-1 through CS-4). In EGPRS an additional set of nine coding schemes, using both GMSK and 8-PSK modulation, were introduced (MCS-1 through MCS-9). As for TCH, each coding scheme offers a different level of error protection (Table 3.6), some are heavily encoded (e.g., MCS-1), whereas others are less protected (e.g., MCS-9). In good channel conditions, where only a little coding is necessary, the highest coding schemes can be used (e.g., MCS-4 and MCS-9). Inversely, in poor channel conditions the stronger coding schemes should be used (e.g., MCS-1) to maintain a low error rate. Throughout a connection, the channel conditions may vary considerably. By dynamically selecting the most appropriate coding scheme for the local channel conditions, link adaptation can optimize the available throughput.

The available data rates run from 8.8 to 59.2 Kbps in a single-slot configuration (Table 3.6). In a multislot configuration the maximum data rate can be brought to 473.6 Kbps. PDTCH can be used by GERAN for any traffic class.

Table 3.6
Packet Data Traffic Channels

PDTCH	Modulation	Code Rate	Data Rate
MCS-9	8-PSK	1.0	59.2 Kbps
MCS-8		0.92	54.4 Kbps
MCS-7		0.76	44.8 Kbps
MCS-6		0.49	29.6 Kbps
MCS-5		0.37	22.4 Kbps
MCS-4	GMSK	1.0	17.6 Kbps
MCS-3		0.85	14.8 Kbps
MCS-2		0.66	11.2 Kbps
MCS-1		0.53	8.8 Kbps
CS-4	GMSK	1.0	21.4 Kbps
CS-3		0.80	15.6 Kbps
CS-2		0.66	13.4 Kbps
CS-1		0.53	9.05 Kbps

3.5.7 Fast Power Control in GERAN

Power control for speech is available in GSM/EDGE through the SACCH, which enables a power control interval of 480 ms. In the downlink, power control commands are sent in every SACCH block, while in the uplink the MS reports the downlink signal quality.

Release 1999 of GSM/EDGE introduced fast power control for ECSD. FPC signaling is made in-band, allowing a control interval of 20 ms, which is 24 times faster than through the SACCH.

Release 5 of GERAN introduces fast power control for speech services on dedicated basic physical subchannels, referred to as *enhanced power control* (EPC). EPC signaling is mapped onto every SACCH burst, allowing a control interval of 120 ms (four times faster than through the SACCH). Unlike in-band signaling, it can be used on any speech traffic channel (both 8-PSK and GMSK modulated) and does not impact the speech channel coding.

The advantages of fast power control are twofold. Not only is signal quality enhanced, but the level of interference to other users is reduced. Fast power control enhances the overall spectral efficiency.

3.6 Security in GERAN

Security is provided in GERAN by means of ciphering and integrity protection. Ciphering is performed in GERAN on RLC and MAC sublayers using the same algorithm (Kasumi f8) as that defined in UTRAN in order to comply with the 3G security requirements. Similarly, integrity protection of RRC messages is performed in GERAN using the same algorithm (Kasumi f9) as that defined in UTRAN. This 3G security alignment creates a harmonized security level between the two systems [30].

3.6.1 Ciphering

This section describes the principle of ciphering used in GERAN and in UTRAN (Figure 3.26). Ciphering is intended to avoid any unauthorized acquisition of data from a third party.

The procedure for ciphering is as follows. A mask M is generated (Kasumi f8 algorithm) that is applied to the incoming data (plaintext) P to yield the ciphered data C, according to (3.1):

$$C = M \oplus P \tag{3.1}$$

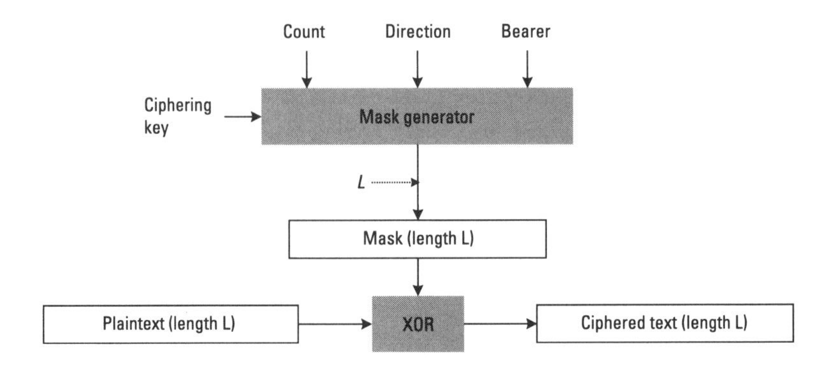

Figure 3.26 Principle of ciphering.

Several parameters (*key, counter, direction, data identifier,* and *bearer*) are needed for generating the mask, among which the most important is the ciphering key (128 bits), the other ones being to apply different ciphering processes (masks) between the blocks of one or more data flows so as to avoid using the same mask twice on two different data blocks. In fact, this would lead otherwise to a significant security loss because the ciphered and nonciphered data would then be linked without the mask:

$$\begin{array}{r} P_1 \oplus M = C_1 \\ \oplus\, P_2 \oplus M = C_2 \\ \hline P_1 \oplus P_2 = C_1 \oplus C_2 \end{array}$$

The length of the mask has to be identical to the one L of the incoming data due to the XOR process applied between the mask and the plaintext.

The different parameters are as follows:

Key: ciphering key;
Count: incoming data block number or frame number;
Direction: uplink or downlink;
Bearer: identifies the data flow.

3.6.2 Integrity Protection

Integrity protection consists of the protection of data from accidental or intentional change, deletion, or substitution (OSI). The principle of integrity protection is depicted in Figure 3.27.

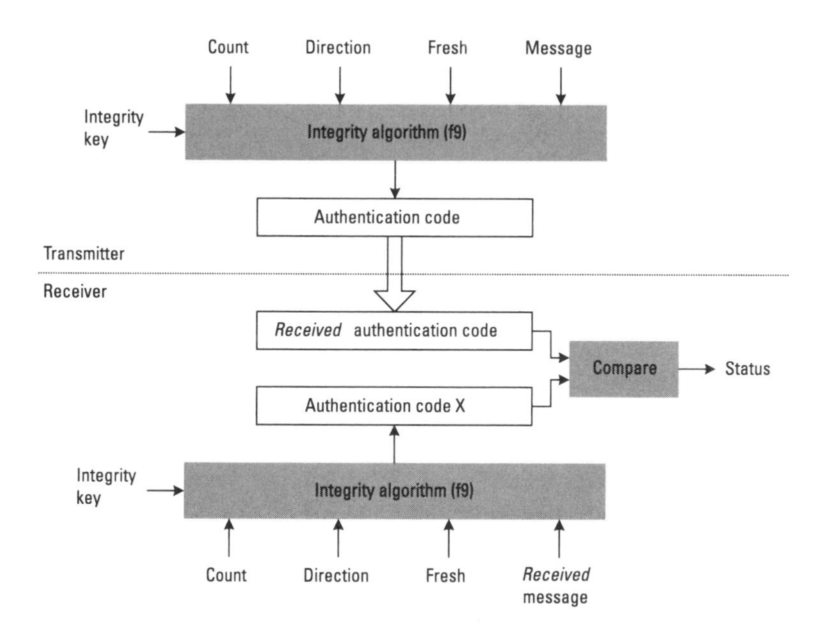

Figure 3.27 Principle of integrity protection.

The input parameters to the integrity algorithm are the *integrity key* (128 bits), the integrity sequence number (*Count*), a random value generated by the network side (*Fresh*) to protect the network against replay of signaling messages by the user, the direction, and the signaling data (*Message*). Based on these parameters, the message authentication code for data integrity is computed using the Kasumi f9 algorithm. This authentication code is then appended to the message before being sent over the air interface. With the received message, the receiver computes an authentication code X using the same other input parameters as those used on the transmitter side and verifies the data integrity of the message by comparison with the received authentication code.

3.7 Conclusion

During the last 10 years, GSM has evolved from a basic voice service to a true 3G multimedia system. Along the way several major steps have been taken including new modulation, new services, and other enhancements. The latest development, under release 5 of 3GPP, aligns the services and the functional split of GERAN with UTRAN, while keeping the backward compatibility. The existing procedures and services are supported via the A and Gb interfaces toward the 2G core network. Toward the 3G core network, the Iu interface is adopted, resulting in a new protocol architecture and various changes to existing Radio Interference Protocols. This harmonization changes in a fundamental way the tasks and behavior of the GSM/EDGE BSS. It allows GERAN to be connected to the same 3G core network and to support the same 3G services as UTRAN. In short, it allows GERAN to be part of UMTS.

References

[1] 3GPP TS 23.107, V5.2.0 (Release 5), *QoS Concept and Architecture,* Oct. 2001.

[2] 3GPP TS 23.121, V3.5.1 (Release 99), *Architecture Requirements for Release 99,* Dec. 2001.

[3] 3GPP TS 04.65, V8.2.0 (Release 99), *GPRS Subnetwork Dependent Convergence Protocol,* Sept. 2001.

[4] 3GPP TS 04.64, V8.6.0 (Release 99), *LLC Specification GPRS,* Dec. 2001.

[5] 3GPP TS 08.18, V8.9.0 (Release 99), *BSS GPRS Protocol,* Oct. 2001.

[6] 3GPP TS 08.16, V8.0.0 (Release 99), *GPRS Gb Interface Network Service,* June 2000.

[7] RFC 791, *Internet Protocol,* Sept. 1981; available at http://www.ietf.org/rfc/rfc0791.txt.

[8] 3GPP TS 44.060, V4.3.0 (Release 4), *GPRS RLC/MAC Protocol,* Sept. 2001.

[9] 3GPP TS 29.060, V4.2.0 (Release 4), *GPRS Tunnelling Protocol (GTP) Across the Gn and Gp Interface,* Sept. 2001.

[10] RFC 768, *User Datagram Protocol,* Aug. 1980; available at http://www.ietf.org/rfc/rfc0792.txt.

[11] 3GPP TS 25.323, V4.2.0 (Release 4), *PDCP Specification,* Sept. 2001.

[12] 3GPP TS 24.022, V4.0.0 (Release 4), *RLP for Data and Telematic Services on the MS-BSS Interface and the BSS-MSC Interface,* March 2001.

[13] 3GPP TS 04.21, V8.3.0 (Release 99), *Rate Adaptation on the MS-BSS Interface,* Dec. 2001.

[14] 3GPP TS 08.20, V8.3.0 (Release 99), *Rate Adaptation on the BSS-MSC Interface,* March 2001.

[15] 3GPP TS 25.413, V4.2.0 (Release 4), *UTRAN Iu Interface RANAP Signalling,* Sept. 2001.

[16] 3GPP TS 48.008, V5.2.0 (Release 5), *BSS-MSC Layer 3 Specification,* Sept. 2001.

[17] 3GPP TS 25.423, V4.2.0 (Release 4), *UTRAN Iur Interface RNSAP Signalling,* Sept. 2001.

[18] 3GPP TS 23.110, V4.0.0 (Release 4), *UMTS Access Stratum—Services and Functions,* April 2001.

[19] 3GPP TS 44.018, V5.2.0 (Release 5), *Radio Resource Control Protocol*, Sept. 2001.

[20] 3GPP TS 25.331, V4.2.0 (Release 4), *Radio Resource Control (RRC) Protocol Specification*, Sept. 2001.

[21] 3GPP TS 45.001, V5.1.0 (Release 5), *Physical Layer on the Radio Path*, Aug. 2001.

[22] 3GPP TS 04.04, V8.1.1 (Release 99), *Layer 1—General Requirements*, Sept. 2001.

[23] 3GPP TS 45.008, V8.0.0 (Release 5), *Radio Subsystem Link Control*, Aug. 2001.

[24] Bellier, T., et al., "Quarter Rate AMR Channels—A Speech Capacity Booster in GSM/EDGE," *Proc. 3rd Int. Symp. Wireless Personal Multimedia Communications*, Bangkok, Thailand, Nov. 12–15, 2000, pp. 1033–1038.

[25] GSM TS 06.75, V7.2.0 (Release 1998), *Performance Characterization of the GSM.*

[26] 3GPP TS 45.002, V5.2.0 (Release 5), *Multiplexing and Multiple Access on the Radio Path*, April 2001.

[27] 3GPP TS 05.10, V8.8.0 (Release 99), *Radio Subsystem Synchronization*, March 2001.

[28] 3GPP TS 43.022, V4.3.0 (Release 4), *Functions Related to MS in Idle Mode and Group Receive Mode*, June 2001.

[29] 3GPP TS 45.003, V5.2.0 (Release 5), *Channel Coding*, Aug. 2001.

[30] 3GPP TS 33.102, V4.2.0 (Release 4), *3G Security—Security Architecture*, Sept. 2001.

Appendix 3A: Radio Access Bearer Realization

The purpose of this appendix is to describe how GERAN can configure the user plane protocol stack to support the desired radio access bearer classes (conversational, streaming, interactive, and background). Only traffic over the Iu-ps interface is considered.

3A.1 Conversational

Table 3A.1 describes how GERAN can configure the protocol stack of the user plane in order to meet the QoS requirements of the conversational traffic class. Generally, the MAC should be in either the dedicated or dual transfer mode state and the RLC in either transparent or unacknowledged mode in order to guarantee low delays.

The optimized support of speech is available for the speech codecs of which the channel coding is optimized (GSM full rate, GSM half rate, enhanced full rate, AMR, and AMR-WB). If the speech codec of the radio bearer is one of them, a speech traffic channel can be used on the physical layer. That requires header removal on PDCP, a transparent RLC, and an MAC in either the dedicated or dual transfer mode state. This case is referred to as the *optimized speech* case. As a matter of fact, from a quality and capacity

Table 3A.1
Conversational Radio Access Bearer

PDCP	RLC	MAC State	Physical Layer BPSCH	Physical Layer Traffic Channel
Nontransparent (header removal)	Transparent	Dedicated or dual transfer mode	DBPSCH/F DBPSCH/H	TCH/F for speech O-TCH/F for speech TCH/H for speech O-TCH/H for speech
Nontransparent (header compression or no adaptation) or transparent	UNACK	Dedicated or dual transfer mode	DBPSCH/F DBPSCH/H	PDTCH/F TCH/F for data E-TCH/F PDTCH/H

perspective, it is the most optimized way to provide speech services in GERAN.

For other codecs, or when full IP transparency is required, PDTCH or data TCH is used on the physical layer. This case is referred to as *generic conversational*.

3A.2 Streaming

Table 3A.2 describes how GERAN can configure the protocol stack of the user plane in order to meet the QoS requirements of the streaming traffic class. The RLC can be either acknowledged or unacknowledged depending on the need for error-free delivery. The MAC can be in the shared, dedicated, or dual transfer mode state. When in the dedicated or dual transfer mode state, the circuit-switched data traffic channels can be used on DBPSCH, thanks to the relaxed delay requirements inherent to streaming services.

3A.3 Interactive and Background

Table 3A.3 describes how GERAN can configure the protocol stack of the user plane in order to meet the QoS requirements of the interactive and background traffic classes. To guarantee error-free delivery, the RLC is set in the acknowledged mode. Delay and multiplexing constraints require PDTCH to be used on the physical layer.

Table 3A.2
Streaming Radio Access Bearer

PDCP	RLC	MAC State	Physical Layer BPSCH	Physical Layer Traffic Channel
Nontransparent (header compression or no adaptation) or transparent	UNACK	Dedicated or dual transfer mode	DBPSCH/F	PDTCH/F TCH/F for data
			DBPSCH/H	E-TCH/F PDTCH/H
	ACK	Dedicated or dual transfer mode	DBPSCH/F	PDTCH/F TCH/F for data
			DBPSCH/H	E-TCH/F PDTCH/H
	UNACK	Shared or dual transfer mode	SBPSCH/F	PDTCH/F
		Dual transfer mode	SBPSCH/H	PDTCH/H
	ACK	Shared or dual transfer mode	SBPSCH/F	PDTCH/F
		Dual transfer mode	SBPSCH/H	PDTCH/H

Table 3A.3
Interactive and Background Radio Access Bearer

PDCP	RLC	MAC State	Physical Layer BPSCH	Physical Layer Traffic Channel
Nontransparent (header compression or no adaptation) or transparent	ACK	Dedicated or dual transfer mode	DBPSCH/F	PDTCH/F
			DBPSCH/H	PDTCH/H
	ACK	Shared or dual transfer mode Dual transfer mode	SBPSCH/F	PDTCH/F
			SBPSCH/H	PDTCH/H

4

CDMA2000 High-Rate Packet Data System

Qiang Wu and Eduardo Esteves

4.1 Introduction

CDMA technology was introduced in cellular systems in the early 1990s with the development of the IS-95 standard. Since then, the technology has been widely deployed throughout the world, reaching the 90-million subscriber mark in early 2001. The experience gained through commercial deployment of IS-95-based networks has shown that on average a 10-times increase in voice capacity versus *first-generation* (1G) analog systems is achievable. In addition, the IS-95 system has significantly evolved and matured in the last decade resulting in the enhanced revisions IS-95-A and IS-95-B in 1994 and 1998, respectively.

The benefits and success of the CDMA technology have been well accepted by the wireless industry such that the two major 3G standards are both CDMA-based technologies. One of these standards, CDMA2000, was developed as a backward-compatible evolution of IS-95 systems to further improve voice services capacity while providing higher data rates for data services. As a result of its enhanced capabilities and simplified migration path, commercial deployment of CDMA2000-based 3G systems began as early as 2000 in South Korea.

Before describing the technical details of the CDMA2000 high-rate packet data system, we first review the evolution of the several CDMA-based standards and their relation to the harmonized 3G mobile communication systems endorsed by the *International Telecommunication Union* (ITU) and known as IMT-2000. In an area overwhelmed by acronyms and rich in technology diversification, this is very important so that the reader will have an overall understanding of the different components that form the 3G CDMA technology.

In Figure 4.1, the timeline of the major CDMA-based standards is shown as they combine to form the basis of the CDMA group within IMT-2000. Both *time-division duplex* (TDD) and *frequency-division duplex* (FDD) modes are supported by IMT-2000. The more common FDD mode uses different frequency bands for the uplink and downlink carriers separated by a fixed frequency, while TDD systems use the same frequency band for both uplink and downlink. Most spectrum allocations for 3G systems are based on paired frequencies (FDD); however, the TDD mode may be used in unlicensed bands and those portions of spectrum for which an FDD allocation is impossible. This mode is accomplished by a TDD variation of WCDMA with an optional spreading rate of 1.28 Mbps and by a mode based on synchronous CDMA (TD-SCDMA). The reader is referred to [1] for further information on the TDD mode of the IMT-2000 CDMA system. The FDD mode consists of the WCDMA and CDMA2000 systems. WCDMA uses direct-sequence spread spectrum with a chip rate of 3.84 Mbps and occupying 5 MHz of spectrum in each uplink and downlink carrier.

The WCDMA standardization process evolved from the first release by the end of 1999 to an updated version (release 4) in early 2001. Release 5, which was expected to occur in 2002, will contain major revisions and additions including a high-speed data packet access mode that relies on similar techniques introduced in the CDMA2000 high-rate packet data (IS-856) system. The other FDD component of IMT-2000 is the family of standards commonly referred to as CDMA2000. The timeline evolution of the major standards that are supported in CDMA2000 is described in Figure 4.1. IS-95-A and its PCS version, J-STD-008, appeared around 1995. After a few years of deployment experience, IS-95-B was introduced and included enhancements that improved access and handoff procedures in addition to other features such as support of medium-rate data services, position location, and global roaming. The IS-95-A/B and several related standards form the basis of the 2G cellular technology known as cdmaOne.

The 3G evolution of cdmaOne consists of a family of standards, known as CDMA2000, which first appeared with the publication of the IS-2000 release 0 in 1999. Shortly after, the release A version of IS-2000 was published

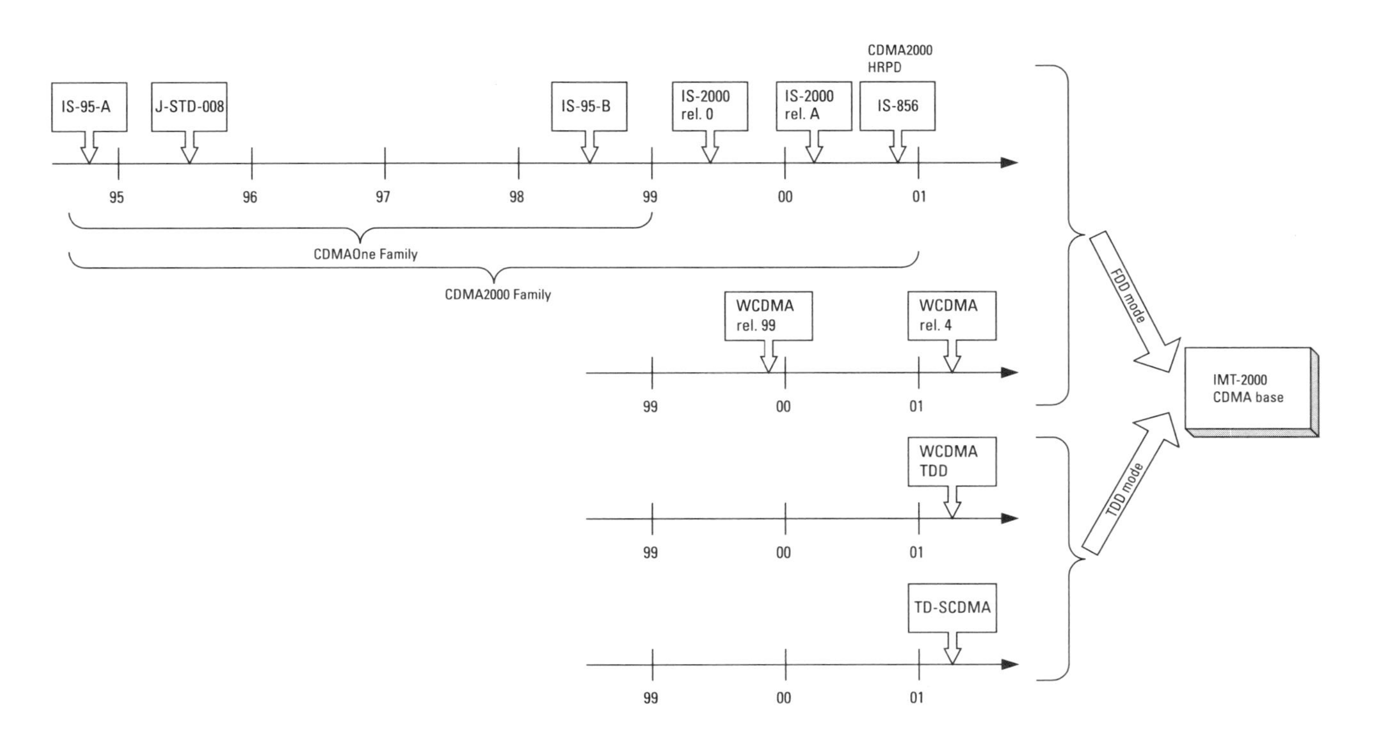

Figure 4.1 Evolution of the CDMA standardization toward 3G.

in 2000 with the inclusion of additional signaling support for features such as new common channels, QoS negotiation, enhanced authentication, encryption, and concurrent services. As a result of the harmonization work with the WCDMA proposal in 1999, the wideband direct-sequence mode of IS-2000 using the 3.6864-Mbps chip rate was withdrawn from the standard in release A. On the other hand, the standard includes both the narrowband (1×) and the wideband (3×) multicarrier modes occupying 1.25 and 3.75 MHz, respectively. In addition, the reverse link of the 3× multicarrier mode can optionally be transmitted using either a 1.2288- or 3.6864-Mbps chip rate. The hybrid 3× forward/1× reverse mode leverages existing BS receiver and mobile transmitter designs while increasing peak rates for asymmetric packet data services. The CDMA2000 system was designed to be backward compatible with existing cdmaOne networks and voice terminals. This characteristic provides a graceful migration path for existing 2G CDMA networks toward high-data-rate, multiservice 3G systems.

The IS-2000 standard [2] introduces several new features as compared to 2G wireless systems. Among those, the introduction of fast forward power control, QPSK modulation, lower code rates, powerful turbo coding, pilot-aided coherent reverse link, and support for transmit diversity are considered the major capacity-enhancing features in IS-2000. The combination of all of the above techniques results in a factor of 2 increase in voice capacity as compared to IS-95. In addition, high data rates can be achieved in IS-2000 by aggregating three standard 1.25-MHz carriers in a multicarrier forward link signal. Another important feature introduced in CDMA2000 is the support of both IS-41 (native to IS-95) and *GSM-Mobile Application Part* (GSM-MAP) network connectivity. This means that while the radio interface is handled according to the CDMA2000 specification, call control, mobility management, and other network aspects can also be handled as they are in the GSM Signaling Protocol. This facilitates the worldwide roaming capabilities of GSM-MAP–enabled terminals.

Even though the IS-2000 standard introduced new features that significantly improve voice capacity and data services, the design was not optimized for high-speed IP traffic. For example, the highest transmit rate on the forward link is 307.2 Kbps using the 1× spreading version (1.25 MHz). A new radio technology was designed to provide efficient high-rate packet data services without the constraints of supporting legacy modes in IS-95.[1] As a result, a major addition to CDMA2000 was accomplished by the introduction

1. Although RF characteristics similar to those of cdmaOne are maintained in order to leverage cdmaOne transceiver designs.

of the high-rate packet data system (IS-856) [3] by the end of 2000. As we will see in the next section and throughout this chapter, the CDMA2000 high-rate packet data standard (IS-856) achieves these goals by introducing new features such as virtual SHO, rate adaptation through closed-loop rate control and hybrid-ARQ, and multiuser diversity scheduling.

4.2 Overview of the IS-856 Basic Concepts

The IS-856 standard is optimized for wireless high-speed packet data services. Because of the typical asymmetric characteristics of IP traffic, the forward link is the more critical of the two links [4]. Thus, several techniques were introduced in IS-856 so as to optimize the forward link throughput.

The IS-856 forward link uses a *time-division-multiplexed* (TDM) waveform, which eliminates power sharing among active users by allocating full sector power and all code channels to a single user at any instant. This is in contrast to *code-division-multiplexed* (CDM) waveform on the IS-95 forward link, where there is always an unused margin of transmit power depending on the number of active users and power allocated to each user. This margin is used to account for large variations of the required per-user transmit power in fading channels in order to guarantee a given target frame error rate. Figure 4.2 shows the sector power usage of the IS-95 and IS-856 forward links. Note that each channel (Pilot, Sync, Paging, and Traffic channels) in IS-95 is transmitted the entire time with a certain fraction of the total sector power, whereas the equivalent channel in IS-856 is transmitted, at full power, only during a certain fraction of time. This efficient use of the sector power resource in IS-856 not only improves cell coverage, but also improves the SINR for noise-limited users.

The IS-856 system consists of a network of servers (BSs) such that a terminal may maintain radio connection with one or more servers, which constitute the active set for the terminal. The concept of and maintenance procedure of the active set in IS-856 are identical to those in IS-95. First, the access terminal reports to the access network the strongest forward link pilots it can measure. In turn, the network selects an active set for each terminal. Each sector in the terminal's active set maintains a connection (an assigned forward traffic channel and reverse traffic channel demodulation resource) with the terminal. For a terminal, its active set of sectors is also the set of power controlling sectors for its reverse link. However, instead of transmitting equal power on all forward traffic channels in the active set as

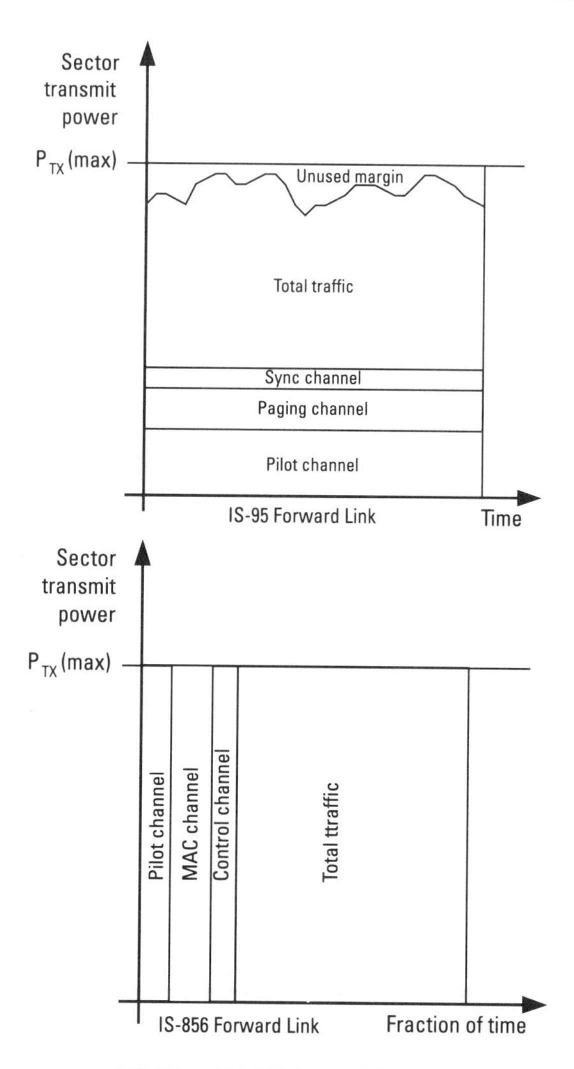

Figure 4.2 Power usage of IS-95 and IS-856 forward links.

adopted in IS-95, the IS-856 network only transmits on the best link and allocates no power on the others.

To accomplish this procedure, a terminal monitors the SINR of all sectors in its active set and informs the network, via a feedback channel, of the identity of the selected serving sector. Such a scheme, called virtual SHO, eliminates the SHO overhead on the forward link. In addition, one can show that this is the optimal strategy for maximizing the received SINR normalized by the total transmit power. Figure 4.3 shows an example of

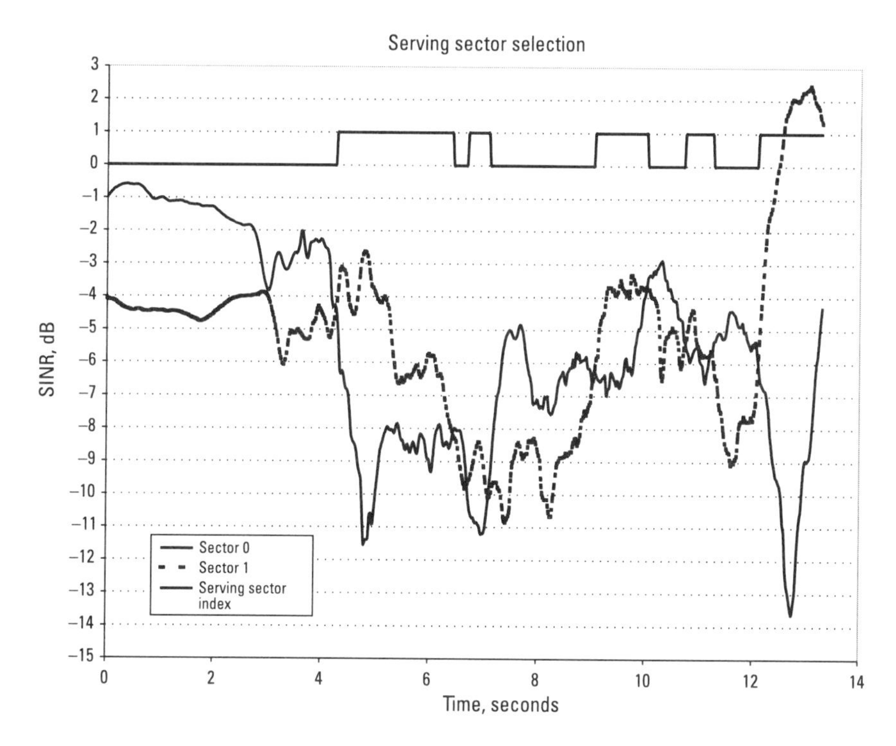

Figure 4.3 Serving sector selection for virtual SHO.

serving sector selection between two sectors using the virtual SHO. A change in the serving sector index indicates a virtual SHO event. As observed from the figure, the terminal attempts to choose the serving sector with the highest SINR in order to achieve selection diversity gain. Thus, virtual SHO is sometimes referred to as (cell) SSTD. It should be clear, however, that the concept is applicable to selection among any active sectors, not just among cell sites.

Due to the TDM waveform of the IS-856 forward link, a terminal is allocated the full sector power whenever it is served; thus, no power adaptation is needed. Rather, rate adaptation is used on the IS-856 forward link. The analysis in Section 4.5.1.2 shows that rate adaptation has an advantage over power adaptation for wireless packet data systems where a guaranteed quality of service is not required. Here, we focus only on the techniques that IS-856 employs to achieve rate adaptation. In general, the highest data rate that can be transmitted to each terminal is a function of the received SINR from the serving sector. This is typically a time-varying quantity, especially for mobile users. To achieve the highest data rate at each time of transmission, each terminal predicts the channel condition over the next packet for its serving sector based on the correlation of the channel states. It selects the highest data rate that can be reliably decoded based on the predicted SINR, and then informs the serving sector its selected rate over the reverse link feedback channel. Whenever the network decides to serve a terminal, it transmits at the most recent selected rate fed back from the terminal. This procedure is referred to as closed-loop rate control.

Figure 4.4 shows the closed-loop rate control performance for a pedestrian user (one-path Rayleigh, Doppler frequency of 5 Hz, and average SINR at 0 dB). Each dot in the plot represents a rate selection event, with the x coordinate showing the selected data rate and the y coordinate showing the measured SINR during the packet transmission. The pair of bars, drawn for each data rate, shows the required SINR thresholds for the rate and the next higher rate. Ideally, the actual received SINR during each packet transmission should always fall between the two thresholds. If a dot falls out of the range of the corresponding pair of bars, then the rate selection is either too conservative if it is higher than the upper bar or it is too aggressive if it is lower than the lower bar.

From Figure 4.4 we can draw the following conclusions for slow fading channels: First, the rate selection for high data rates is quite accurate because

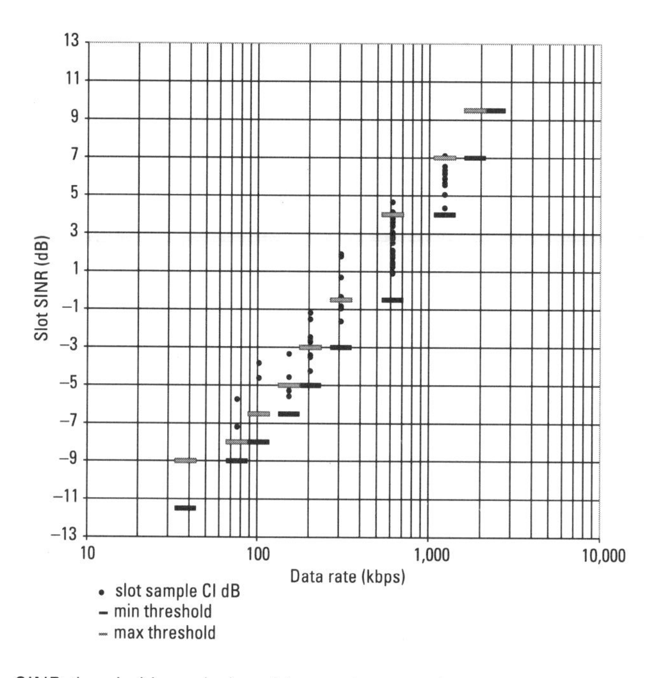

Figure 4.4 SINR thresholds and closed-loop rate control.

the majority of the dots fall into the range of the corresponding pairs of bars. This is because high data rates employ short packet lengths, and the channel state over the next packet is highly correlated with the present and past channel states. Consequently, the terminal can accurately predict the maximum data rate for the next packet. Second, the rate selection for low data rates is quite conservative as evidenced by the many dots located above the corresponding upper bars. This is because low data rates employ long packet lengths and the channel state over the next packet is not so correlated with the present and past channel states. Thus, the terminal has to be conservative when selecting a data rate so as to maintain a satisfying level of packet error rate. For slow fading channels, hybrid-ARQ and the multiuser diversity scheduler will minimize the impact of the conservative rate selection at low data rates. As further explained later in this section, hybrid-ARQ terminates multislot packet transmissions whenever a packet can be correctly decoded while a capacity-enhancing scheduler attempts to serve a user only at its local peaks (higher data rates).

As we have discussed, when the channel states are highly correlated, the access terminal obtains an accurate prediction of the future channel. The terminal can thus select data rates that are well adapted to the varying channel conditions. As the channel becomes less correlated due to the increased vehicle speed, packet length, or unpredictable traffic load in the adjacent interfering sectors, the data rates requested by the terminal must necessarily become more conservative. The conservative rate selection can result in a waste of transmitting power. To reduce the excess power, IS-856 defines a hybrid-ARQ mechanism that can terminate the transmission of a multislot packet as soon as it can be correctly decoded. To accomplish this, a terminal attempts to decode a packet whenever it receives a new portion of the packet (a new slot), and it informs the network to stop transmitting when the packet is correctly decoded. When the network receives an ACK indication, it will not transmit the remaining portion of the packet, effectively reducing the excess transmit power. As discussed in Section 4.4.1, to allow time for decoding the packet and informing the access network, multislot packets are transmitted with a four-slot interlaced timing. Figure 4.5 shows a typical reduction of the excess transmitting power due to ARQ for the fast vehicular channel at 120 km/hr.

Because a sector transmits traffic data to a single user at any instant of time, a scheduling algorithm is implemented in each sector in order to fairly allocate the available time slots among the active users and, at the same time, maximize the capacity by exploiting the channel dynamics.

Because different users experience independent fading processes, it is unlikely that all users' SINR will fall into deep fades at the same time. Most

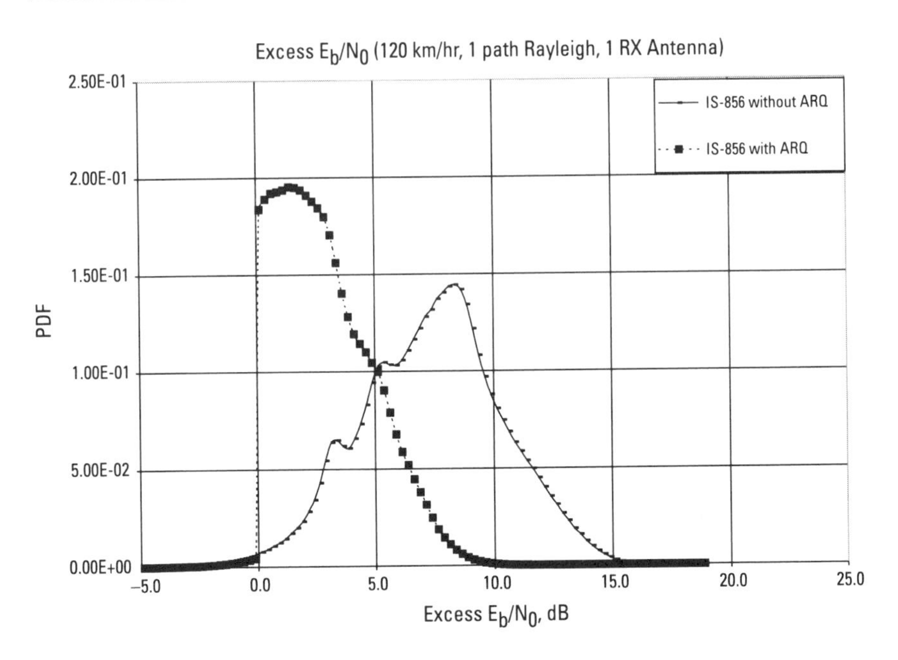

Figure 4.5 Excess transmitting power reductions due to ARQ.

likely, when some users experience a deep fade, others reach peaks in their received signal strength. Contrary to voice services, IP traffic can tolerate relatively longer and variable time delays. A smart scheduler will attempt to serve an active user near its peak SINR while maintaining a certain degree of fairness. The IS-856 standard does not specify the data scheduler; therefore, manufacturers may include their own efficient algorithms.

The proportional fair scheduling algorithm, which is further investigated in Section 4.5.1.3, is a good baseline given that it incorporates the two important features of a capacity enhancing scheduler: multiuser diversity gain and fairness. The algorithm selects the terminal based on a metric equal to the ratio of the instantaneous channel state to the long-term average of the served throughput. Thus, it attempts to serve each terminal at its local peak of channel conditions and maintain higher average throughput when the terminal is experiencing better coverage.

Figure 4.6 shows the histogram of the served rates of an equal time round-robin scheduler and a proportional fair scheduler for 16 single-antenna pedestrian users (one-path Rayleigh, 3 km/hr) obtained from a network simulation. We can observe that the proportional fair scheduler increases the serving probability of the higher data rates. In this example, the proportional fair scheduler attempts to serve each terminal for approximately the

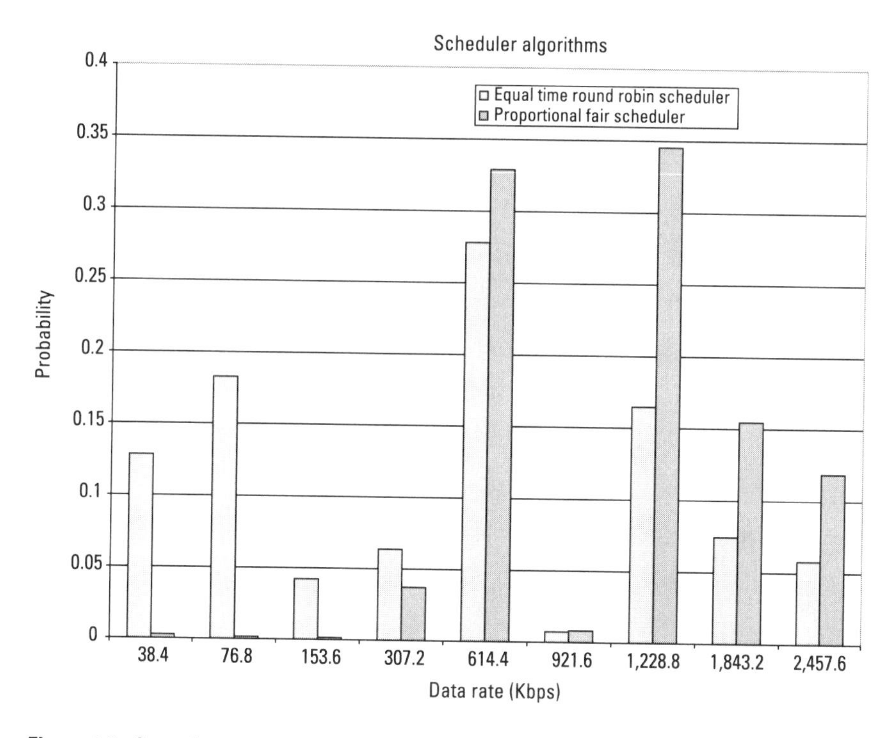

Figure 4.6 Served rates of different scheduling algorithms.

same amount of time. Thus, a proportional fair scheduler is a more efficient equal-time scheduler that takes advantage of the variability of the mobile channel experienced by different terminals. The gain in throughput is referred to as scheduler gain or multiuser diversity gain, which is analyzed further in Section 4.4.

In Sections 4.4, 4.5, and 4.6 we describe the structure and analyze the performance of the IS-856 system in much more detail.

4.3 Overlaying IS-2000 and IS-856 Systems

A hybrid access terminal is defined as an MS that can operate while on both IS-2000 and IS-856 networks capable of handing off between the two systems. This hybrid terminal is capable of receiving voice, short messaging, and dedicated data services on IS-2000 networks and high-speed packet data on IS-856 networks.

IS-2000 and IS-856 systems can be overlaid with a frequency multiplexing scheme similar to the one shown in Figure 4.7.

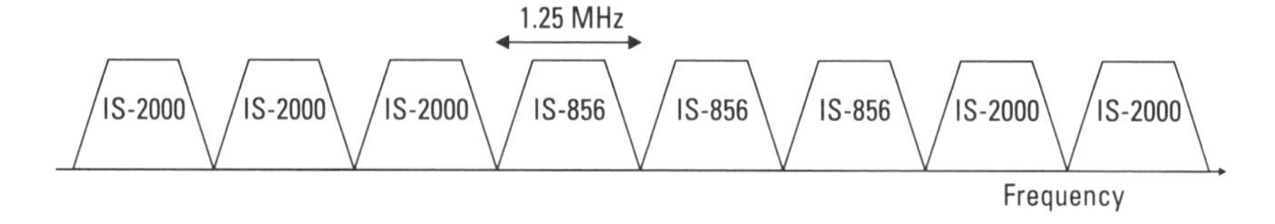

Figure 4.7 Frequency allocation for overlaying IS-2000 and IS-856 systems.

While idle for a long period on either system, a hybrid terminal will perform slotted operation, which optimizes standby time by monitoring the IS-2000 common channels (i.e., F-QPCH, F-PCH, or F-CCCH/F-BCCH, whichever is applicable) or IS-856 control channel in accordance with the slot cycle. The hybrid terminal monitors the IS-2000 common channels to receive incoming pages for voice and SMS.

If a hybrid terminal has an active connection with the IS-856 system, it periodically tunes to the frequency of the IS-2000 system to monitor the IS-2000 common channels, and then tunes back to the IS-856 frequency to resume the active data session. The assigned IS-856 control channel cycle can be selected such that it does not overlap with the assigned IS-2000 paging slot.

If a hybrid terminal becomes idle on the IS-856 system, it monitors the IS-856 control channel and IS-2000 paging channel sequentially. This allows the hybrid terminal to resume high-speed packet data service on the IS-856 system and to receive incoming voice and SMS on the IS-2000 system. To reduce the amount of time that a hybrid terminal performs concurrent slotted operation on both the IS-2000 and IS-856 systems, the terminal may transfer its IS-856 data session to the IS-2000 system and cease monitoring the IS-856 control channel. The reduction in the amount of time that a hybrid terminal must perform concurrent slotted operation results in an increase in the battery life of the terminal. Transferring the terminal's data session to the IS-2000 system ensures that the packet data serving node sends the packets destined for the hybrid terminal to the IS-2000 system and, therefore, the hybrid terminal does not miss packets that are destined for it as a result of not monitoring the IS-856 control channel. When idle on the IS-2000 system only, a hybrid terminal may perform periodic off-frequency searches in order to discover existing IS-856 systems. Also, once tuned to an IS-2000 system, the hybrid terminal may use the preferred roaming list in order to search for an IS-856 system that exists within the same geographical location.

4.4 IS-856 Physical Layer Description

4.4.1 Forward Link

The IS-856 forward link consists of the following TDM channels: the pilot channel, the MAC channel, the forward traffic channel, and the control channel. The MAC channel consists of three subchannels: the reverse activity channel, the *data rate control* (DRC) Lock channel, and the reverse power control channel. Figure 4.8 shows the forward channel hierarchies.

4.4.1.1 The Forward Link Waveform

An IS-856 forward link carrier is allocated 1.25 MHz of bandwidth and is direct-sequence spread at the rate of 1.2288 Mbps. The forward link transmission consists of time slots having a length of 2,048 chips (1.66 ms). Groups of 16 slots, referred to as frames, are aligned to the CDMA system time. Within each slot, the pilot, MAC, and traffic or control channels are time-division multiplexed as shown in Figure 4.9 and are transmitted at the same power level. A slot during which no traffic or control data are transmitted is referred to as an idle slot. During an idle slot, the sector transmits the pilot and MAC channels only, thus reducing interference to other sectors.

The overall channel structure of the forward link physical layer is shown in Figure 4.10 and is briefly described in the following sections.

4.4.1.2 Pilot Channel

The pilot channel transmits an unmodulated signal with full sector power during the 96 chips at the center of every half slot (see Figure 4.9). The

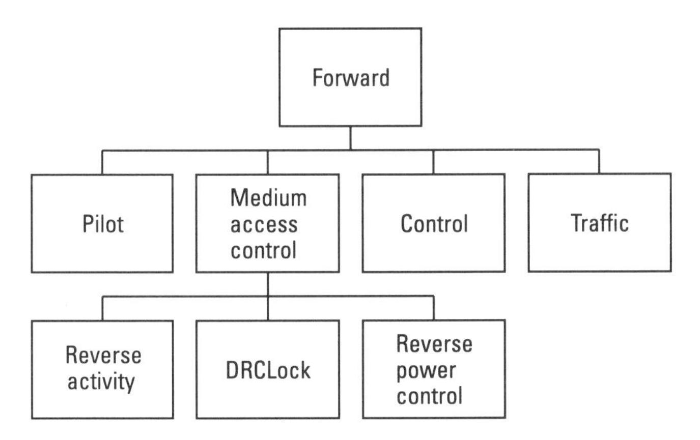

Figure 4.8 Forward channel hierarchies.

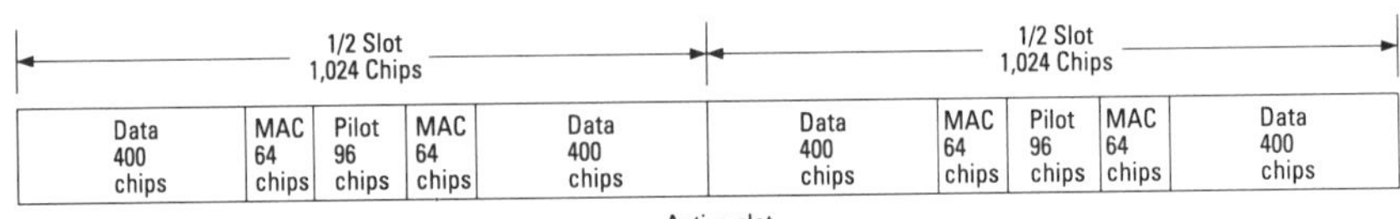

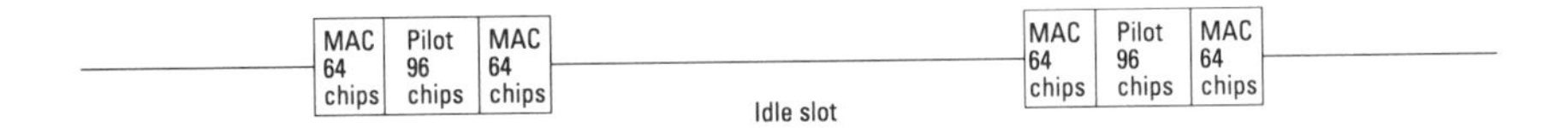

Figure 4.9 Forward link slot structure.

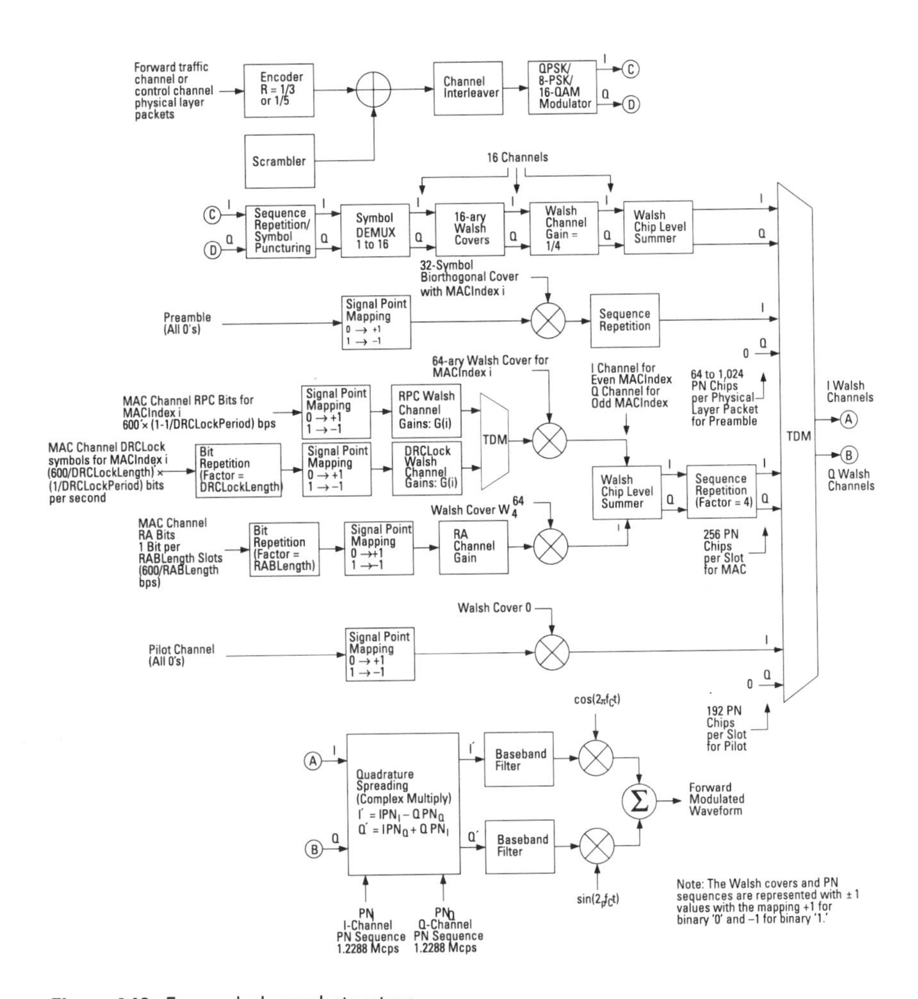

Figure 4.10 Forward channel structure.

unmodulated signal is multiplexed on the I channel with Walsh cover 0 (see Figure 4.10). The pilot channel of each sector is distinguished by the *pseudonoise* (PN) offset in increments of 64 chips. Concerning the pilot channel, the only difference between IS-95 and IS-856 is that the former transmits a continuous pilot signal while the latter transmits a gated pilot signal. The access terminal uses the pilot for initial acquisition, phase recovery, timing recovery, and symbol combining. An additional function of the pilot in IS-856 is to provide a channel estimate for the purpose of rate adaptation.

4.4.1.3 Forward MAC Channel

The MAC channel is transmitted in the 256 chips surrounding the two pilot bursts every slot as depicted in Figure 4.9. It consists of Walsh channels that are orthogonally covered and *binary phase shift-key* (BPSK)–modulated on a particular phase of the carrier (either in-phase or quadrature phase). A Walsh channel is identified by a MACIndex value that is between 0 and 63, which defines a unique 64-ary Walsh cover and a unique carrier phase as follows:

- For MACIndex i the assigned Walsh function shall be:

$$W_{i/2}^{64} \quad \text{for } i = 0, 2, \ldots, 62$$

$$\overline{W_{(i-1)/2+32}^{64}} \quad \text{for } i = 1, 3, \ldots, 63$$

  where W_i^N is the ith row of an N by N Walsh-Hadamard matrix.

- The Walsh channels with even-numbered MACIndex values are assigned to the in-phase modulation phase, while those with odd-numbered indexes are assigned to the quadrature modulation phase.

Table 4.1 specifies MAC channel use versus MACIndex.

As discussed in Section 4.5.2, the reverse activity channel is used by the reverse MAC algorithm to control the total interference level received in a given sector. The reverse activity channel transmits the *reverse activity bit* (RAB) stream over the Walsh channel with MACIndex 4. The RAB is transmitted over a certain number of successive slots, RABLength. Thus, it is transmitted 4*RABLength times to fit in the allocated 256*RABLength chips. The transmission of each RAB starts in a slot that satisfies

```
T mod RABLength = RABOffset
```

Table 4.1
MAC Channel and Preamble Use Versus MACIndex

MACIndex	MAC Channel Use	Preamble Use
0 and 1	Not used	Not used
2	Not used	76.8-Kbps control channel
3	Not used	38.4-Kbps control channel
4	RA channel	Not used
5–63	Available for reverse power control and DRCLock channel transmissions	Available for forward traffic channel transmissions

where T is the system time in slots and RABLength and RABOffset are user-specific parameters specified by the access network.

The DRCLock channel for an access terminal with an open connection is assigned to an available Walsh channel with a MACIndex between 5 and 63. The MACIndex used for the DRCLock channel is the same as that used for the reverse power control channel for a particular terminal. It is used for the transmission of the DRCLock bit stream destined to that terminal for the purpose of indicating whether or not the sector can reliably decode the DRC sent by the terminal. The DRCLock bit is transmitted over DRCLockPeriod*DRCLockLength slots. DRCLockPeriod specifies the number of slots between transmissions of two consecutive DRCLock symbols. DRCLockLength specifies the number of consecutive DRCLock symbols that are combined by the terminal to determine a single DRCLock bit. Thus, a DRCLock bit is repeated 4*DRCLockLength times to fit in the allocated 256*DRCLockLength chips. A DRCLock bit is transmitted in slots T such that

```
(T-FrameOffset) mod DRCLockPeriod = 0
```

The transmission of each DRCLock bit starts in a slot that satisfies

```
(T-FrameOffset) mod DRCLockPeriod*DRCLockLength = 0
```

where FrameOffset, DRCLockPeriod, and DRCLockLength are user-specific parameters specified by the access network.

The reverse power control channel for an access terminal with an open connection is assigned to an available Walsh channel with a MACIndex between 5 and 63. The MACIndex used for the reverse power control channel is the same as that used for the reverse power control channel for a particular terminal. It is used for the transmission of the reverse power control bit stream destined to that terminal for the purpose of reverse link transmitting power control. The reverse power control bit is transmitted over a single slot. Thus, it is repeated four times to fit in the allocated 256 chips. A reverse-power-control bit is transmitted in slots T such that

```
(T-FrameOffset) mod DRCLockPeriod ≠ 0
```

Figure 4.11 shows an example of how the DRCLock and reverse-power-control channels for a particular terminal are time-division multiplexed on an assigned Walsh channel.

The DRCLock and reverse-power-control channels are power controlled to achieve the required levels of performance. Thus, the transmitting power allocated to each Walsh channel can vary, but the total power allocated to the entire MAC channel must equal to that of the pilot channel. An active terminal demodulates the reverse-activity, DRCLock, and reverse-power-control channels from all the sectors in the active set. The DRCLock symbols and reverse-power-control symbols received from sectors that belong to a common cell are soft combined, respectively, prior to binary decision. The reverse-power-control decisions from different cells are combined such that if a down command is received from one of the cells, the terminal will reduce the transmit power. Only if all reverse-power-control decisions are up commands will the transmit power be increased.

Because the DRCLock channel can be considered as being punctured into the reverse-power-control channel for a given terminal, some reverse link performance degradation occurs when compared to the case where no puncturing is present. The configuration of the DRCLock channel should be such that DRCLockPeriod and DRCLockLength are chosen to be as small as possible while introducing a negligible impact on the reverse link

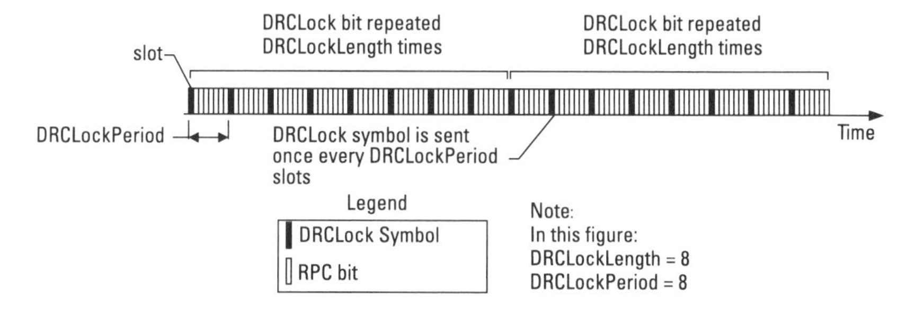

Figure 4.11 DRCLock and reverse-power-control channel (RPC) timing example.

performance and ensuring reliable transmission of the DRCLock bit. In this way, the DRCLock channel can provide the terminal with an update of its DRC channel reliability while causing minimal impact on the terminal's reverse link performance.

4.4.1.4 Forward Traffic Channel

The forward traffic channel is a shared medium that carries user physical layer packets. Because it transmits to a single user at a time, a preamble sequence is transmitted to indicate the presence and starting point of the packet, also indicating the intended receiving terminal.

The preamble sequence consists of all 0 symbols on the I channel and is time multiplexed into the forward traffic channel as described in Figure 4.10. It is covered by a biorthogonal sequence, which is determined by the MACIndex of the desired terminal as follows:

$$\begin{array}{ll} W_{i/2}^{32} & \text{for } i = 0, 2, \ldots, 62 \\ \overline{W_{(i-1)/2}^{32}} & \text{for } i = 1, 3, \ldots, 63 \end{array}$$

where $i = 0, 1, \ldots, 63$ is the MACIndex value and $\overline{W_i^{32}}$ is the bit-by-bit complement of the 32-chip Walsh function of order i. The preamble sequence is repeated as needed depending on the preamble length required by the physical layer packet.

Each physical layer packet is encoded with a turbo encoder, scrambled,[2] permuted with a channel interleaver, modulated using QPSK/8-PSK/16-QAM, subjected to sequence repetition or truncation, demultiplexed into 16 streams, Walsh covered, gain controlled, summed to form a single in-phase stream and a single quadrature stream, and finally time multiplexed with the preamble and pilot/MAC channels.

Packet Types

To support closed-loop rate control, IS-856 defines a set of packet types for the forward traffic channel physical layer. Each packet type is uniquely specified by its nominal data rate and nominal packet length, which can be regarded as the primary attributes of the packet type [5]. Table 4.2 shows the fundamental parameters associated with each packet type. These packet types used for closed-loop rate control are referred to as complete packet types. Hybrid-ARQ allows multislot packets to be terminated early, thus

2. Randomize the data prior to modulation to reduce the peak-to-average ratio of the forward link waveform.

Table 4.2
Complete Packet Types for the Forward Traffic Channel Physical Layer

Complete Packet Type	Nominal Data Rate (Kbps)	Nominal Packet Length (slots)	Packet Size (bits)
38K	38.4	16	1,024
76K	76.8	8	1,024
153K	153.6	4	1,024
307K-2S	307.2	2	1,024
307K-4S	307.2	4	2,048
614K-1S	614.4	1	1,024
614K-2S	614.4	2	2,048
921K	921.6	2	3,072
1.2M-1S	1,228.8	1	2,048
1.2M-2S	1,228.8	2	4,096
1.8M	1,843.2	1	3,072
2.4M	2,457.6	1	4,096

providing an additional set of truncated packet types. Some truncated packet types provide new data rates such as 204.8 and 409.6 Kbps, while the majority of the truncated packet types results in a data rate that coincides with one of the complete packet types.

Note that the additional set of truncated packet types does not add more complexity to the forward link transmitter, since the truncated packet types are naturally derived from the complete packet types through early termination (hybrid-ARQ). We will see in the following sections that the construction of the complete packet types (including code, modulation, and interleaving) is carefully designed such that early termination of the complete packet types leads to near optimal performance.

It is evident from Table 4.2 that packet types with lower data rates generally span longer time durations. This is due to the fact that packet sizes are chosen to lie between 1,024 and 4,096 bits. The packet sizes need to be large enough to exploit the coding gain offered by turbo codes, while minimizing the impact of data encapsulation overhead at higher layers. On the other hand, the packet sizes need to be small enough to avoid excessive transmission delay and packing inefficiency for small data payloads.

Code and Modulation

Each slot contains 2,048 chips at the rate of 1.2288 Mcps. As shown in Figure 4.9, 1,600 chips are allocated to the forward traffic channel/control channel in each slot, excluding the 192 chips for the pilot channel and 256 chips for the MAC channel. The preamble accounts for 64 chips per slot for the complete packet types, which leaves 1,536 chips per slot for the data payload. Consequently, the total number of data chips allocated to a complete packet type is 1,536*N*, where *N* stands for the nominal packet length (slots). The spectral efficiency (bits/chip) of a packet type is defined to be the ratio of the packet size to the total data chips. Table 4.3 shows the spectral efficiency (bits/chip) for complete packet types.

The spectral efficiency of a packet type may also be thought of as the product of its code rate and modulation rate, divided by the repetition factor. The repetition factor can be considered as the inverse of the repetition code rate.

Table 4.4 shows how the spectral efficiency of each complete packet type is distributed over its code rate, modulation rate, and repetition factor.

The code and modulation rates are chosen to be consistent with the required spectral efficiency. We limit the lowest code rate to 1/5 because very low code rates increase decoding complexity, while providing only a small increase in the coding gain. A rate 1/5 code with QPSK modulation achieves a spectral efficiency of 2/5 bits/chip. For packet types that require a spectral efficiency lower than 2/5 bits/chip, the rate 1/5 code with QPSK modulation is used. Complete or partial sequence repetition of the modulation symbols is employed to achieve the desired spectral efficiency. For packet types that require a spectral efficiency higher than 2/5 bits/chip, higher rate

Table 4.3
Spectral Efficiency for Complete Packet Types

Complete Packet Type	Packet Size (bits)	Preamble (chips)	Data (chips)	Spectral Efficiency (bits/chip)
38K	1,024	1024	24,576	1/24
76K	1,024	512	12,288	1/12
153K	1,024	256	6,144	1/6
307K-2S	1,024	128	3,072	1/3
307K-4S	2,048	128	6,272	16/49
614K-1S	1,024	64	1,536	2/3
614K-2S	2,048	64	3,136	32/49
921K	3,072	64	3,136	48/49
1.2M-1S	2,048	64	1,536	4/3
1.2M-2S	4,096	64	3,136	64/49
1.8M	3,072	64	1,536	2
2.4M	4,096	64	1,536	8/3

Table 4.4
Code and Modulation for Complete Packet Types

Complete Packet Type	Spectral Efficiency (bits/chip)	Modulation Rate	Code Rate	Repetition Rate
38K	1/24	2 (QPSK)	1/5	48/5 = 9.6
76K	1/12	2 (QPSK)	1/5	24/5 = 4.8
153K	1/6	2 (QPSK)	1/5	12/5 = 2.4
307K-2S	1/3	2 (QPSK)	1/5	6/5 = 1.2
307K-4S	16/49	2 (QPSK)	16/49	2
614K-1S	2/3	2 (QPSK)	1/3	1
614K-2S	32/49	2 (QPSK)	16/49	1
921K	48/49	3 (8-PSK)	16/49	1
1.2M-1S	4/3	2 (QPSK)	2/3	1
1.2M-2S	64/49	4 (16-QAM)	16/49	1
1.8M	2	3 (8-QAM)	2/3	1
2.4M	8/3	4 (16-QAM)	2/3	1

codes and possibly higher order modulations are used. Higher rate codes are obtained by indirectly puncturing the basic rate 1/5 turbo code. To do so, the output of the rate 1/5 turbo code is first interleaved, and then simple truncation of the interleaver output leads to good puncture patterns of the basic code. This also facilitates hybrid-ARQ, because early termination of a packet achieves the same effect as puncturing the underlying code. Such a design provides high-performance adaptive rate coding for both complete and truncated packet types while maintaining a simple encoder structure.

In a few cases, relatively higher-order modulations than those needed to meet the required spectral efficiency are used to construct packet types. In particular, packet types of 921K and 1.2M-2S are constructed as code extensions (less truncations of the basic code) of 1.8M and 2.4M packet types, respectively. This provides the opportunity for hybrid-ARQ to improve the effective data rate of these packet types in unpredictable channel conditions.

Channel Interleaver

The output of the turbo encoder is permuted by the channel interleaver with the following two objectives:

- To randomize code symbol errors at the encoder output, or equivalently, at the decoder input. Most decoders deal better with random error than burst errors.

- To ensure that a simple truncation of the interleaver output leads to a good puncture pattern of the basic code. This simplifies the hybrid-ARQ procedure since early termination of a packet achieves the same effect as puncturing the underlying code.

IS-856 employs a basic rate 1/5 turbo code that is also used in IS-2000 systems [2]. It consists of two identical rate 1/3 systematic, recursive convolutional codes separated by a code interleaver. The systematic part of the second convolutional code is discarded, resulting in a rate 1/5 code. Let $U(D)$ denote the input binary sequence to the encoder, and $U'(D)$ denote the turbo-code-interleaved input sequence. The output of the first constituent encoder is given by

$$[V_0(D) \quad V_1(D)] = U(D)\left[\frac{1 + D + D^3}{1 + D^2 + D^3} \quad \frac{1 + D + D^2 + D^3}{1 + D^2 + D^3}\right]$$

The output of the second constituent encoder is given by

$$[V_0'(D) \quad V_1'(D)] = U'(D)\left[\frac{1 + D + D^3}{1 + D^2 + D^3} \quad \frac{1 + D + D^2 + D^3}{1 + D^2 + D^3}\right]$$

The channel interleaver structure is described in Figure 4.12. The code symbols are randomized by the following three steps:

1. *Row-wise write, column-wise aggregate in rectangular buffers:* The output of the turbo encoder is written row-wise, into three rectangular buffers, labeled U, (V_0 / V_0'), and (V_1 / V_1'). The U buffer is filled with the $U(D)$ sequence from the turbo encoder. The (V_0 / V_0') buffer is filled with the $V_0(D)$ followed by the $V_0'(D)$ sequence. Similarly, the (V_1 / V_1') buffer is filled with the $V_1(D)$ followed by the $V_1'(D)$ sequence. The number of rows in each buffer is equal to the modulation rate (two for QPSK, three for 8-PSK, and four for 16-QAM). The code symbols are aggregated column-wise to form modulation symbols, which are read out sequentially, first from the U buffer, then from the (V_0 / V_0') buffer, and finally from the (V_1 / V_1') buffer, after end-around shift and bit-reversal interleaving. This step ensures that the code symbols that go into a given modulation symbol are well separated at the encoder output, and

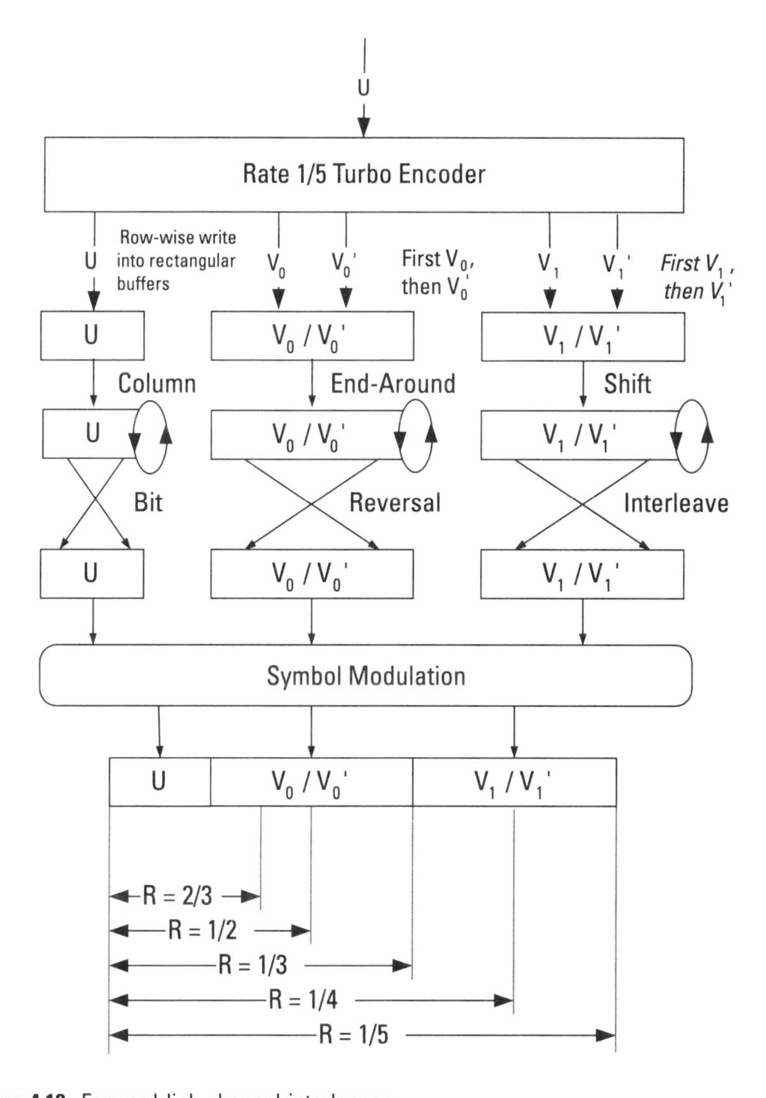

Figure 4.12 Forward link channel interleaver.

thus transforms a modulation symbol error into random errors in the code symbol sequence.

2. *End-around shift:* Within each buffer, the symbols in the jth column are cyclically shifted by an amount $[j/4]$. This is referred to as end-around shift. The symbols in a given column are eventually aggregated to form a modulation symbol. The column-wise end-around shift operation is intended to randomize the unequal error

protection offered to different bit positions by the 8-PSK and 16-QAM constellations with gray code mapping. Figures 4.13, 4.14, and 4.15 show the signal constellations for the QPSK, 8-PSK, and 16-QAM modulations with gray code mapping, respectively. Note that for a given modulation symbol, bit position s_i maps to the code symbol at row i in the corresponding column of interleaving

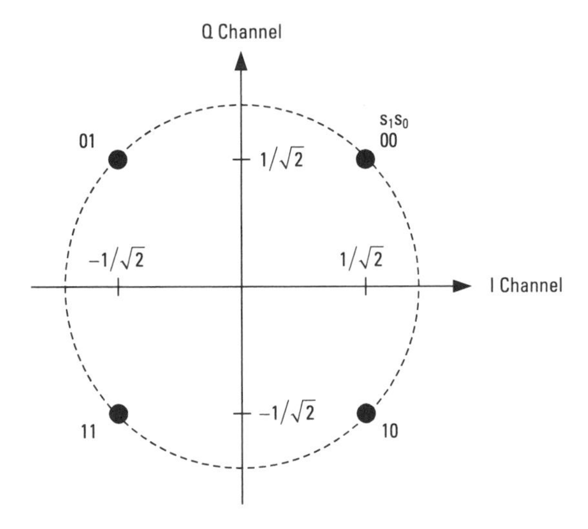

Figure 4.13 Signal constellation for QPSK modulation.

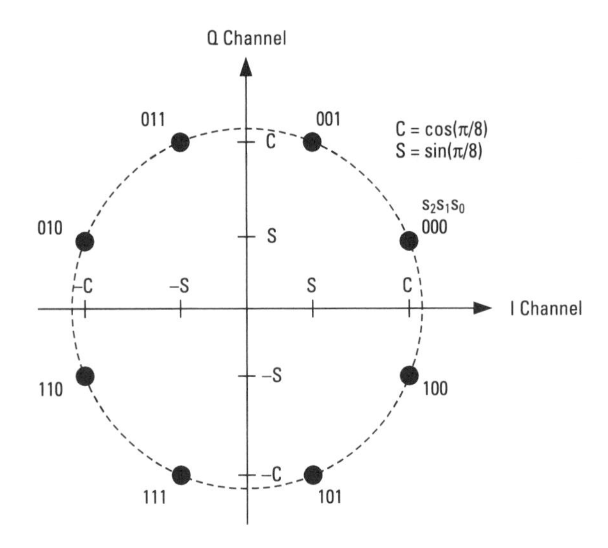

Figure 4.14 Signal constellation of 8-PSK modulation.

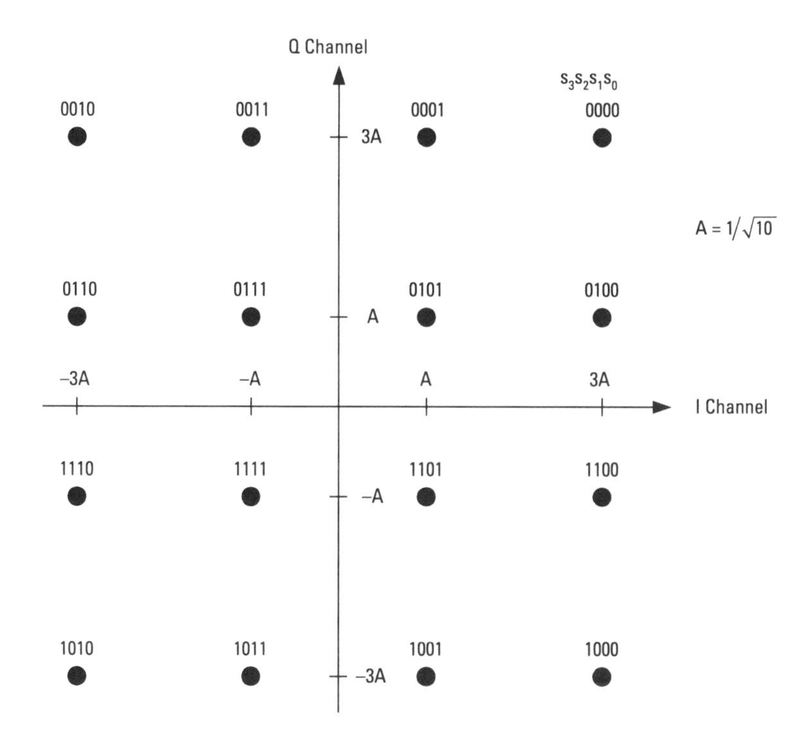

Figure 4.15 Signal constellation of 16-QAM modulation.

buffer. For 8-PSK, it is apparent that the bit-position s_0 is the worst protected position. Similarly, for 16-QAM, the bit positions s_0 and s_2 are the worst protected positions. This step prevents any contiguous sub-block of code symbols at the encoder output from clustering in less protected bit positions. Such a clustering would be detrimental to the turbo code performance.

3. *Bit-reversal interleaver:* The column-wise aggregated code symbols are row-wise permuted using a bit-reversal interleaver. Code symbols that are adjacent to one another at the encoder output are moved to columns that are as far away as possible, thus transforming burst modulation symbol errors at channel interleaver output into random code symbol errors at the encoder output. Bit-reversal interleaving also ensures that simple sequence truncation after interleaving leads to a regular puncture pattern at the encoder output, which provides near-optimal performance for all packet types (complete and truncated) in IS-856. Figure 4.12 also shows the extent of sequence truncation needed to generate code rates of 1/4, 1/3, 1/2, 2/3, and so on.

4.4.1.5 Forward Control Channel

The control channel transmits broadcast messages and access-terminal-directed messages. The control channel messages are transmitted at a data rate of 76.8 or 38.4 Kbps. The modulation characteristics of the control channel physical layer packet are the same as those of the forward traffic channel at the corresponding data rates.

An access terminal attempts to detect the preamble of a control channel packet at both 76.8 and 38.4 Kbps. Thus, the terminal is able to receive control channel packets transmitted at either rate.

4.4.1.6 Quadratic Spreading and Filtering

The forward traffic channel or control channel data chips (including preamble) are time-division multiplexed with the pilot channel and MAC channel chips as described in Figure 4.10. The combined sequence is then quadrature spread by the pilot PN sequence of length 2^{15}. After the quadrature spreading operation, the I and Q impulses are applied to lowpass pulse-shaping filters and then upconverted to the carrier frequency to produce the forward link waveform.

4.4.1.7 Physical Layer Packet Interlacing

The forward traffic channel and control channel transmit multislot physical layer packets with four-slot interlacing. That is, the transmit slots of a physical layer packet are separated by three intervening slots, which can be used to transmit other physical layer packets. Such a timing structure is important to support the early termination mechanism on the forward link. It allows time for the receiving access terminal to attempt to decode the partially received packet and to return an indication to the transmitting sector via the ACK channel on the reverse link. If a positive acknowledgment is received before all of the allocated slots have been transmitted, the remaining slots are not transmitted and the next allocated slot may be used for the first slot of the next physical layer packet.

This procedure is exemplified in Figure 4.16. A data rate of 153.6 Kbps is requested via the DRC channel and the corresponding four-slot packet is transmitted. First, the terminal sends out two *negative acknowledgment* (NAK) responses indicating that the decoding attempts after receiving the first slot and after receiving the second slot both failed. Later, the terminal transmits an ACK response on the ACK channel after the third slot is received indicating that it has correctly received the packet. In this case, the request of a 153.6-Kbps packet results in an effective transmission rate of 204.8 Kbps.

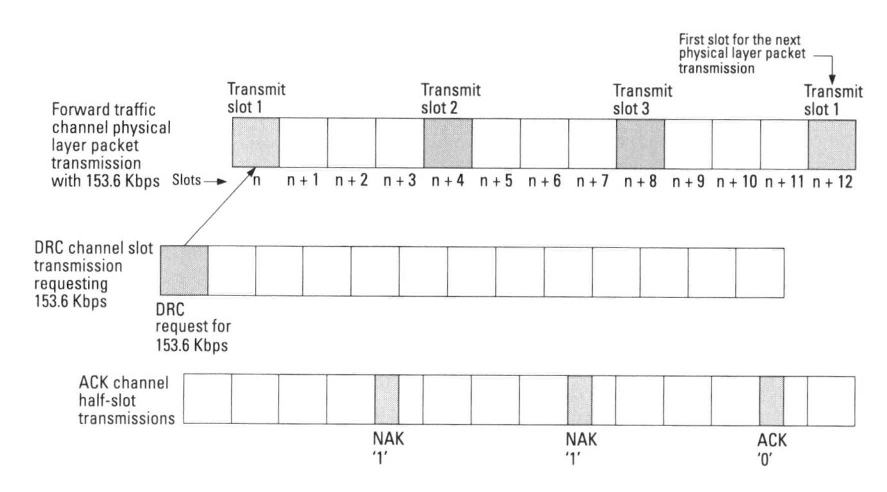

Figure 4.16 Multislot physical layer packet with early termination.

Note that each sector may serve a single user with all four interlaced slot sequences on the forward traffic channel. Thus, the delay involved in the packet decoding, NAK indication, and retransmission neither increases the delay of transmitting packets nor reduces the user throughput.

4.4.1.8 Traffic and Control Channel Performance

The *additive white Gaussian noise* (AWGN) channel packet error rate performance of the forward link complete packet types is shown in Figure 4.17. The result accounts for pilot/MAC/preamble puncturing on the forward link waveform, as well as channel estimation errors. We can see from the figures that the set of data rates, with appropriate modulation and coding schemes, achieves reasonable PER performance in a wide range of SINR values typical for cellular systems.

Table 4.5 shows the SINR thresholds of the complete packet types for closed-loop rate control. Note that the construction of the complete packet types (including code, modulation, and interleaving) is designed such that early termination of the complete packet types leads to a near-optimal code and modulation. Thus, the performance of a truncated packet type is close to that of the corresponding complete packet type with the same data rate.

4.4.2 Reverse Link

The IS-856 reverse channel structure, as described in Figure 4.18, consists of the access channel and reverse traffic channel. The access channel, which further consists of pilot and data channels, is used by an access terminal that

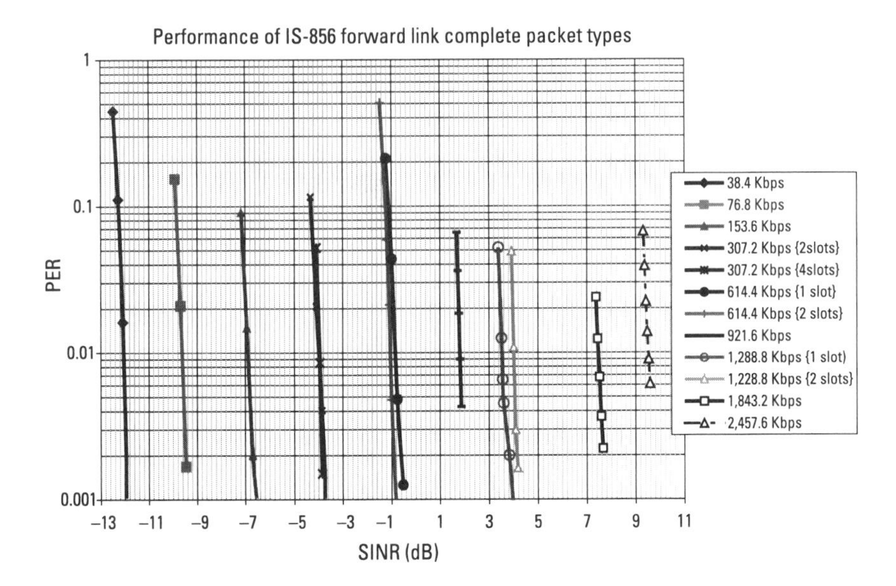

Figure 4.17 Traffic channel performance for data rates from 76.8 to 307.2 Kbps.

Table 4.5
SINR Thresholds of Complete Packet Types for Closed-Loop Rate Control

Complete Packet Type	SINR Threshold (dB)
2.4M	10.3
1.8M	8.0
1.2M-2S	4.0
1.2M-1S	3.9
921K	2.2
614K-2S	−0.6
614K-1S	−0.5
307K-4S	−3.5
307K-2S	−3.5
153K	−6.5
76K	−9.2
38K	−11.5

is not in the connected state to send signaling messages to the access network. In the connected state, the access terminal transmits on the reverse traffic channel, which contains a pilot channel, a reverse rate indicator channel, a DRC channel, an ACK channel, and a data channel. The reverse rate indicator

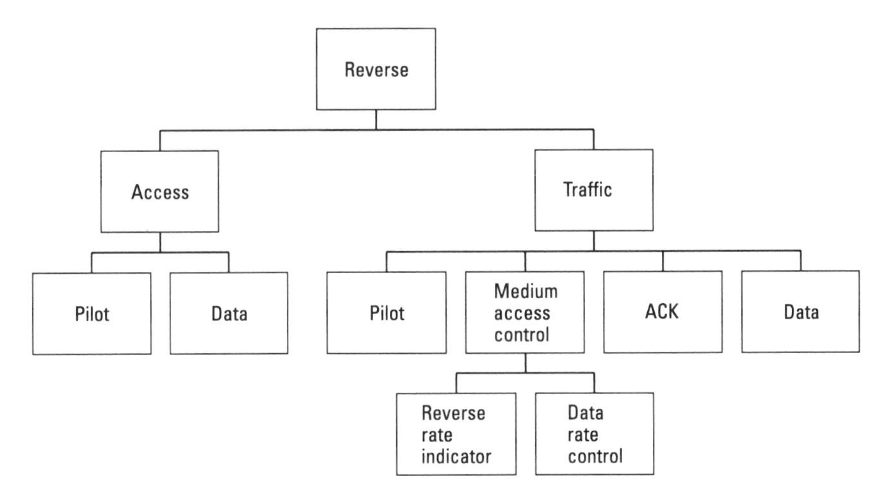

Figure 4.18 Reverse HDR channel structure.

channel is used to indicate whether or not the data channel is being transmitted on the reverse traffic channel and its associated data rate. Thus, complex rate determination algorithms can be avoided in the IS-856 system. The DRC channel is used to indicate to the access network the supportable forward traffic channel data rate and the best serving sector for the forward link. The ACK channel is used to inform the access network whether a data packet transmitted on the forward traffic channel has been received successfully.

4.4.2.1 Reverse Link Waveform

Figures 4.19 and 4.20 show the reverse traffic channel structure of the IS-856 standard. It has four orthogonal CDM channels. As shown in Figure 4.21, the pilot/reverse rate indicator channel is time multiplexed so that the reverse rate indicator channel is transmitted during 256 chips at the beginning of every slot (1.66 ms). The 3-bit reverse rate indicator symbol transmitted every frame (16 slots) is encoded using a 7-bit simplex codeword as described in Table 4.6. Each codeword is repeated 37 times over the duration of the frame, while the last three code symbols are not transmitted.

The DRC symbols (4 bits indicating the desired rate) are encoded using 16-ary biorthogonal code as specified in Table 4.7. Each code symbol is further spread by one of the 8-ary Walsh functions in order to indicate the desired transmitting sector on the forward link. The DRC message is transmitted in a half-slot offset manner with respect to a slot boundary. This is done to minimize prediction delay while providing enough time for

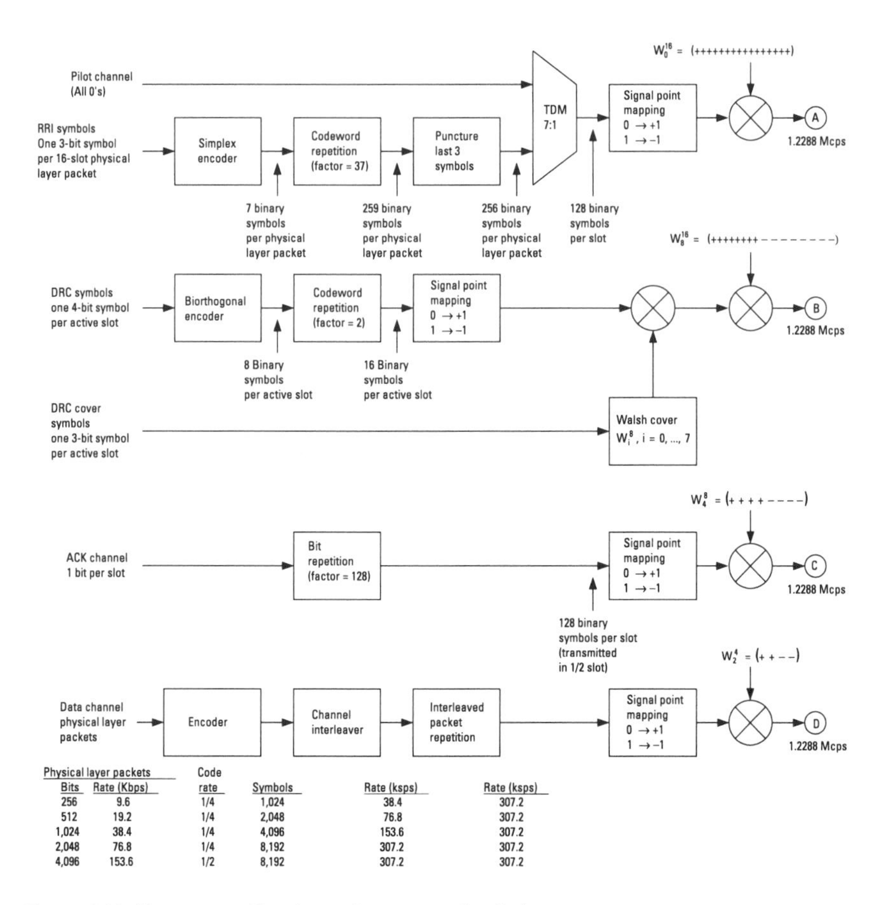

Figure 4.19 Reverse traffic channel structure (1 of 2).

processing at the desired sector before transmission on the forward link starts on the next slot. A DRC message indicating the desired forward link data rate and transmitting sector may be repeated over DRCLength slots, a user-specific parameter set by the access network. The DRC transmission timing is shown in Figure 4.22.

The ACK channel is BPSK modulated in the first half-slot (1,024 chips) of an active slot. A 0 bit is transmitted on the ACK channel if a data packet has been successfully received on the forward traffic channel, otherwise a 1 bit is transmitted. Transmissions on the ACK channel only occur if the access terminal detects a data packet directed to it on the forward traffic channel. For a forward traffic channel data packet transmitted in slot n, the corresponding ACK channel bit is transmitted in slot $n + 3$ on the reverse traffic channel. The three slots of delay allow the terminal to demodulate and decode the received packet before transmitting on the ACK channel.

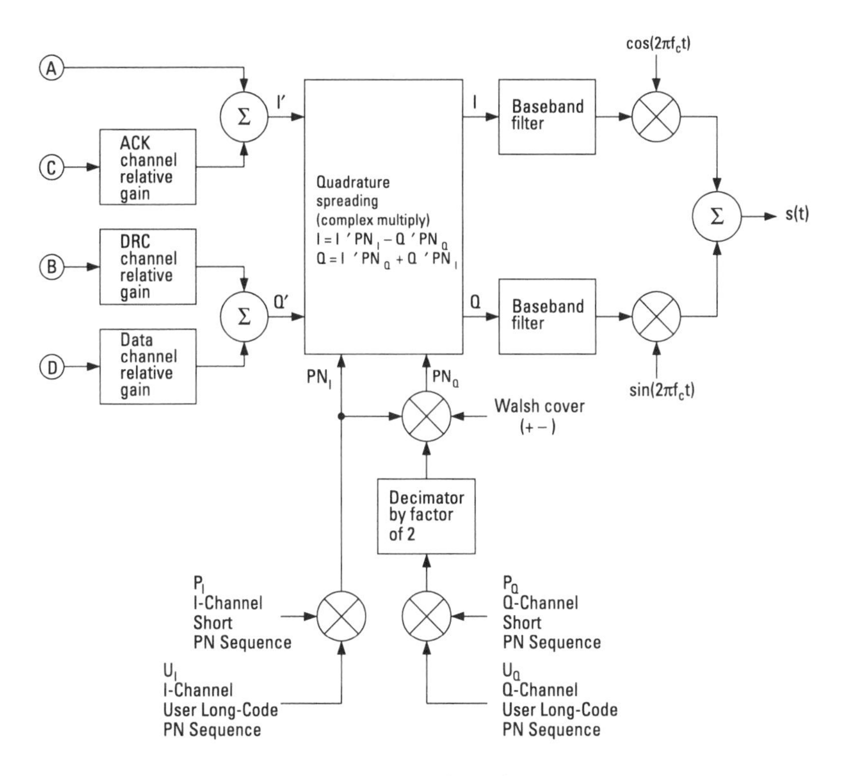

Figure 4.20 Reverse traffic channel structure (2 of 2).

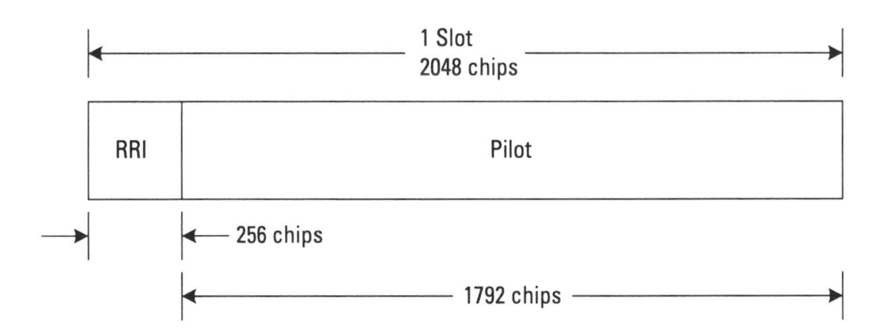

Figure 4.21 Pilot and reverse rate indicator channel TDM allocations.

The data channel supports data rates from 9.6 to 153.6 Kbps with 16-slot packets (26.66 ms). The packet is encoded using either a rate 1/2 or rate 1/4 parallel turbo code as specified in IS-856. The parameters of the reverse link encoder for different data rates are described in Table 4.8. The code symbols are bit-reversal interleaved and block repeated to achieve the 307.2 Ksps modulation symbol rate.

Table 4.6
Simplex Encoding for Reverse Rate Indicator Channel

Reverse Link Data Rate (Kbps)	Reverse Rate Indicator Symbol	Reverse Rate Indicator Codeword
0	000	0000000
9.6	001	1010101
19.2	010	0110011
38.4	011	1100110
76.8	100	0001111
153.6	101	1011010

Table 4.7
DRC Codeword

Forward Link Data Rate (Kbps)	Number of Slots	DRC Symbol	DRC Codeword
0	—	0000	00000000
38.4	16	0001	11111111
76.8	8	0010	01010101
153.4	4	0011	10101010
307.2	2	0100	00110011
307.2	4	0101	11001100
614.4	1	0110	01100110
614.4	2	0111	10011001
921.6	2	1000	00001111
1,228.8	1	1001	11110000
1,228.8	2	1010	01011010
1,843.2	1	1011	10100101
2,457.6	1	1100	00111100

The pilot/reverse rate indicator, DRC, ACK, and data channel modulation symbols are each spread by an appropriate orthogonal Walsh function as shown in Figure 4.19. Before quadrature spreading (see Figure 4.20), the pilot/reverse rate indicator and ACK channels are scaled and combined to form the in-phase component. Similarly, the data and DRC channels are scaled and combined to form the quadrature component of the baseband signal. Reverse link power control (both open and closed loops) is applied to the pilot/reverse rate indicator channel only. The powers allocated to the DRC, ACK, and data channels are adjusted by a fixed gain relative to the pilot/reverse rate indicator channel in order to guarantee the desired

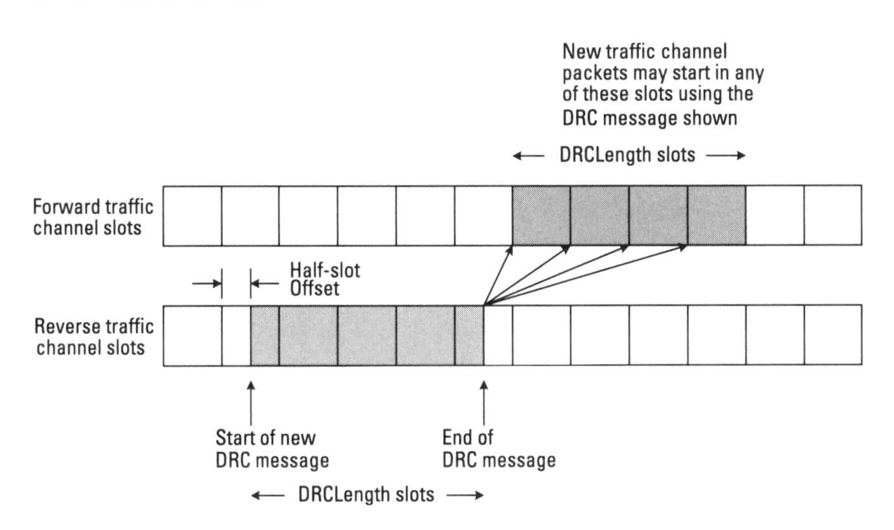

Figure 4.22 DRC timing as a function of DRCLength.

Table 4.8
Parameters for the Reverse Link Encoder

Data rate (Kbps)	9.6	19.2	38.4	76.8	153.6
Reverse rate index	1	2	3	4	5
Bits per physical layer packet	256	512	1,024	2,048	4,096
Number of turbo encoder input symbols	250	506	1,018	2,042	4,090
Turbo encoder code rate	1/4	1/4	1/4	1/4	1/2
Encoder output block length (code symbols)	1,024	2,048	4,096	8,192	8,192

performance of these channels. For example, the relative gain of the data channel increases with the data rate so that the received E_b/N_t is adjusted to achieve the required packet error rate.

4.4.2.2 Reverse Link Performance

The performance of the reverse traffic channel was simulated under different channel scenarios and with power control. In particular, for the data channel, we consider two independent and equal-strength paths, each being affected by Rayleigh fading channel with a classic Doppler spectrum at vehicular speeds of 0, 3, 30, and 120 km/hr (1.9-GHz carrier frequency). The 0 km/hr speed corresponds to a simple AWGN channel. A typical packet error rate performance of the reverse data channel is shown in Figure 4.23 for the 76.8-Kbps packets. The performance is shown as a function of total average pilot E_c/N_t.

To evaluate the required E_b/N_t for each data rate, one has to include the effect of the pilot overhead and processing gain as follows:

$$\frac{E_b}{N_t}\,(\text{dB}) = \text{Pilot}\,\frac{E_c}{N_t}\,(\text{dB}) + 10\log_{10}\left(\frac{R_b}{W}\right) + 10\log_{10}\left(1 + 10^{\text{DataChannelGain}/10}\right)$$

where the second term on the right-hand side of the equation corresponds to the processing gain. The parameter DataChannelGain (in decibels) depends on the data rate and was assumed to be one of the default values provided in the IS-856 standard as described in Table 4.9. Similarly, the preceding equation may be modified to include the effects of DRC and ACK channel overheads.

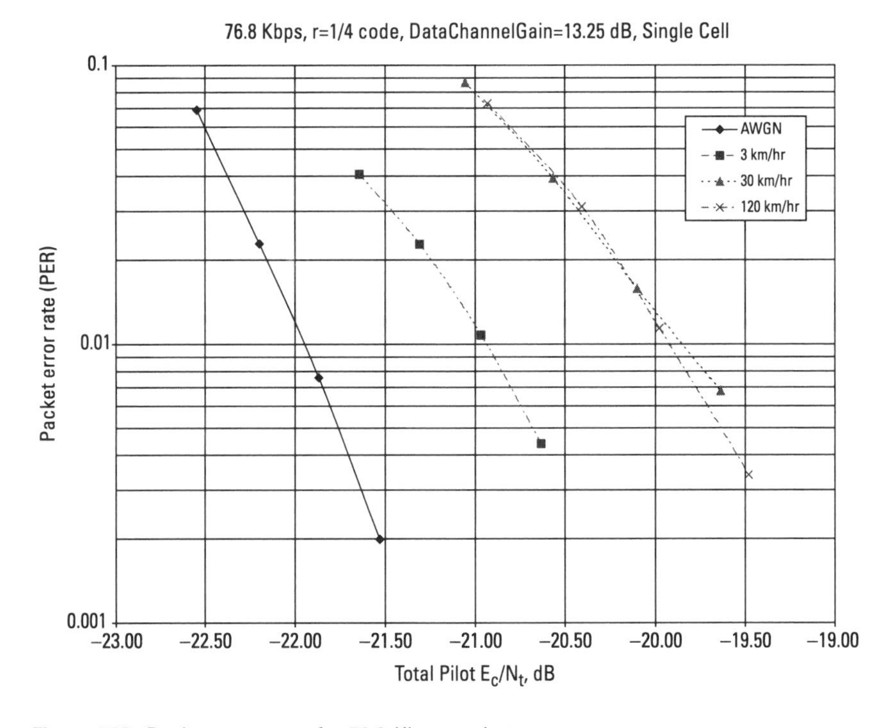

Figure 4.23 Packet error rate for 76.8-Kbps packets.

Table 4.9
Default Data Channel Gains

Data Rate (Kbps)	Data Channel Gain (dB)
9.6	3.75
19.2	6.75
38.4	9.75
76.8	13.25
153.6	18.5

The required total pilot E_c/N_t for 1% packet error rate is shown in Figure 4.24 for all data rates as a function of the vehicular speed. For all code rate 1/4 packets, the performance is very similar requiring a total pilot E_c/N_t between −21.8 and −19.8 dB to achieve 1% packet error rate. Also in this case, we can conclude that there is a small difference (2.0 dB) in performance between stationary versus mobile users, which is a desirable feature for adjusting the power control setpoint as the mobile environment changes.

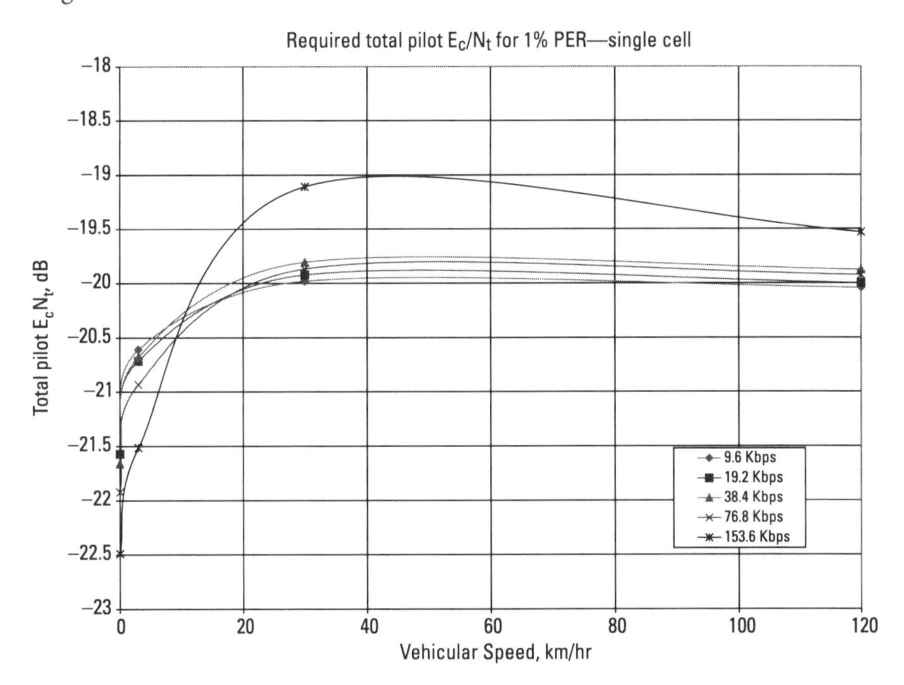

Figure 4.24 Required pilot E_c/N_t for 1% packet error rate.

For the code rate 1/2 packet (153.6 Kbps), the spread in required pilot E_c/N_t is about 3.5 dB. For a 153.6-Kbps data rate, the default data channel gain of 18.5 dB was chosen in order to "equalize" the required pilot E_c/N_t across the range of speeds. It is clear that, at low speeds, the selected gain makes the occasional transition to 153.6 Kbps conservative. On the other hand, such transitions will achieve a slightly higher packet error rate at high vehicular speeds. Of course, for long periods of 153.6-Kbps transmission, the outer loop power control will eventually converge to the correct setpoint.

Next we examine the performance of the DRC channel, which is important since the IS-856 system relies on DRC not only to provide accurate data rate information to the forward link adaptive scheduler but also to perform virtual soft handoff (sector selection). Due to delay constraints, the DRC channel is trivially encoded resulting in a high required E_s/N_t level for acceptable performance. Increasing the DRC channel gain relative to the pilot improves the performance of the channel. However, because the DRC channel is an overhead channel on the reverse link, the tradeoff between reverse link capacity and a more reliable DRC channel (i.e., increased forward link capacity) is an important issue and therefore we discuss it in more detail.

The IS-856 standard provides several parameters that can be adjusted for optimal system deployment. Such parameters include DRC length (one, two, four, or eight slots), DRC channel gain and DRC gated/continuous mode. In addition, a typical receiver at the base station may erase unreliable DRC decisions in order to guarantee a low DRC symbol error rate even if the given BS is not the one power controlling the reverse link. The tradeoff between DRC erasure rate and symbol error rate is possible because these events have a different impact on forward link performance. If an erroneous DRC causes the serving BS to schedule a packet at the wrong data rate, the forward link packet error rate and throughput is directly impacted. On the other hand, a DRC erasure only prevents the scheduler from selecting a given terminal during a certain time. If the number of active users is high, this will have a negligible effect on the sector throughput, because other terminals will likely be available for scheduling. The rule of thumb is that if only a small portion of the active users in the sector have erasure rates as high as 50%, neither the sector nor the respective users will have a significant reduction in their throughputs.

In Figure 4.25 we present the results for a single power controlling base station with a target data channel packet error rate of 1% in a two-path Rayleigh fading channel at 30 km/hr. The plots are shown as a function of the equivalent pilot E_c/N_t used to derive the appropriate erasure threshold. The threshold was scaled by the DRCLength (two slots) and the square root

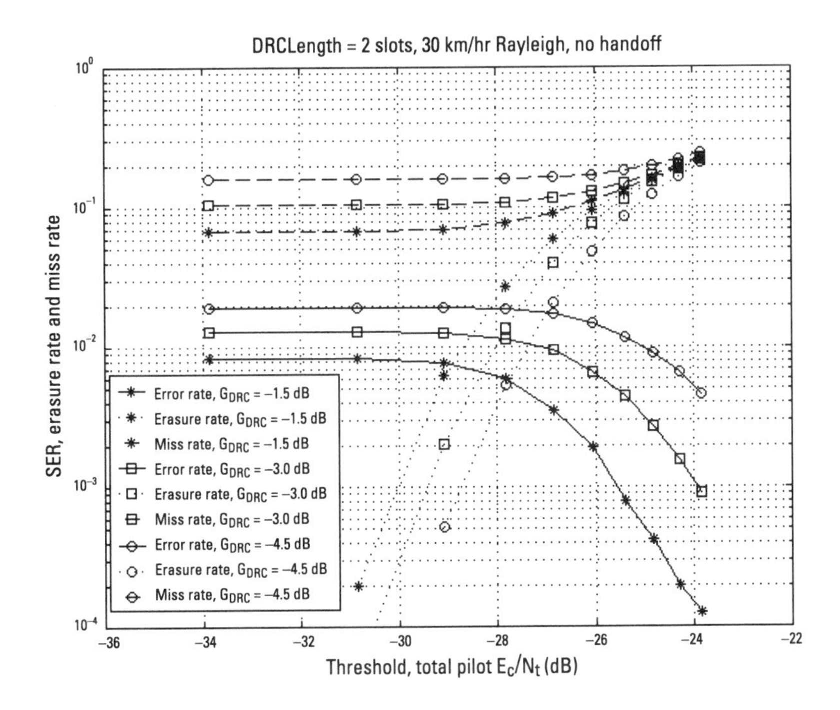

Figure 4.25 DRC channel performance at 30 km/hr.

of the DRC gain (converted to linear). The error rate curves correspond to DRC decisions that had the correct Walsh cover but wrong DRC codeword (data rate request). This is the type of error that directly generates a packet error on the forward traffic channel. The miss rate curves correspond to DRC decisions that are either erased or do not contain the correct Walsh cover associated with the serving sector. A DRC miss only limits the scheduler's ability to serve a particular terminal at the optimal times and has negligible effects on throughput as discussed before.

The erasure rate is also shown as a function of the DRC erasure threshold. We consider a target DRC symbol error rate of 10^{-3}. The figure shows we can achieve this level of performance with, for example, DRC gain = −3.0 dB and erasure rate = 20% or DRC gain = −1.5 dB and erasure rate = 12%. It is always possible to improve the DRC channel reliability at the expense of reverse link capacity.

Next we examine a typical performance curve of the ACK channel in SHO. The two metrics of interest are the probability of a false ACK being sent and the probability of an ACK miss. These are shown in Figure 4.26

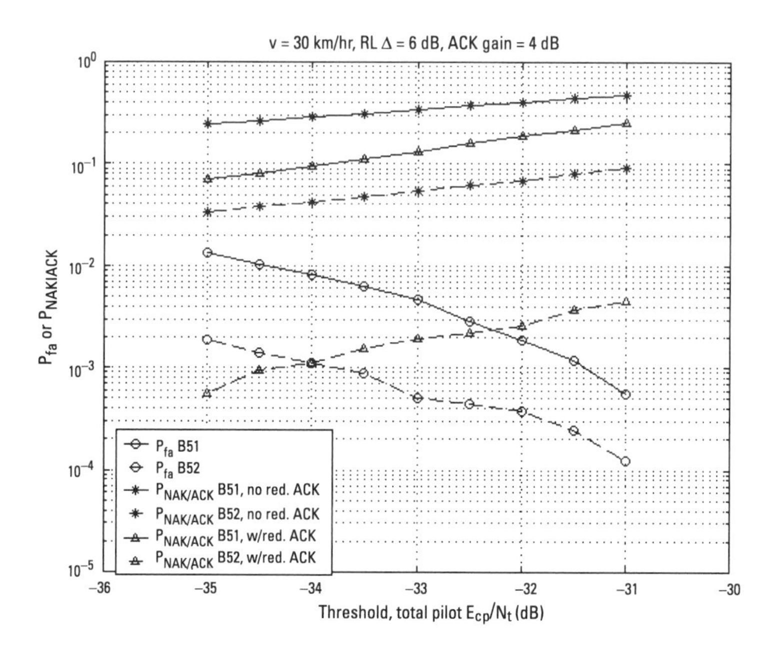

Figure 4.26 ACK channel performance at 30 km/hr in a two-way SHO.

for a two-path Rayleigh channel at 30 km/hr and an ACK channel gain of 4 dB. The scenario is such that two BSs are power controlling the reverse link such that BS 1 has an average propagation loss that is 6 dB greater than that of BS 2. The power control setpoint is chosen to provide a 1% data channel packet error rate after selection combining.

In this case we also run a receiver with a threshold that biases the decisions toward ACK misses. The reason is that ACK misses only cause the sector to transmit unnecessary slots, thus reducing the achievable hybrid-ARQ gains. On the other hand, ACK false alarms cause packet errors, hence increasing the forward link packet error rate and eventually reducing the throughput perceived by the application in more than one way.

We also show the effect of the redundant ACK on the ACK miss probability. According to the IS-856 standard, the redundant ACK is an extra ACK bit that the terminal may send four slots after the first if no new packet is being demodulated on the same interlacing phase. Again, the desired ACK false alarm is 10^{-3}. We can observe that this goal can easily be achieved on the stronger reverse link (BS 2) with an ACK miss rate of only 4% (first ACK) or 0.1% (with redundant ACK transmission). However, for the weaker reverse link (BS 1), the erasure threshold has to be increased significantly in order to achieve 10^{-3} false alarm rate resulting in 40% and 20% miss rates for the first and redundant ACK transmissions, respectively.

Because there are at most two access terminals transmitting on the ACK channel at any given slot, increasing the ACK channel gain would not significantly impact RL capacity. However, it may cause peak-to-average power constraints on the terminal's power amplifier. Thus, the final goal is to use the smallest ACK channel gain as possible while maintaining most of the hybrid-ARQ benefits.

4.5 IS-856 MAC Layer Description

The IS-856 MAC layer consists of forward control channel MAC, forward traffic channel MAC, reverse access channel MAC, and reverse traffic channel MAC protocols. To better understand the HDR concept of IS-856 systems, we focus here only on the traffic channel MAC.

4.5.1 Forward Traffic Channel MAC

The IS-856 forward traffic channel provides high-rate packet transmissions and is a shared medium among all the connected access terminals in the sector. Addressing on the shared medium is achieved by a MAC index that identifies a particular access terminal in the sector. This is accomplished by covering the packet preamble with a 32-chip biorthogonal sequence determined by the MAC index of the desired access terminal. The access terminal attempts to decode a traffic channel packet only if it contains the appropriate preamble. To maximize the performance in a variety of mobile conditions, three basic mechanisms exist to control access to the forward traffic channel:

1. DRC messages are sent by each access terminal containing a requested data rate and a serving sector indication.
2. A fast feedback ACK channel allows the data rate of a packet to be effectively increased beyond the requested DRC if the channel conditions experienced by the access terminal are improved.
3. An adaptive data scheduler takes into account the fairness and channel conditions provided by the DRCs.

4.5.1.1 Virtual SHO

When communication is established between the access terminal and a sector, the terminal monitors the signal strength of measurable pilot signals transmitted from various sectors. When a pilot's signal strength exceeds a predetermined threshold level, the pilot is added to a candidate set at the terminal. The terminal communicates a message to the access network identifying the new pilot and its signal strength.

The network then uses this message to decide whether to add this pilot to the terminal's active set. Should the network decide to include the pilot, a message is transmitted to the terminal. This message identifies the active set pilots that correspond to sectors through which the terminal is able to communicate. The network also instructs each sector in the active set to establish a connection with the terminal. That is, each sector in the terminal's active set transmits power control bits on the allocated reverse power control channel, and demodulates the reverse data channel for selection diversity and the DRC channel in order to determine if the terminal is requesting forward link traffic transmissions from the sector.

In the conventional CDMA cellular telephone system, forward traffic information is routed through all sectors in the mobile's active set to improve reliability, especially during handoffs. This is due to the stringent and fixed delay requirement for voice services. Because data services can tolerate larger and variable delay, reliability can be achieved more efficiently through retransmissions. Thus, in IS-856 systems, forward traffic is routed to only one sector selected by the terminal from its active set. Because the terminal receives forward traffic from only one sector at a time, no SHO throughput losses occur in IS-856 systems. On the other hand, the terminal can quickly reselect its serving sector to adapt to new channel conditions.

Ideally, the serving sector selected from the active set should be the one that maximizes the forward link throughput as perceived by the terminal. Thus, a sector with the strongest pilot strength is generally preferred. However, the best pilot measurement can toggle between two or more sectors over successive time slots if the terminal observes approximately equal signal strength from these sectors. Because some overhead time is always associated with the switching mechanism, fast cell site selection can result in efficiency loss or even connection failure. The excessive switching among competing sectors can be addressed by the use of hysteresis through the combination of the signal level and timing. In addition, the reliability of the reverse link DRC and ACK channels also affects the forward link throughput. Because power control is dominated by the sector with the minimum reverse link attenuation, the strongest forward link sector may not receive enough power from the terminal to reliably demodulate the DRC and ACK channels. Such a situation is generally referred to as an imbalance scenario. Under severe imbalance, the terminal should either select the sector with the strongest pilot but in fixed-rate state[3] or select a different sector with reliable DRC and ACK channels from the active set. A terminal can obtain information about the reverse link reliability of all its active sectors from the DRCLock channels as discussed in Section 4.4.1.3.

To accomplish virtual SHO, a terminal in the connected state monitors the pilot channels from all sectors in the active set and also collects feedback information about the quality of the reverse link channel seen by each sector. The terminal covers its DRC transmissions with a particular Walsh function to indicate its decision about the selected serving sector. Only the sector whose assigned Walsh function matches the DRC cover will serve the terminal with forward traffic channel packets at the requested rate (DRC value).

If the terminal decides to transition to the fixed-rate state, it will indicate its decision by transmitting a control message indicating the selected serving sector and requested data rate. Subsequently, the terminal will cover its DRC with a reserved Walsh function that is never assigned to any sector. The network does the assignment of the Walsh functions for each terminal's active set, and updates the assignment whenever there is a change in the active set.

4.5.1.2 Link Adaptation

The quality of a wireless channel varies with time due to the path loss and fading. When the transmitter is provided with channel state information, it may, for example, adapt its transmitter power and data rates. For voice-oriented systems like IS-95, a certain guaranteed grade of service is required. Power adaptation, which adapts transmit power to the fading channel environment, is needed to ensure a fixed information bit rate and frame error rate at all times. However, for packet data-oriented systems like IS-856, a guaranteed minimum bit rate may not be required. In this case rate adaptation, which adapts the data rate to the fading channel environment, is a better link adaptation scheme in the sense that it achieves higher throughput (average data rate) under the constraint of equal average transmitter power.

The following analysis is based on the assumption of perfect link adaptation. Even though it relies on a simplified model, it provides insights into the behavior of the two link adaptation schemes.

We first assume the required E_b/I_0 is a constant for all data rates, where E_b is the energy per information bit and I_0 is the total interference

---

3. In this state, the access network transmits the forward traffic channel to the access terminal from a selected sector at a fixed data rate.

and noise power spectrum density. Moreover, the achievable data rate is proportional to SINR, that is,

$$R = \frac{W}{(E_b/I_0)_{req}} \cdot \frac{S}{I} = k_0 \cdot \frac{S}{I} = k_0 \cdot \frac{P}{I} \cdot g \tag{4.1}$$

where R is the data rate, k_0 is a constant determined by the bandwidth and receiver performance, I is the total interference and noise power, S is the receiver signal power, P is the transmitter power, and g is the fading factor. The perfect rate adaptation scheme with a fixed transmitter power P_0, achieves an average data rate (or throughput) of,

$$\overline{R} = k_0 E\left[\frac{P_0}{I} g\right] = k_0 \cdot E[\text{SINR}]$$

where SINR denotes the receiver SINR given the fixed transmitter power P_0. For the perfect power adaptation scheme, in order to achieve a fixed data rate of $\overline{R}$, the transmitter power P has to be adapted according to the following equation:

$$P = \frac{\overline{R}}{k_0} \cdot \frac{I}{g}$$

Thus, the average transmitter power required to maintain the data rate of $\overline{R}$ is given by

$$\overline{P} = \frac{\overline{R}}{k_0} \cdot E\left[\frac{I}{g}\right]$$

The gain of rate adaptation over power adaptation, under the constraint of equal effective data rate, can be characterized as the ratio of the average transmitter power for the two link adaptation schemes:

$$\frac{\overline{P}}{P_0} = E[\text{SINR}] \cdot E\left[\frac{I}{P_0 \cdot g}\right] = E[\text{SINR}] \cdot E\left[\frac{1}{\text{SINR}}\right] \tag{4.2}$$

It can be shown that the above gain is always greater or equal to 1. The equality is satisfied only when the SINR is a constant corresponding to static channel conditions.

According to Shannon's limit in AWGN channels, the assumption of a common required E_b/I_0 for all data rates is reasonable when the spectral efficiency in bits per second per hertz is less than one. Such an assumption leads to (4.2), which provides insight into the two link adaptation schemes. Typically, however, the required E_b/I_0 increases as spectral efficiency increases above unity. For a HDR system like IS-856, it is more appropriate to assume the following:

$$R = f\left(\frac{S}{I}\right) = f\left(\frac{P}{I} \cdot g\right) \tag{4.3}$$

where $f(x)$ is a function determined by the modem performance and is subject to limitations imposed by the Shannon capacity. Typically, $f(x)$ can be expressed as

$$f(x) = W \log_2 (1 + \eta x), \qquad 0 \le \eta < 1 \tag{4.4}$$

where η defines the modem performance loss relative to the Shannon limit. In this case, the perfect rate adaptation scheme with a fixed transmitter power P_0 achieves a throughput of

$$\overline{R} = E[f(\text{SINR})]$$

Note that $f(x)$ is a nondecreasing concave function. Applying Jensen's inequality, we have

$$\overline{R} \le f(E[\text{SINR}])$$

For the perfect power adaptation scheme the transmitter power P has to be adapted according to the following equation:

$$P = f^{-1}(\overline{R}) \cdot \frac{I}{g}$$

Thus, the average transmitter power to maintain the data rate of $\overline{R}$ is

$$\overline{P} = f^{-1}(\overline{R}) \cdot E\left[\frac{I}{g}\right]$$

The gain of rate adaptation over power adaptation can be represented as

$$\frac{\overline{P}}{P_0} = f^{-1}(\overline{R}) \cdot E\left[\frac{I}{P_0 \cdot g}\right] = f^{-1}(\overline{R}) \cdot E\left[\frac{1}{\text{SINR}}\right]$$

Since $f^{-1}(x)$ is also a nondecreasing function, we have

$$\frac{\overline{P}}{P_0} \leq f^{-1}(f(E[\text{SINR}])) \cdot E\left[\frac{1}{\text{SINR}}\right] = E[\text{SINR}] \cdot E\left[\frac{1}{\text{SINR}}\right] \tag{4.5}$$

Note that under the more realistic modem performance model given by (4.3) and (4.4), the gain of rate adaptation over power adaptation is bounded by the result obtained from the simplified model given in (4.1). The "less" concave $f(x)$ is, the tighter the bound is. Given a certain range of SINR values in which we are interested, η defines the degree of "concaveness" of $f(x)$.

In reality, for the power adaptation scheme, the transmitter does not have unlimited power to compensate for deep fades. On the other hand, for the rate adaptation scheme, the transmitter only has a limited range of data rates that can be used. This imposes certain constraints on the link adaptation capability for both schemes.

Assuming a one-path Rayleigh channel with white Gaussian interference and noise, SINR is characterized by the central chi-square distribution with $N = 2$ degrees of freedom. We impose a limited SINR adaptation range of [−12, 10] dB for both the perfect rate adaptation and power adaptation schemes. That is, for the perfect rate adaptation scheme, the data rate becomes 0 when SINR is below −12 dB and saturates to a maximum rate when SINR is above 10 dB. For the perfect power adaptation scheme, the transmitter stops transmitting when SINR is below −12 dB and transmits a minimum power level when SINR is above 10 dB. Such a system is simulated over the average SINR range of [−5, 9] dB. This corresponds to a typical range of average SINR in an existing CDMA network with single-antenna terminals.

The gains of perfect rate adaptation over perfect power adaptation for the linear model (4.1) and nonlinear model (4.3) are shown Figure 4.27. The gain ranges from approximately 0.5 to 5.5 dB, with an average around 4 dB.

As analyzed above, for a packet data-oriented system, perfect rate adaptation is a more efficient link adaptation scheme than perfect power adaptation.

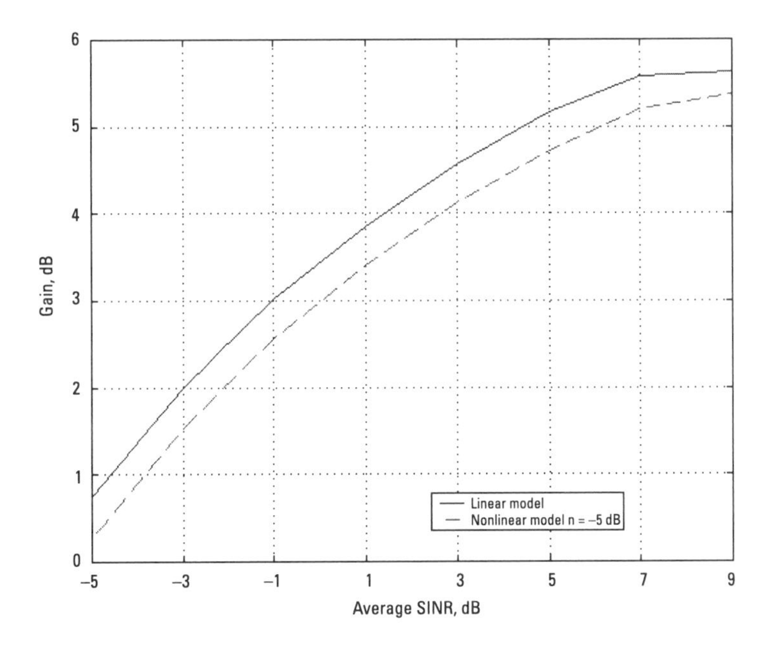

Figure 4.27 Gain of the perfect rate adaptation over the perfect power adaptation.

As discussed in the following sections, IS-856 employs closed-loop rate control and hybrid-ARQ mechanisms in an attempt to achieve the perfect rate adaptation.

Closed-Loop Rate Control

In IS-856, forward traffic channel transmissions occur at full sector power and, thus, there is no need for forward power control. To adapt the preamble length and coding/modulation scheme to the varying channel, forward closed-loop rate control is employed such that the access terminal can request the highest data rate subjected to a certain packet error rate. The requested data rate is also used by the data scheduler to achieve multiuser diversity gains.

Similarly to the reverse closed-loop power control the forward link closed-loop rate control also contains two loops: the inner loop and the outer loop.

The inner loop generates the terminal's best estimate of the supportable data rate, providing it to the access network through the DRC channel. Accordingly, the terminal continuously measures the pilot SINR from the serving sector and predicts the channel condition over the next packet based on the correlation of the channel. The inner loop selects the highest data rate (DRC) whose SINR threshold is lower than the predicted SINR. Whenever the network decides to serve the terminal with traffic data, it transmits at the rate indicated by the most recently received from the terminal.

The outer loop adjusts the SINR thresholds of the data rates based on the packet error rate of the forward traffic channel packets. If the packet error rate is higher than the target value, the outer loop increases the SINR thresholds of the data rates. However, if the packet error rate is lower than the target value, the outer loop decreases the SINR thresholds.

The procedure is illustrated in Figure 4.28. The supportable data rate is defined so that overall system efficiency is optimized. That is, forward link throughput is maximized subjected to a reasonable allocation of resources for reliable DRC, preamble detection, and ACK/NAK detection.[4]

Hybrid-ARQ

Hybrid-ARQ is the coding scheme wherein the information blocks are encoded for forward error correction, while retransmissions requested by the receiver attempt to correct the detectable errors that cannot be corrected by forward error correction. In IS-856 hybrid-ARQ is used on the forward traffic channel to further adapt the code rate (turbo code rate and repetition code rate) and the data rate to the mobile channel. IS-856 employs a basic

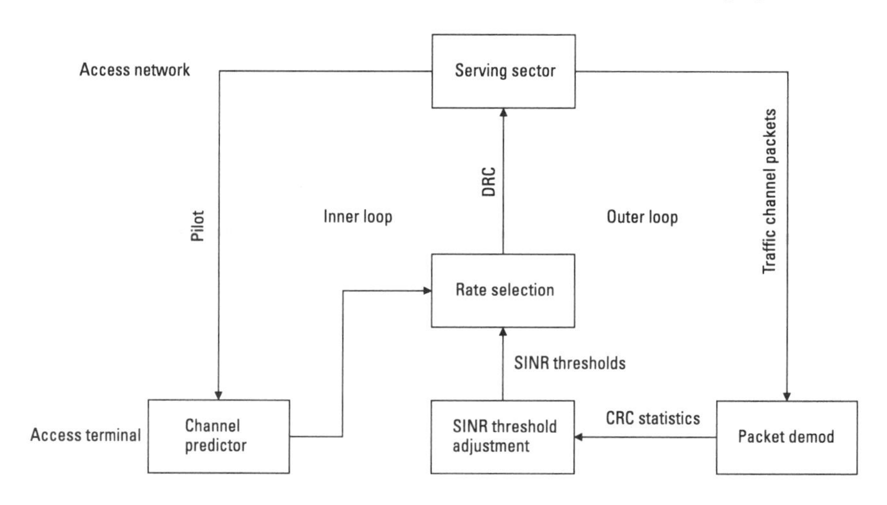

Figure 4.28 Closed-loop rate control.

4. The channel prediction and rate selection functions are not specified in the IS-856 standard, allowing manufacturers to include the latest and more efficient algorithms when they become available without the need for revision of the standard.

rate 1/5 turbo code that is also used in IS-2000 systems [2]. As discussed in Section 4.4.1.4, it consists of two identical rate 1/3 systematic, recursive convolutional codes separated by a code interleaver. The systematic part of the second convolutional code is discarded, resulting in a rate 1/5 code. For packet types (with low data rates) that require spectral efficiency lower than 2/5 bits/symbol, the rate 1/5 code with QPSK modulation is used.

Complete or partial sequence repetition of the modulation symbols is employed to achieve the desired spectral efficiency. For packet types (with high data rates) that require spectral efficiency higher than 2/5 bits/symbol, higher rate codes and possibly higher order modulations are used. Higher rate codes are obtained by indirectly puncturing the basic rate 1/5 turbo code. Accordingly, the output of the rate 1/5 turbo code is first interleaved, followed by a simple truncation of the interleaver output. This design facilitates hybrid-ARQ and provides near-optimal performance for truncated packet types, because early termination of a packet achieves the same effect as puncturing the underlying code.

As indicated earlier, the forward traffic channel adapts its forward error correction code (turbo code and repetition code) and modulation based on the DRC sent by the terminal. However, DRC is determined by the predicted SINR of the forward link based on pilot measurements. To ensure that the packet error rate is maintained below a certain level, the terminal tends to request conservative data rates due to the effects of pilot measurement noise, prediction delay, and inactivity of interfering sectors. It is desirable to transmit a packet that best adapts its code rate and thus data rate to the channel condition at the time of transmission. Thus, a two-step rate adaptation mechanism is utilized in IS-856: Closed-loop rate control adapts a data rate based on the a priori information of the channel state, whereas hybrid-ARQ further adapts the data rate based on the a posteriori information of the channel state.

For slow fading channels there is essentially no difference between the a priori and a posteriori information of the channel states. However, for the more unpredictable fast fading channels, the DRC might be much lower than the rate that could have been supported. In this case, closed-loop rate control naturally selects multislot packets since they introduce more time diversity,[5] giving the terminal an opportunity to terminate the transmission

5. Such time diversity reduces the dynamic range of the average channel state, thus improving the prediction of the packet average SINR. So, for fast fading channels, selecting the longer version of each data rate packet leads to less conservative rate selection while maintaining a desired packet error rate level.

early. As discussed in Section 4.4, hybrid-ARQ employs the ACK channel to terminate a multislot forward traffic packet if it is decoded correctly in fewer slots than the nominal packet length indicated by the DRC. The effective transmitted data rate is increased by the ratio of the nominal packet length (in slots) of the DRC to the number of slots transmitted before the ACK is received.

4.5.1.3 Scheduler

In Section 4.5.1.2 we argued that for packet data service where a guaranteed grade of service is not required, rate adaptation is a more efficient link adaptation scheme than power adaptation. In other words, it is more efficient to allocate to each user a certain fraction of power while adapting the data rate, constellation size, and coding scheme to the link quality. In this section, we study the optimum power allocation strategy that maximizes total throughput under a certain fairness constraint.

Forward Link Model

We consider the forward link of a sector embedded in a network. The base station transmits to N users, which are multiplexed on different orthogonal code channels. Suppose P_T is the total sector power, L_i paths of the desired signal are received by user i. Assuming that a fraction ξ_i, $0 \le \xi_i \le 1$ of the total power is allocated to user i, then its SINR can be written as

$$\mathrm{SINR}_i = \sum_{l=1}^{L_i} \frac{\xi_i \cdot P_T \cdot g_{i,l}}{\sum_{k=1, k \neq l}^{L_i} P_T \cdot g_{i,k} + I_i}$$

where $g_{i,l}$ is the link gain of path l for user i, and I_i is the other cell interference and noise power received by user i. Let γ_i denote the SINR for user i assuming the total sector power is being allocated to user i,

$$\gamma_i = \sum_{l=1}^{L_i} \frac{P_T \cdot g_{i,l}}{\sum_{k=1, k \neq l}^{L_i} P_T \cdot g_{i,k} + I_i}$$

Moreover, γ_i represents the channel state information that is fed back to the network by user i. Then we have

$$\mathrm{SINR}_i = \xi_i \cdot \gamma_i \tag{4.6}$$

The maximum rate at which data can be transmitted to user i is a function of the user's SINR and the fraction of code space allocated to the user. Similar to the model developed in Section 4.5.1.2, we may write

$$R_i = \mu_i W \log_2\left(1 + \eta \cdot \frac{1}{\mu_i} \cdot \mathrm{SINR}_i\right) = \mu_i f\left(\frac{\xi_i}{\mu_i} \cdot \gamma_i\right) \quad (4.7)$$

where

$$f(x) = W \log_2(1 + \eta x), \qquad 0 \le \eta < 1$$

and R_i is the data rate for user i, W is the forward link bandwidth, η defines the modem performance loss relative to the Shannon limit, and μ_i is the fraction of the orthogonal code space allocated to user i. In this section, as the strategy of power allocation among multiple users is studied, it is necessary to include the orthogonal code allocation in the model.

Optimal Orthogonal Code Allocation

The optimal orthogonal code allocation is defined to be the code allocation scheme that achieves the desired broadcast channel capacity given that the received SINR for each user is known. We first study a code allocation scheme that assigns

$$\mu_i = \frac{\mathrm{SINR}_i}{\sum_{i=1}^{N} \mathrm{SINR}_i} \quad (4.8)$$

In this scheme, the fraction of orthogonal code space assigned to a user is proportional to the user's SINR. Then the maximum data rate for user i is

$$R_i = \mu_i W \log_2\left(1 + \eta \cdot \sum_{i=1}^{N} \mathrm{SINR}_i\right) \quad (4.9)$$

In addition, the total throughput achieved by the forward link is given by

$$\sum_{i=1}^{N} R_i = W \log_2\left(1 + \eta \cdot \sum_{i=1}^{N} \mathrm{SINR}_i\right) \quad (4.10)$$

Equation (4.10) is actually the capacity that can be achieved by the broadcast channel modeled in the preceding subsection when orthogonal division signaling is used for the users.[6] Thus, (4.8) defines the optimal orthogonal code allocation for the CDMA forward link. From (4.8) to (4.10), we can also derive the following:

$$\frac{R_i}{\sum_{i=1}^{N} R_i} = \frac{\text{SINR}_i}{\sum_{i=1}^{N} \text{SINR}_i} \tag{4.11}$$

which shows that allocating code space proportional to SINR is equivalent to allocating code space proportional to the data rate. In the following sections, we investigate the optimal power allocation strategy (scheduler algorithm) that maximizes the total utility function defined for the user fairness requirement under the assumption that the optimal code allocation (4.8) is used.

Problem Definition and Constraints

Let $b_i(n)$ denote the total number of bits delivered to user i over time interval $[n - M, n]T$, and let $U_i(b)$ denote the utility function for delivering b bits to user i. Here, T is the time granule of packet transmission; MT is the length of scheduler window or fairness window. A certain set of utility functions $\{U_i(b)\}_{1 \le i \le N}$ defines the degree of fairness (or grade of service) over the scheduler window. At the time of $(n - 1)T$, the goal of the optimal scheduler is to form a power fraction vector $\xi(n)$ for time interval $[n - 1, n]T$ so as to maximize the total utility that can be offered by the BS at time nT:

$$\Psi(n) = \sum_{i=1}^{N} U_i(b_i(n))$$

The total number of bits delivered to user i over the time interval $[n - M, n]T$ can be expressed as

$$b_i(n) = \int_{(n-M)T}^{nT} R_i(t)\,dt = \sum_{k=n-M+1}^{n} R_i(k) \cdot T$$

6. Orthogonal division user signaling cannot achieve the capacity region of the broadcast channel. However, here orthogonal user code division is assumed for the forward link.

The data rate $R_i(k)$ over $[k-1, k]T$ can be derived based on (4.7) and (4.8), that is,

$$R_i(k) = \frac{\xi_i(k) \cdot \gamma_i(k)}{\sum_j \xi_j(k) \cdot \gamma_j(k)} f\left(\sum_j \xi_j(k) \cdot \gamma_j(k)\right) \tag{4.12}$$

where we defined the average SINR during the interval $[k-1, k]T$ to be

$$\gamma_i(k) = \frac{1}{T} \int_{(k-1)T}^{kT} \gamma_i(t)\, dt$$

Note that (4.12) assumes perfect rate adaptation to the forward link channel state $\gamma_i(k)$, which is a quite reasonable assumption due to the two-step rate adaptation mechanism in IS-856.

Moreover, at time $(n-1)T$, the power fraction vector $\xi(n)$ determined by the optimal scheduler has to satisfy the following constraints:

$$\sum_{i=1}^{N} \xi_i(n) = 1 \qquad \text{for } \xi_i(n) \geq 0,\ 1 \leq i \leq N$$

Optimal Scheduler

For a given power fraction vector $\xi(n)$ determined at time $(n-1)T$, the total utility offered by the BS at time nT can be written as

$$\Psi(n) = \sum_{i=1}^{N} U_i(b_i(n)) = \sum_{i=1}^{N} U_i(b_i(n-1) + \Delta b_i(n))$$

where

$$\Delta b_i(n) = b_i(n) - b_i(n-1) = [R_i(n) - R_i(n-M)] \cdot T$$

We assume that $\Delta b_i(n)$ is small enough so that $U_i(b)$ over $[b_i(n-1), b_i(n-1) + \Delta b_i(n)]$ can be approximated as a linear function. We then have

$$\Psi(n) \cong \Psi(n-1) - T \cdot \sum_i U_i'(b_i(n-1)) \cdot R_i(n-M) + T \cdot \sum_i U_i'(b_i(n-1)) \cdot R_i(n)$$

Only the last term of the above equation is related to $\xi(n)$, so we can reformulate our optimization problem by simplifying the notations. Hence, maximize

$$\phi(\xi) = \sum_{i=1}^{N} u_i \cdot R_i = \sum_{i=1}^{N} u_i \cdot \frac{\xi_i \cdot \gamma_i}{\sum_j \xi_j \cdot \gamma_j} f\left(\sum_j \xi_j \cdot \gamma_j\right) \quad (4.13)$$

Subject to

$$\sum_{i=1}^{N} \xi_i = 1$$

and

$$\xi_i \geq 0, \ 1 \leq i \leq N,$$

where $u_i(n) = U_i'(b_i(n-1))$.

From (4.7) we know that $f(x)$ is a concave function and $f(0) = 0$. Thus, for any $0 \leq \theta \leq 1$, we have

$$f(\theta \cdot x) \geq \theta \cdot f(x) \quad (4.14)$$

The objective function ϕ can be reformulated as

$$\phi = \frac{\lambda}{\lambda} \cdot \frac{\sum_i u_i \cdot R_i}{\sum_i R_i} \cdot \sum_i R_i = \frac{1}{\lambda} \cdot \frac{\sum_i \lambda \cdot u_i \cdot R_i}{\sum_i R_i} \cdot f\left(\sum_i \mathrm{SINR}_i\right)$$

We assume that the utility function $U_i(b)$ is always a nondecreasing function. Thus, we have $u_i \geq 0$. A constant λ is introduced such that $0 \leq \lambda \cdot u_i \leq 1$ is satisfied for all users. Then by applying (4.14), we have

$$\phi \leq \frac{1}{\lambda} \cdot f\left(\frac{\lambda \cdot \sum_i u_i \cdot R_i}{\sum_i R_i} \cdot \sum_i \mathrm{SINR}_i\right)$$

From (4.11), we can derive that

$$\frac{\sum_i u_i \cdot R_i}{\sum_i R_i} = \frac{\sum_i u_i \cdot \mathrm{SINR}_i}{\sum_i \mathrm{SINR}_i} \quad (4.15)$$

Then we have

$$\phi(\xi) \leq \frac{1}{\lambda} \cdot f\left(\lambda \cdot \sum_{i=1}^{N} u_i \cdot \text{SINR}_i\right) = \frac{1}{\lambda} \cdot f\left(\lambda \cdot \sum_{i=1}^{N} \xi_i \cdot u_i \cdot \gamma_i\right) \tag{4.16}$$

which shows the upper bound for the objective function. Obviously the bound is maximized when all the power is allocated to the user

$$k_0 = \underset{1 \leq i \leq N}{\arg\max}\,[u_i \cdot \gamma_i]$$

Even though maximizing the upper bound does not necessarily indicate the maximization of the objective function, this result still effectively justifies the efficiency of the TDM waveform in IS-856 forward link.

As the BS can only have knowledge about an estimated discrete channel state information $\tilde{\gamma}_i(n)$, the corresponding practical scheduling algorithm is to set

$$\xi_{k_0}(n) = 1,\ \xi_{\substack{1 \leq i \leq N \\ i \neq k_0}}(n) = 0 \qquad \text{for } k_0 = \underset{1 \leq i \leq N}{\arg\max}\ u_i(n) \cdot \tilde{\gamma}_i(n) \tag{4.17}$$

In contrast, if the sector power were adaptively distributed among multiple users, the scheduled data rate would be unknown to each user unless additional message exchange or signal processing is introduced for each user. In addition, due to the limited set of data rates and user orthogonal codes, allocating code space proportional to the data rate among multiple users might result in an unused portion of the user codes. Such a problem which further reduces the efficiency of the forward link waveform, exists in IS-2000 and WCDMA systems.

Proportional Fair Scheduler

The proportional fairness is defined by the following utility function:

$$U_i(b) = \ln(b/MT) \tag{4.18}$$

Thus, the scheduler defined in (4.5) through (4.17) can be reformulated accordingly as the following equation:

$$\xi_{k_0}(n) = 1, \ \xi_{\substack{1 \le i \le N \\ i \ne k_0}}(n) = 0 \qquad \text{for } k_0 = \underset{1 \le i \le N}{\arg\max} \frac{\gamma_i(n)}{T_i(n-1)} \tag{4.19}$$

where

$$T_i(n-1) = \frac{b_i(n-1)}{MT}$$

where $T_i(n)$ denotes the average throughput for user i over the time interval $[n-M, n]T$. More specifically, in IS-856, the discrete channel state feedback for each user is the DRC information, which is the estimate of the maximum data rate that can be supported assuming all the sector power is allocated to the user. The proportional fair scheduler defined in (4.19) can be equivalently implemented as

$$\xi_{k_0}(n) = 1, \ \xi_{\substack{1 \le i \le N \\ i \ne k_0}}(n) = 0 \qquad \text{for } k_0 = \underset{1 \le i \le N}{\arg\max} \frac{\mathrm{DRC}_i(n)}{T_i(n-1)} \tag{4.20}$$

In addition, the average throughput received by each user can be updated by the following IIR filter:

$$T_i(n+1) = (1-\alpha) \cdot T_i(n) + \alpha \cdot R_i(n) \tag{4.21}$$

where $R_i(n)$ is the data rate served to user i during slot n. If user i is not being served, then $R_i(n)$ is set to zero.

The scheduling algorithm specified in (4.20) and (4.21) attempts to serve each user at the peak of its channel condition, thus exploring the multiuser diversity gain and increasing the throughput of the sector. Fairness (or maximum packet delay) is controlled by selecting the averaging filter time constant, $\approx 1/\alpha$, which basically sets the tradeoff between throughput optimization and the amount of latency a user experiences when its channel condition degrades abruptly. Note that to fully explore the capacity-enhancing potential of this scheduler, the DRC values need to be updated fast enough to allow the scheduler to take advantage of channel variations experienced by different users.

This algorithm incorporates the two important features of a capacity enhancing scheduler: multiuser diversity and fairness. This will be further exemplified by simulations presented in Section 4.5.1.4.

4.5.1.4 Simulation Results

The IS-856 forward link capacity is simulated based on a network of 37 tri-sectored cells, a total of 111 sectors arranged in a hexagonal grid as shown in Figure 4.29. The capacity is obtained from one of the central embedded sectors [6]. The pedestrian and vehicular path-loss models and the Type A and B multipath models are based on the well-known ITU-R M. 1225, which is summarized in Table 4.10. The network simulation parameters are shown in Table 4.11.

For a given number of N terminals in the embedded sector, the simulation is run over a series of random location sets. For each set of the random locations, the simulation is run over 30 seconds. A complex channel coefficient vector is generated at every half slot for each path and antenna. The coefficient includes a static attenuation due to path loss and antenna directive

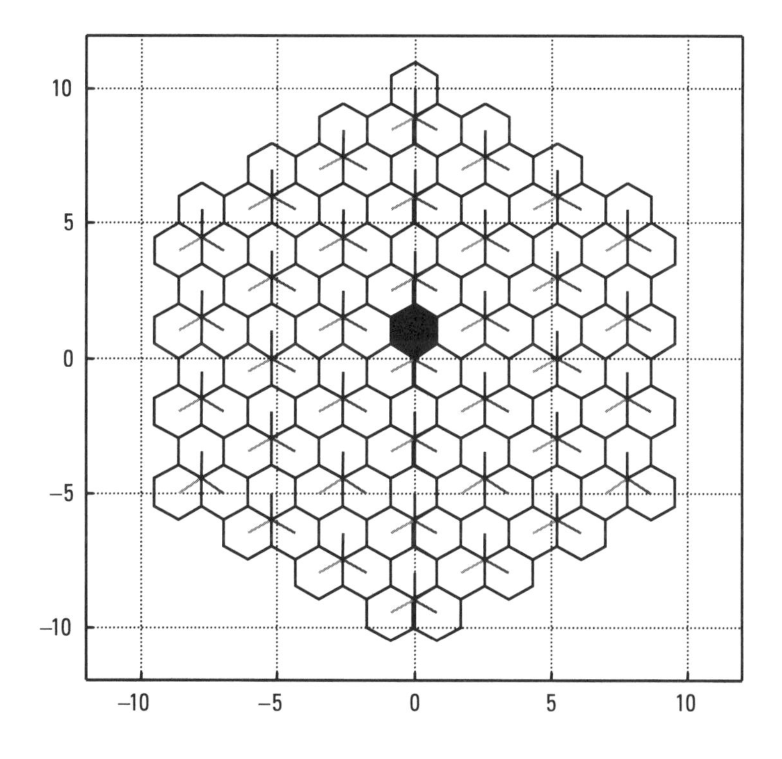

Figure 4.29 Three-tier network and embedded sectors.

Table 4.10
Channel Model

Pedestrian (3 km/hr) **Parameter**	**Value**
Path loss (dB)	$L = 30.0 \log_{10}(f_{cMHz}) + 49.0 + 40.0 \log_{10}(d_{km})$
Number of fingers per antenna per sector	1 for type A, 3 for type B
Captured multipath linear power gains (type A)	{1.0} Uncaptured power = 0.0
Captured multipath linear power gains (type B)	{0.71, 0.165, 0.068} Uncaptured power = 0.057 represented as a separate path
Vehicle (120 km/hr) **Parameter**	**Value**
Path loss (dB)	$L = -18.0 \log_{10}(H_{BS,m}) + 21.0 \log_{10}(f_{c,MHz}) + 80.0 + 40.0(1.0 - 0.004H_{BS,m}) \log_{10}(d_{km})$
Number of fingers per antenna per sector	2 for Type A, 3 for Type B
Captured multipath linear power gains (type A)	{0.784, 0.09} Uncaptured power = 0.126 represented as a separate path
Captured multipath linear power gains (type B)	{0.808, 0.03, 0.058} Uncaptured power = 0.104 represented as a separate path

gain, time-varying shadowing attenuation, and a short-term complex fading component.

The average sector loading effect on the cochannel interference from the adjacent sector is modeled by adjusting the relative power of the traffic relative to pilot burst. For example, the interference due to 50% loading is modeled by a traffic-to-pilot power ratio of −3 dB.

Thermal noise is set by the receiver noise figure. Post-AGC receiver self-noise equals −17.24 dB of the total receiver power. Waveform distortion induced ISI is set to be −15 dB relative to the path. These terms limit the maximum SINR at the demodulator for an AWGN channel to 13 dB for one antenna and 13.9 dB for dual antennas. These values are chosen consistent with existing IS-95 components for both transmitters and receivers.

The receiver has a RAKE structure with spatial-temporal signal combining. Temporal paths (multipaths) are combined using MRC assuming uncor-

Table 4.11
Network Level Simulation Parameters

Parameter	Value
Network topology	Hexagonal grid, 3 tiers (37 cells)
Sectorization	3 sectors/cell
Carrier frequency	1,960 MHz
BS antenna height	15m
MS antenna height	1.5m
Log-normal shadow standard deviation	10 dB
Site-to-site shadow correlation coefficient	0.5
Building penetration loss	12 dB for pedestrian, 0 dB for vehicular
Access point (AP) TX	1 antenna/sector
Access terminal (AT) RX	1 antenna/2 antennas
TX power (dBm) + TX antenna gain (dBi) – cable loss (dB) – body loss (dB) + RX antenna gain (dBi)	41.8 dBm + 17.0 dBi – 0.3 dB – 3.0 dB + 1.5 dBi = 57.0 dBm
Thermal noise figure	9 dB (= –104 dBm at 290K and assuming a 1.25-MHz bandwidth)
Cell radius	Set by pedestrian path loss model

related interference. Spatial paths (dual antenna) are combined using minimum mean square error combining given the correlated interference between antennas.

The half-slot SINR is calculated for the best serving sector selected by the terminal from the active set of candidate serving sectors. The model of active set management is consistent with the handoff rules defined in IS-856.

Rate adaptation and the proportional fair scheduler are simulated. A reverse link model with power-controlled soft/softer handoff is included to simulate the impact of ACK and DRC symbol errors and erasures on forward link performance. The differential path loss for the reverse link is equal to the forward link differential static attenuation and shadowing, while the forward and reverse link short-term fading processes are assumed to be independent. The reverse link power control setpoint is adjusted by an outer loop algorithm such that 1% packet error rate is achieved for the reverse traffic channel.

The imbalance issue reduces the reliability of the DRC and ACK channels. The practical parameters of the DRC and ACK channels are summarized in Table 4.12. These values are chosen for robust performance in a SHO scenario with 6 dB of imbalance. These are typical settings that

Table 4.12
DRC/ACK Channel Parameters

Parameter	One Way SHO	More Than One Way SHO
DRC length	Two slots	Four slots
DRC-to-pilot ratio	–1.5 dB	–3.0 dB
ACK-to-pilot ratio	4.0 dB	4.0 dB

would ensure reliability of the DRC and ACK channels with minimal impact on both reverse link capacity and forward link adaptation in slow fading channels. The number of ways in SHO is equal to the number of distinct cells in the active set.

Figures 4.30 and 4.31 show the sector throughput for single-antenna and dual-antenna cases, respectively. In each figure, four ITU channel models are presented: pedestrians A and B at 3 km/hr, and Vehicles A and B at 120 km/hr. The increase in throughput with the number of users manifests the multiuser diversity gain. The multiuser diversity gain expressed in the throughput improvement from 1 to 16 users is summarized in Table 4.13, from which we can derive the following conclusions:

1. For a given channel model, the multiuser diversity gain for a single antenna is higher than that for dual antennas. This is because antenna diversity significantly improves the distribution of the fading process

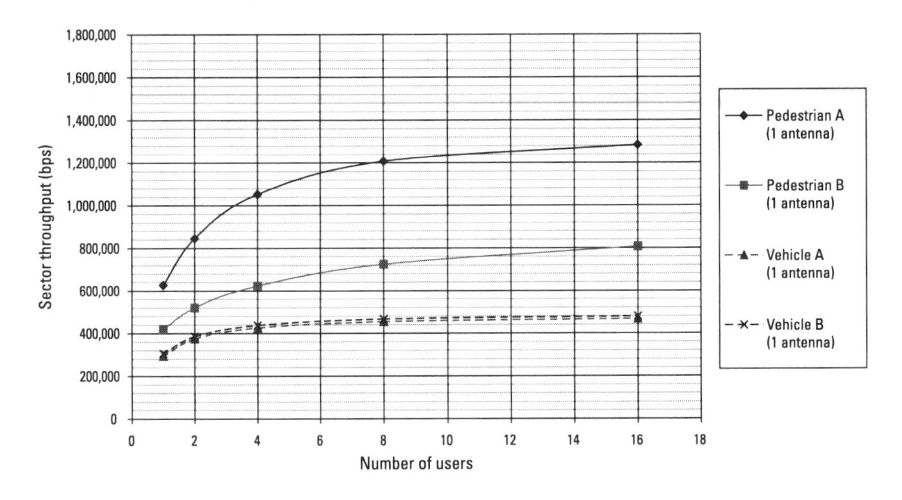

Figure 4.30 Single-antenna forward link capacity.

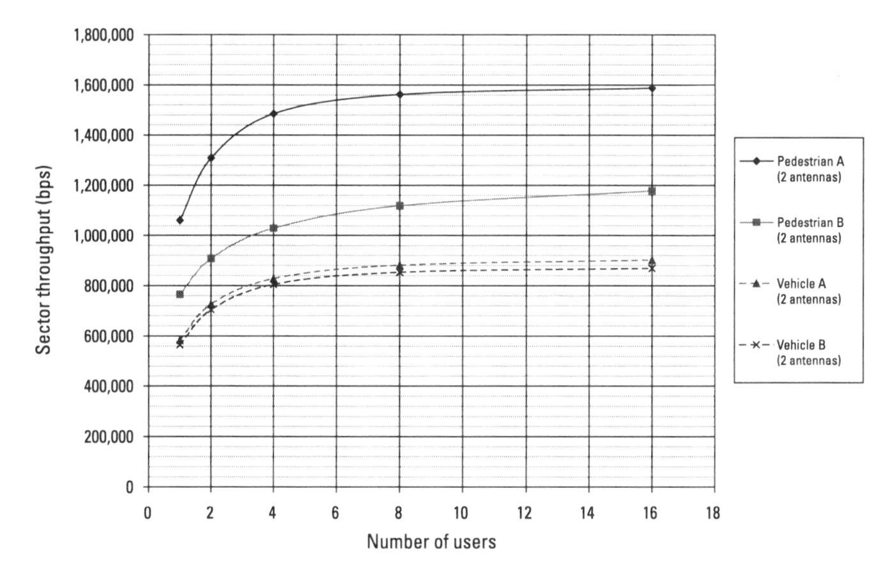

Figure 4.31 Dual-antenna forward link capacity.

Table 4.13
Sector Throughput and 16-User Multiuser Diversity Gain

	Single Antenna		**Dual Antenna**	
Channel Model	**Sector Throughput (Kbps)**	**Multiuser Diversity Gain (dB)**	**Sector Throughput (Kbps)**	**Multiuser Diversity Gain (dB)**
Pedestrian A	1,284	3.11	1,587	1.75
Pedestrian B	807	2.85	1,178	1.87
Vehicular A	467	2.01	902	1.90
Vehicular B	479	1.95	869	1.87

and, thus, reduces the channel dynamics. Because multiuser diversity is a form of selection diversity that exploits local peaks in the SINR, reduced channel dynamics leads to reduced selection diversity gain.

2. For the single-antenna case, multiuser diversity gain for slow fading is higher than that for fast fading. For slow fading channels, closed-loop rate control provides reliable channel state information (DRC) to the scheduler, and thus makes it possible for the scheduler to achieve higher throughput gains.

Figure 4.32 shows the histogram of the scheduled (initial DRC) and served data rates for a single antenna user with the ITU Vehicle A channel

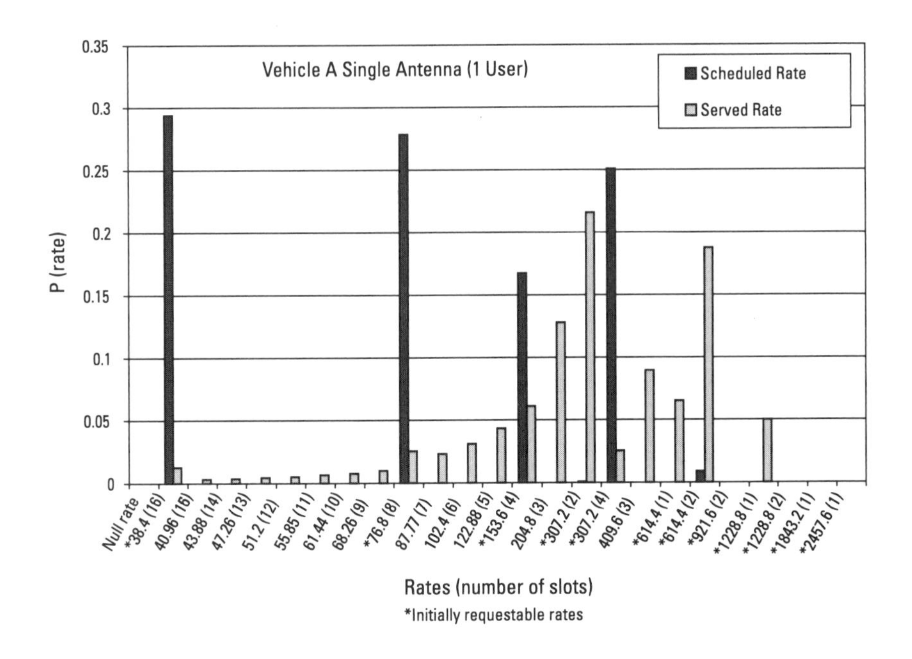

Figure 4.32 Histogram of the scheduled and served rates.

model. It gives insight into the effectiveness of hybrid-ARQ for a highly unpredictable, fast fading channel. In fact, for a slow fading channel, the ARQ gain is small due to the fact that the adjacent channel states are correlated and rate control (DRC) can accurately predict future channel conditions. Thus, the terminal is able to provide a good estimate of the highest supportable data rate for the next packet. For the fast fading channel, however, rate control is unable to accurately predict the uncorrelated channel states, and thus conservative data rates are requested to ensure low packet error rate. In this case, the ARQ gain becomes significant because it allows the terminal to decode and terminate the packets early before the end of the full packet transmission. In this way, closed-loop rate control and hybrid-ARQ cooperate to achieve the best link adaptation in all channel conditions.

Figure 4.33 shows how hybrid-ARQ increases the sector throughput by taking advantage of the cell loading. Because the pilot channel is always transmitted from each sector, a channel estimate obtained from the pilot burst contains no information about the actual cell loading. In other words, the channel state estimate obtained from the pilot burst reflects the channel condition, assuming that all interfering cells are fully loaded. Such an assumption results in a conservative rate request. The hybrid-ARQ mechanism

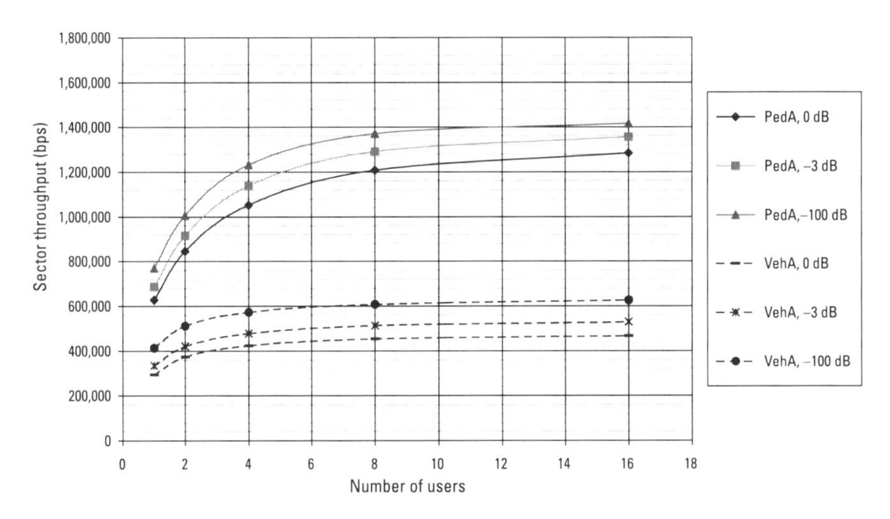

Figure 4.33 Single-antenna forward link capacity showing hybrid-ARQ gain from cell loading.

allows the terminal to decode and terminate the packet prior to its complete transmission based on the actual channel states experienced by the terminal during the packet transmission. This therefore allows the actual served rate to be adapted to the cell loading.

The ARQ gains for network loads of −3 and −100 dB relative to a fully loaded network are summarized in Table 4.14. Note that the cell-loading-induced ARQ gain is more pronounced for the vehicular case than for pedestrian case. This occurs for two reasons. First, for a highly unpredictable, fast fading channel, multislot packets are more likely to be requested, providing more opportunities to terminate the transmission early. Second, given the ITU path-loss models, vehicular users have a higher receiver power and are more interference limited than noise limited. The CDFs of the receiver power for pedestrian and vehicular users are shown in Figure 4.34.

Table 4.14
Network-Loading-Induced ARQ Gain

Network Load	Pedestrian A (1 user)	Pedestrian B (16 users)	Vehicular A (1 user)	Vehicular B (16 users)
−3 dB	0.39 dB	0.23 dB	0.58 dB	0.54 dB
−100 dB	0.89 dB	0.42 dB	1.49 dB	1.28 dB

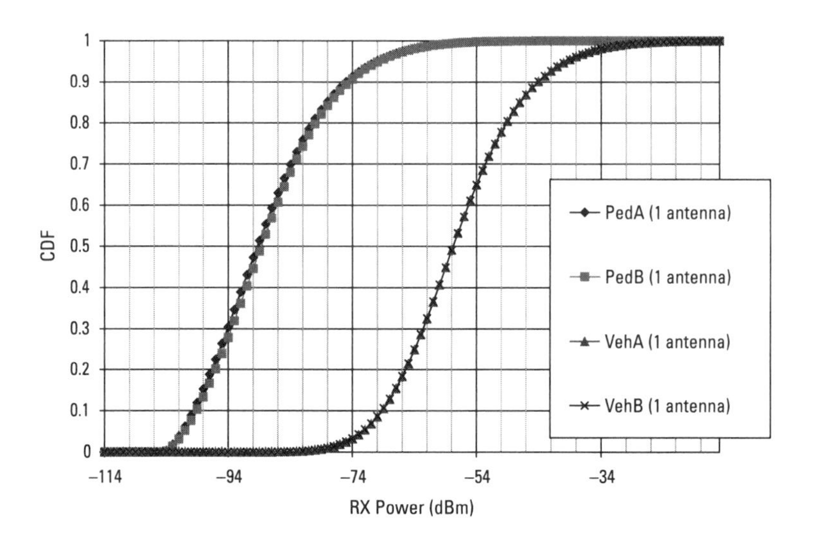

Figure 4.34 CDF of the received power of ITU path loss models.

As a result, the variation of interference from adjacent sectors has a more pronounced impact on the vehicular model.

The results in this section indicate that for a mix of mobile and portable users, the sector spectral efficiency is approximately 0.92 bps/Hz, which translates to a sector throughput of 1.1 Mbps. Moreover, the system supports a high-user peak-to-average throughput that further enhances the user experience.

4.5.2 Reverse Traffic Channel MAC

The reverse traffic channel MAC protocol in IS-856 defines the rules used by the access terminal to determine its transmission data rate on the reverse data channel [7, 8]. Two mechanisms allow the access network to control the transmission data rate of a given terminal. First, the terminal may receive a message from the network indicating the maximum data rate (RateLimit) it is allowed to transmit on the reverse link. Second, the terminal receives a reverse activity bit from each sector in its active set, indicating whether the total reverse traffic channel interference received at the sector is above a certain value.

This information determines whether the terminal may increase or decrease its data rate. The terminal increases or decreases its data rate in a probabilistic manner. The probabilities governing these transitions are specified by the network for each terminal, allowing the network to differentiate the behavior of the reverse traffic channel rate control algorithm among different terminals. Prior to the beginning of a new packet on the reverse traffic channel, the terminal uses the RateLimit, the reverse activity bits, and information related to the amount of data to send and available transmitting power to determine its reverse data rate. The complete procedure is described as follows.

Let CurrentRate be the terminal's current transmission data rate on the reverse link (it may be 0 if the terminal is not transmitting data on the reverse link). Let CombinedRABit be the logical OR operation on the most recent RA bits received from all sectors in the active set.

1. If CombinedRABit equals 1, then set MaxRate = max(9.6, CurrentRate/2) with probability p or MaxRate = max(9.6, CurrentRate) with probability $(1 - p)$. The transition probability p is a function of CurrentRate. For the results shown in this section, we consider the transition probabilities specified in Table 4.15.
2. If CombinedRABit equals 0, then set MaxRate = min(153.6, CurrentRate*2) with probability q or MaxRate = min(153.6, CurrentRate) with probability $(1 - q)$. The probability q is a function of CurrentRate. Table 4.15 shows the values of q that are used in the rest of this section.
3. The new transmission data rate may not exceed the limit given by the network, thus, set NewRate = min(MaxRate, RateLimit).
4. If the transmitting power available at the access terminal is not sufficient to support transmission at NewRate, then the access terminal decreases NewRate to the highest data rate that can be accommodated by the available transmit power.
5. In addition, if the amount of data to transmit is less than the payload size of NewRate, then NewRate is decreased to the lowest data rate

Table 4.15
Transition Probabilities and Total-to-Pilot Transmit Power Ratios

	Data Rate (Kbps)					
	0	**9.6**	**19.2**	**38.4**	**76.8**	**153.6**
q	1	3/16	1/16	1/32	1/32	0
p	0	0	1/16	1/16	1/8	1
Total to pilot	2	4.37	6.73	11.44	23.13	72.79

for which its payload size is large enough to send all the available data.

6. The terminal transmits at NewRate.

4.5.2.1 Simulation Results

Despite the fact that the IS-856 air interface specifies which algorithm an access terminal must use to determine its transmitting rate, it does not specify the algorithm the access network must use to set the rate limit, transition probability parameters, or the RABs transmitted by the sectors. This allows flexibility for both network optimization and grade-of-service selection.

The algorithm used in our simulations sets the rate limit of each access terminal to 153.6 Kbps, the highest rate allowed, and sets the transition probabilities as specified in Table 4.15. In addition, each sector determines the value of the RAB that it transmits to all terminals within its coverage area as follows [8]:

$$X = \sum_{k \in C(S)} (E_c)_k / I_0 = \sum_{k \in C(S)} ((E_{cp})_k / I_0)(\mathrm{TTP})_k$$

where $C(S)$ is the set of all access terminals that have the sector S in their active set, $(E_c)k$ is the total received energy per chip from access terminal k, and I_0 is the total received power spectral density in the sector. In the IS-856 system, the total received chip energy can be seen as product of two terms: the term $(\mathrm{TTP})_k$ represents the ratio of the total power in the traffic channel to the power in the pilot portion of the traffic channel transmitted by access terminal k, and the term $(E_{cp})_k$ represents the pilot chip energy. The term $(\mathrm{TTP})_k$ is, in general, a function of the data rate. A typical set of values is given in Table 4.15. Sector S then compares its load X against a threshold T. If $X > T$, it sets the RAB to 1. If $X \leq T$, it sets the RAB to 0.

A simulation of the rate allocation algorithm was carried out for a seven-cell network with typical link budget parameters, which included the COST-231 path-loss model, log-normal shadow with 8-dB standard deviation, and fast Rayleigh fading process at 30 km/hr. Closed-loop power control was run to achieve the desired 1% packet error rate after selection combining from all sectors in the active set. We present results for 4, 8, and 16 terminals per sector, where the link budget was calculated for I_0 / N_0 = 5 dB[7] and a maximum transmitting power of 23 dBm. The load X and RAB determination were updated every 26.67 ms with a threshold T = 0.65625.

7. N_0 is the thermal noise power spectral density.

Sector throughput is calculated by averaging more than 10 runs. The throughput was determined with four and eight terminals per sector to demonstrate the robustness of selected parameters to varying number of users.

Table 4.16 presents the average and standard deviation of the time-averaged throughput and load at the central sector. It shows that the sector throughput is approximately 180 Kbps for four and eight terminals per sector and goes down to approximately 125 Kbps when each sector has 16 terminals. This is because the number of pilot and DRC channels increases, contributing to the sector load but increasing data throughput.

The average observed load, however, was not constant when the number of users in the sector was changed. This is related to the choice of the transition probabilities. As the data rates increase, the q (transition probability) values go down, while the p values go up. This choice ensures that the sector does not get overloaded too frequently and if it does, the load is reduced quickly. As a result, when the terminals are operating at a high data rate, as is the case for four terminals per sector, the load of the sector decreases quickly when the RAB is set to 1 but increases slowly when the RAB is set to 0. This causes the average load to be lower than the threshold used to set the RAB.

Next we compare the measured load and I_0/N_0 as a function of time in Figure 4.35 obtained for the eight terminals per sector case. We observe that the resulting I_0/N_0 is approximately (below) 5 dB, which corresponds to the value used in the link budget. This suggests that controlling the sector load results in a desired value of I_0/N_0. Also, as shown in Figure 4.35, both load and I_0/N_0 monotonically change as the number of terminals increases. Finally, note that a tradeoff between reverse link throughput and cell range can be achieved by selecting an appropriate target I_0/N_0. A higher I_0/N_0 allows more active terminals in each sector, thus increasing throughput. However, coverage may be reduced, as terminals at the edge of the cell will

Table 4.16
Average and Standard Deviation of Load, Sector Throughput, and I_0/N_0

Terminals per Sector	Load Average	σ	THPT (Kbps) Average	σ	I_0/N_0 Average	σ
4	0.54	0.10	175	36	3.75	0.46
8	0.65	0.07	187	29	4.83	0.49
16	0.74	0.04	125	11	5.53	0.30

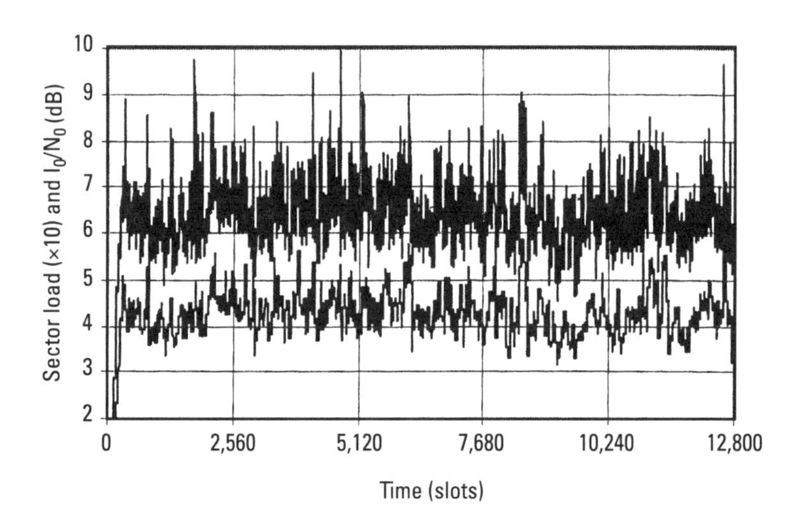

Figure 4.35 Typical sector load (×10) (upper curve) and I_0/N_0 (lower curve) as a function of time.

need to transmit more power to compensate for the additional interference levels. In addition, although not discussed here in detail, IS-856 allows for classes of terminals achieving different average throughputs to be created by assigning different sets of probability pairs (p, q) to different users.

4.6 Link Budget

Improved link budget or cell coverage is an additional advantage of the efficient waveform design in IS-856 compared with other digital cellular standards. Analysis and computer simulation show that, under an equally effective served rate, IS-856 has link budget advantages over IS-95-A of approximately 10 dB on the forward link and 1.5 dB on the reverse link [9].

4.6.1 Link Budget Analysis

A link budget determines the maximum allowable path loss of a communication link. For wireless systems, it is simply the difference between the transmitter EIRP and the receiver sensitivity, plus receiving antenna gain, less fade margin, building penetration loss and body (or cable) loss (all on the decibel scale).

Given a certain transmitter power and antenna gain, one observes a certain maximum radiation intensity. The transmitter EIRP is the effective input power to a hypothetical isotropic antenna that achieves such radiation intensity in any direction. It is a function of transmitter power, transmitter antenna gain, and cable (or body) loss. The transmitter EIRP is calculated as follows:

```
Transmitter EIRP (dBm) =
Transmitter Power (dBm) + Tx Antenna Gain (dBi) - Cable (or Body) Loss
(dB)
```

The receiver sensitivity denotes the minimum signal level that a receiver is able to receive. It is a function of the receiver's thermal noise and the required I_{0r}/N_0 to ensure a certain throughput or data rate under a certain geometry for the worst case fading channel. It is derived to be:

Receiver Sensitivity (dBm) =
Thermal Noise (dBm/Hz) + (I_{0r}/N_0)req(dB) + Bandwidth(dB-Hz)

It is also derived to be:

Receiver Sensitivity (dBm) =
Thermal Noise (dBm/Hz) + (E_b/N_o)req(dB) + Data Rate(dB-Hz)

For a CDM link, of which each channel transmits at an allocated portion of the total transmitter power, it is somewhat conventional in link budget calculations to scale both transmitter EIRP and receiver sensitivity by the same fraction. Obviously, the scaling will not affect the difference between the two quantities (in decibels). However, after scaling, conceptually the transmitter EIRP becomes transmitter EIRP per channel, and receiver sensitivity refers to the minimum signal level required for the channel.

The required I_{or}/N_0 or average E_b/N_0 is a common measure of modem efficiency. We dedicate Section 4.6.2 to the discussion of this topic.

The random fade ς (in decibels) due to shadowing requires that the maximum path loss be reduced by a certain fade margin to guarantee—most of the time—the same signal level that could be received without shadowing. In this document, we assume ς has zero mean and standard deviation of 8 dB. For hard handoff, if we require that 90% of the time adequate coverage be achieved for points on the cell boundary, it follows that the fade margin is 10.3 dB. However, a mobile at the edge of one BS's coverage is very likely to also be covered by a neighboring BS. The IS-856 system would place such a mobile in SHO. Both forward link and reverse link of IS-856 use selection diversity (instead of combined power as on the IS-95-A forward link) when in SHO mode. For the forward link the mobile selects a BS that has the best signal level as its serving BS, while for the reverse link the access network selects the better of the two BSs' receptions of any given packet. For selection diversity SHO, we see that the required fade margin is much reduced to 6.2 dB. Therefore, we have a 4.1-dB SHO gain for the IS-856 link budget.

4.6.2 IS-856 Link Budget Assumptions

4.6.2.1 Forward Traffic Channel

In this section, we discuss the required I_{or}/N_0 to ensure a certain forward link user throughput under a certain geometry for the worst case fading channel.

Forward Link User Throughput

For the IS-95-A CDM approach at the fixed data rate of 9.6 Kbps, the typical maximum traffic channel gain is −12.7 dB (5.3% of the BTS power). Because the forward link budget calculation for IS-95-A assumes two-way power combined SHO, the effective maximum traffic channel gain normalized to a single serving sector is −9.7 dB (10.72% of the BTS power). Assuming 30% pilot, paging, and sync channel overhead, this is equivalent to 15.3% of the available traffic channel resource. Given that the IS-856 forward traffic channel is variable rate, time division multiplexed, and uses selection diversity SHO, a fair link budget comparison would be at the same average user throughput of 9.6 Kbps with the same serving fraction of the traffic channel slots. Assuming a control channel rate of 38.4 Kbps, 15.3% of the available traffic channel slots is equivalent to 14.3% of the total serving slots.

Geometry

A mobile at the edge of one BS's coverage is very likely to be also covered by a neighboring BS. For the purpose of link budget calculation, we assume two-way SHO and equal distance between the two BSs with a third cell interference 6 dB down. The total noise and interference power spectrum density is $N_0 + 1.25I_{or}$. We will run variable-rate single-user link simulations under such geometry and plot sector throughput versus I_{or}/N_0. The throughput is obtained assuming 100% available resources. While deriving the user throughput, the sector throughput will be scaled by 14.3% to reflect a 14.3% serving fraction of the time for the user.

Fading Channel

For the purpose of link budget calculation, only the worst case fading channel that corresponds to the highest required I_{or}/N_0 is of interest. However, to determine the worst case, we have to run link simulations for a complete set of channels: PedA/B, VehA, and Rice.

The very slow log-normal shadow fading is not included in the simulation. Rather, that effect is considered in a separate entry of link budget, that is, fade margin.

Multiuser Diversity Gain

Given that the mobile is served only 14.3% of the time, multiuser diversity gain is realized from the scheduler that should be included. The single-user link simulations assume 100% of the serving time and do not include such gain.

We will run N user network simulations with all the users in the same geometry as in the single-user link simulations and plot sector throughput versus I_{or}/N_0. The sector throughput is obtained assuming 100% available resources for the N users. While deriving the user throughput, the sector throughput shall be scaled by 14.3% to reflect a 14.3% serving fraction of the time for the user. For a given user throughput, the difference between the required I_{or}/N_0 for a single-user link and the required I_{or}/N_0 for N user networks is the multiuser diversity gain. For link budget calculations, multiuser diversity gain is estimated from $N = 4$ simulations.

Figure 4.36 demonstrates the forward link sector throughput versus I_{or}/N_0 for a single-user link with a single receiving antenna. Figure 4.37 demonstrates the forward link sector throughput versus I_{or}/N_0 for an N-user network with a single receiving antenna. From these figures, we notice that the worst case occurs in the PedA 3 km/hr channel for the single-user case and the VehA 30 km/hr channel for the multiuser case under the given user throughput of 9.6 Kbps (which corresponds to a sector throughput of 67.1 Kbps assuming 14.3% serving fraction of time). The required I_{or}/N_0 are −5.0 and −7.0 dB for the single-user link and N-user network, respectively. As a result, the multiuser diversity gain is 2.0 dB at the user throughput of 9.6 Kbps.

4.6.2.2 Reverse Traffic Channel

In this section, we discuss the required average E_b/N_0 to ensure a reliable reverse traffic channel at the fixed data rate of 9.6 Kbps under 50% system loading for the worst case fading channel. The target packet error rate is 2%.

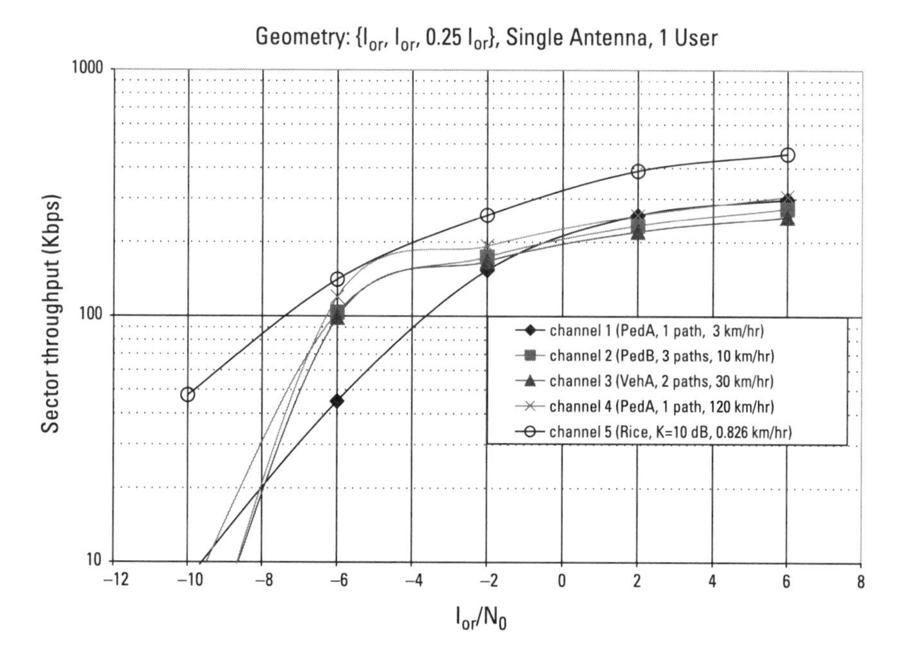

Figure 4.36 Single-user link simulations.

Dual Receiving Antennas

As with IS-95, we assume dual receiving antennas at the BS receiver. For link budget calculation, the required average E_b/N_0 per antenna includes the diversity gain of interest.

Percent Power for Data Channel

The reverse traffic data channel of the IS-856 system transmits at a fraction α of the total transmitter power. For the purpose of link budget calculation, we can scale the transmitter power by α to derive the MS EIRP for the data channel. Alternatively, in this chapter we will scale the data channel $\overline{E}_b/N_0$ per antenna by $1/\alpha$ to obtain the total $\overline{E}_b/N_0$ per antenna. Such $\overline{E}_b/N_0$ can be interpreted as the ratio of the total energy (including pilot, DRC, and ACK) received per antenna from that MS during an information bit to thermal noise *power spectral density* (psd).

System Loading

We will first obtain the required $\overline{E}_b/N_0$ per antenna assuming 0% system loading: $I_0/(N_0 + I_0) = 0$. However, the assumption of 0% system loading is optimistic. As the cell loading increases, the interference in the

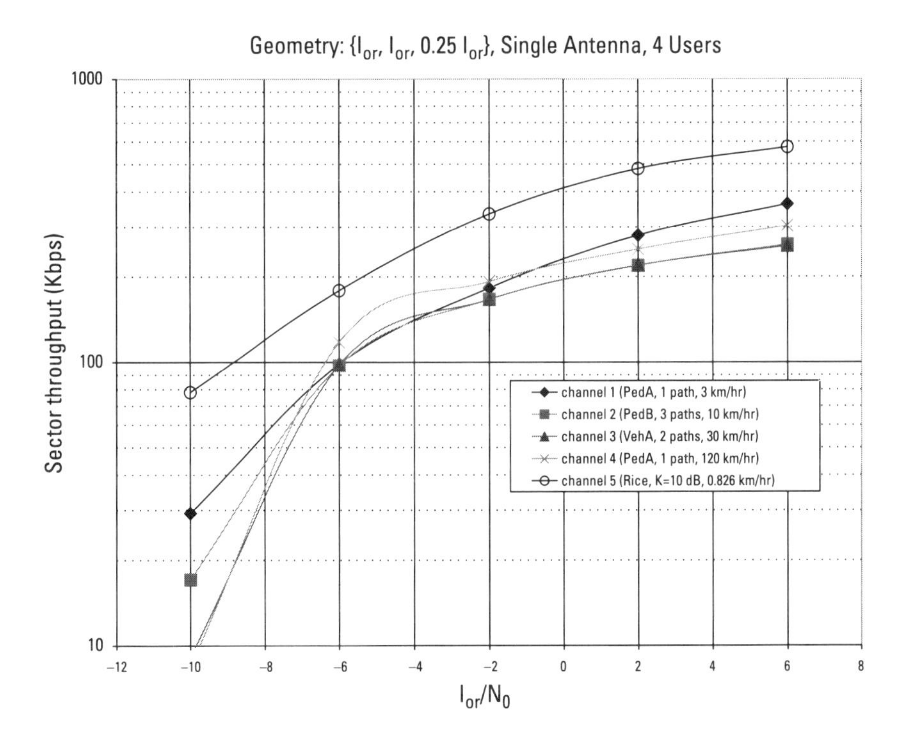

Figure 4.37 *N*-user network simulations (*N* = 4).

system increases. Therefore, a higher average E_b/N_0 per antenna is required to maintain the same QoS. This in turn reduces the maximum path loss and thus the size of a cell. A loading of 50%, that is, $I_0/(N_0 + I_0) = 0.5$, is assumed in the IS-95-A reverse link budget calculation.

To yield a fair comparison, we will make the same assumption for IS-856. Higher system loading can be accommodated at the expense of higher required average E_b/N_0 per antenna. To be flexible, we will create an entry of required $\overline{E}_b/N_0$ per antenna for 0% system loading, while the system loading margin, which is derived to be −10 log10 (one-system loading), is accounted for as a separate entry in the link budget.

Power Control

The link budget is concerned with the study of maximum allowable path loss for which link closure can occur—in which case, the mobile transmits at maximum power. When a mobile transmits at maximum power, there is actually no reverse link power control. Thus, we will run reverse link simulations for this section with power control off.

Two-Way SHO

A mobile at the cell edge is mostly covered by a neighboring BS. An IS-856 system would place such a mobile in the SHO mode. As reverse link selection diversity is performed on a fast frame-by-frame basis, it combats not only shadowing but also multipath fading. A reconstructed frame in the switching center is in error only if both reverse link frames are in error. Thus, the required $\overline{E}_b/N_0$ is determined by noting where the product of the two reverse link FERs is at the specified operating point. As an example, if the overall target packet error rate is 2%, each link can have 14% packet error rate. This provides a significant gain in slow speed conditions.

Figure 4.38 demonstrates the reverse traffic channel packet error rate versus total $\overline{E}_{c,p}/N_t$ per antenna (or $\overline{E}_{c,p}/N_0$ per antenna at 0% loading in which situation $N_t = N_0$). From the figure we notice that the worst case occurs in the PedB 3 km/hr channel condition. The required total $\overline{E}_{c,p}/N_t$ at 0% loading is −22.5 dB for the reverse traffic channel to achieve 14% packet error rate at the fixed data rate of 9.6 Kbps with SHO.

We can derive $\overline{E}_b/N_0$ per antenna as:

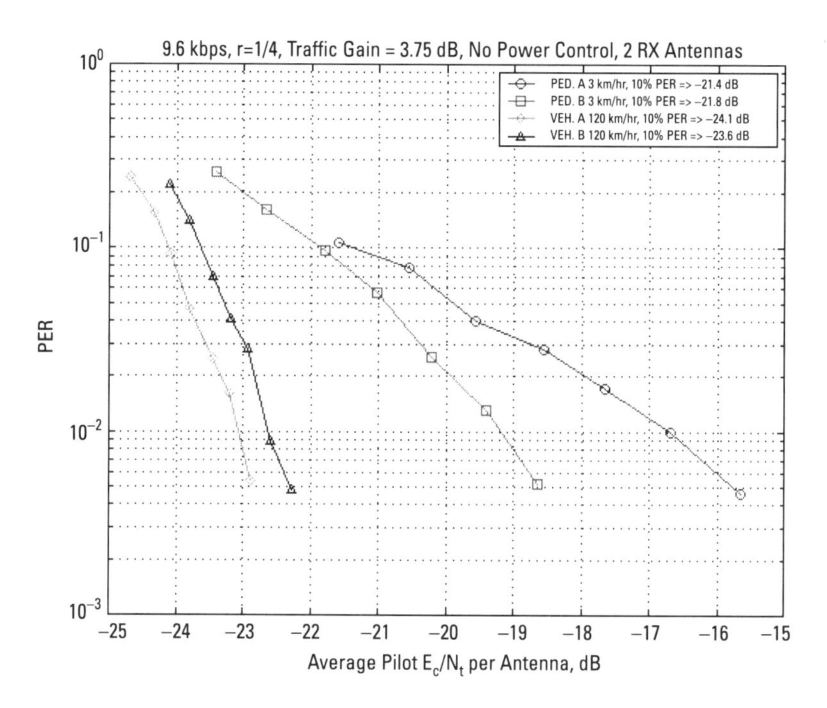

Figure 4.38 Reverse traffic channel packet error rate (PER) at fixed data rate of 9.6 Kbps with power control off.

$$\frac{\overline{E}_b}{N_0} = \frac{\overline{E}_{c,p}}{N_0} + 10\log\left(\frac{W}{R}\right) + 10\log\left(1 + 10^{\text{DataGain}/10} + 10^{\text{DRCGain}/10} + 0.5 \cdot 10^{\text{ACKGain}/10}\right)$$

All the gains in the above equation define the relative power levels of the corresponding channels to the pilot channel. The factor 0.5 for ACKGain accounts for the fact that ACK channel can at most transmit every other half slot. We will assume DataGain = 3.75 dB for 9.6 Kbps, DRCGain = –3.0 dB (with DRCLength = 4 slots under the SHO geometry assumed), and ACKGain = 4.0 dB. Thus, when a terminal is in two-way SHO and transmits at maximum power the required $\overline{E}_b/N_0$ per antenna is 5.67 dB in order to ensure the reverse traffic channel can achieve 2% effective packet error rate at fixed data rate of 9.6 Kbps under 0% system loading.

For all reverse link data rates, due to the carefully selected data to pilot channel gains (DataGain), the packet error rate performance is very similar with respect to the pilot $E_{c,p}/N_t$. Thus, Figure 4.38 can be applied to all reverse link data rates. Table 4.17 shows the required average E_b/N_0 per antenna for each reverse link data rate under 2% packet error rate and 0% system loading.

As observed, higher data rates come with lower relative overhead for the pilot, DRC, and ACK channels, and thus result in lower required "total" average E_b/N_0 under the same required data channel average E_b/N_0. At 153.6 Kbps, higher E_b/N_0 is due to the rate 1/2 versus rate 1/4 code of the other rates.

4.6.3 Link Budget Calculation

Here, we first derive the link budget for the IS-856 forward link and then show that it has better coverage than IS-95-A (under conditions of equal effective served rate). Because service providers may want to deploy IS-856 systems using the same network plans as IS-95-A but on a separate carrier

Table 4.17
Required Average E_b/N_0 at 0% Loading

Data rate (bps)	9,600	19,200	38,400	76,800	153,600
Data gain (dB)	3.75	6.75	9.75	13	17.5
Required average E_b/N_0 per antenna (dB)	5.67	4.31	3.41	3.10	4.26

frequency, it is also interesting to see the IS-856 forward user throughput under the IS-95-A link budget (under conditions of equal path loss).

4.6.3.1 Forward Link Budget

From Table 4.18, for 90% cell edge coverage of a fixed data rate at 9.6 Kbps with a –9.7-dB effective traffic channel gain, the forward link budget for IS-95-A is 138.2 dB.

For terminals with a single receiving antenna, as observed from Table 4.19, for 90% cell edge coverage of user throughput at 9.6 Kbps with 14.3% serving fraction of the time, the link budget for IS-856 is 147.7 dB. Under

Table 4.18
IS-95-A Forward Link Budget

	Traffic	Equation
Data rate (bps)	9,600	
Data rate (dB-Hz)	39.8	A
BTS Tx power (W)	15.0	
BTS Tx power (dBm)	41.8	
Maximum power for each channel (%)	5.3	
Maximum power allocated per channel (W)	0.8	
Maximum power allocated per channel (dBm)	29.0	B
BTS antenna gain (dBi)	17.0	C
BTS cable loss (dB)	3.0	D
BTS EIRP per TCH (dBm)	43.0	$E = B + C - D$
MS Rx antenna gain (dBi)	0.0	F
Body loss (dB)	3.0	G
Noise figure (dB)	9.0	H
Thermal noise (dBm/Hz)	–165.0	$I = -174.0 + G$
Target packet error rate (%)	3	
E_b/N_0 required for single antenna (dB)	12.3*	J
MS receiver sensitivity (dBm)	–112.9	$K = I + J + A$
Log-normal standard deviation (dB)	8	
Log-normal fade margin (dB)	10.3	L
SHO gain (dB)	5.6	M
Building penetration loss (dB)	10.0	N
Maximum path loss (dB)	138.2	$O = E - K + F - G - L + M - N$

*Assume two-way SHO with two received signals having the same psd. No third cell interference is allowed (i.e., $I_{0c} = 0$). So it is better geometry than what we assumed for IS-856 link budget calculation. Also, the required average E_b/N_0 is obtained assuming 3% target FER, a looser constraint than the 2% target packet error rate assumed in IS-856.

Table 4.19
IS-856 Forward Link Budget

	Traffic	Traffic	Equation
Average throughput (or data rate) (bps)	9,600	32,175	
Average burst rate (bps)	67,100	225,000	
Serving time fraction (%)	14.3	14.3	
Bandwidth (Hz)	1,228.8k	1,228.8k	
Bandwidth (dB-Hz)	60.9	60.9	A
BTS Tx power (W)	15.0	15.0	
BTS Tx power (dBm)	41.8	41.8	B
BTS antenna gain (dBi)	17.0	17.0	C
BTS cable loss (dB)	3.0	3.0	D
BTS EIRP (dBm)	55.8	55.8	$E = B + C - D$
MS Rx antenna gain (dBi)	0.0	0.0	F
Body loss (dB)	3.0*	3.0	G
Noise figure (dB)	9.0	9.0	H
Thermal noise (dBm/Hz)	–165.0	–165.0	$I = -174.0 + H$
Target packet error rate (%)	2	2	
I_{0r}/N_0 required per antenna (dB)	–7.0	2.5	J
Multiuser diversity gain (dB)	2.0	1.0	
MS receiver sensitivity (dBm)	–111.1	–101.6	$K = I + J + A$
Log-normal standard deviation (dB)	8	8	
Log-normal fade margin (dB)	10.3	10.3	L
SHO gain (dB)	4.1	4.1	M
Building penetration loss (dB)	10.0	10.0	N
Maximum path loss (dB)	147.7	138.2	$O = E - K + F - G - L + M - N$

*For nonhandheld terminals (e.g., wireless data modems), the body loss can be assumed to be zero.

the IS-95-A link budget, an IS-856 single receiving antenna terminal would achieve a long-term average forward user throughput of 32.2 Kbps or an equivalent short-term average burst rate of 225 Kbps.

IS-856 obtains link budget gain over IS-95-A from the following sources:

- Coding gain of 2.0 dB. IS-856 uses turbo code as opposed to convolutional code in IS-95-A.

- Variable rate gain of 7.0 dB. IS-856 uses DRC and ARQ to achieve fast rate adaptation as opposed to slow power control in IS-95-A forward link.
- Multiuser gain of 2.0 dB. IS-856 uses a proportional fair scheduling algorithm that explores multiuser diversity gain as opposed to the equal grade of service in IS-95-A.

IS-856 loses 1.5 dB in the SHO gain for fade margin because IS-856 uses selection diversity SHO as opposed to the power combined diversity SHO used in IS-95-A. These add up to the 10-dB forward link budget gain for IS-856 over IS-95-A.

4.6.3.2 Reverse Link Budget

From Table 4.20, we conclude that for 90% cell edge coverage of a fixed data rate at 9.6 Kbps, the reverse link budget for IS-95-A is 136.0 dB, whereas for IS-856, as observed from Table 4.20, for 90% cell edge coverage of a fixed data rate at 9.6 and 19.2 Kbps, the reverse link budgets are 138.3 and 136.7 dB, respectively. Thus, operators can plan an IS-856 network for adequate coverage of reverse link data rate at 19.2 Kbps and still achieve a better link budget than that of an IS-95-A network, which only has adequate coverage of the reverse link data rate at 9.6 Kbps.

Acknowledgments

We thank Roberto Padovani, Peter Black, Rajesh Pankaj, Yu-Cheun Jov, and Ramin Rezaiifar for their valuable review of the chapter.

Table 4.20
IS-856 and IS-95-A Reverse Link Budgets

	IS-856	IS-856	IS-95	Equation
Data rate (bps)	9,600	19,200	9,600	
Data rate (dB-Hz)	39.8	42.8	39.8	A
MS Tx power (mW)	200	200	200	
MS Tx power (dBm)	23.0	23.0	23.0	B
MS antenna gain (dBi)	0.0*	0.0	0.0	C
Body loss (dB)	3.0	3.0	3.0	D
MS EIRP for data channel (dBm)	20.0	20.0	20.0	$E = B + C - D$
BTS Rx antenna gain (dBi)	17.0	17.0	17.0	F
BTS cable loss (dB)	3.0	3.0	3.0	G
BTS noise figure (dB)	5.0	5.0	5.0	H
BTS thermal noise (dBm/Hz)	–169.0	–169.0	–169.0	$I = -174.0 + H$
Target packet error rate (%)	2	2	3	
E_b/N_0 required per antenna @ 0% loading (dB)	5.7	4.3	8.0	J
System loading margin (dB)	3.0	3.0	3.0	K
BTS receiver sensitivity (dBm)	–120.5	–118.9	–118.2	$L = I + J + K + A$
Log-normal standard deviation	8	8	8	
Log-normal fade margin (dB)	10.3	10.3	10.3	M
SHO gain (dB)	4.1	4.1	4.1	N
Building penetration loss (dB)	10.0	10.0	10.0	O
Maximum path loss (dB)	138.3	136.7	136.0	$P = E - L + F - G - M + N - O$

*Nonhandheld terminals may yield higher antenna gain. The comparison here assumes a mobile handset.

References

[1] 3GPP TS 25.221, v3.1.0, *Physical Channels and Mapping of Transport Channels onto Physical Channels (TDD),* 2001.

[2] 3GPP2, *Physical Layer Standard for CDMA2000 Spread Spectrum System,* Technical Report C.S002-A-1, Dec. 2001.

[3] Third-Generation Partnership Project 2, *CDMA2000 High Rate Packet Data Air Interface Specification,* Technical Report C.S20024, version 2.0, Oct. 2000.

[4] Bender, P., et al., "CDMA/HDR: A Bandwidth Efficient High Speed Wireless Data Service for Nomadic Users," *IEEE Communications Magazine,* Vol. 38, No. 7, July 2000, pp. 70–77.

[5] Black, P., and N. Sindhushayana, "Forward Link Coding and Modulation Design for IS-856," unpublished manuscript.

[6] Black, P., and M. Gurelli, "Capacity Simulation of CDMA2000 1xEV Wireless Internet Access System," *Proc. IEEE MWCN 2001,* Aug. 2001.

[7] Esteves, E., "The High Data Rate Evolution of the CDMA2000 Cellular Systems," in Stuber, G. L., and B. Jabbari (eds.), *Multiaccess, Mobility and Teletraffic for Wireless Communications,* Vol. 5, Norwell, MA: Kluwer Academic Publishers, 2000, pp. 61–72.

[8] Chakravarty, S., R. Pankaj, and E. Esteves, "An Algorithm for Reverse Traffic Channel Rate Control for CDMA2000 High Rate Packet Data Systems," *Proc. IEEE Globecom 2001,* San Antonio, TX, Nov. 2001.

[9] Black, P., and Q. Wu, "Link Budget of CDMA200 1xEV Wireless Internet Access System," unpublished manuscript.

5

Peak-to-Average Ratio of CDMA Systems

Vincent Lau

The *peak-to-average ratio* (PAR) of a signal is an important parameter in CDMA systems because it determines the *backoff* factor that needs to be applied to the power amplifier in order to avoid clipping of the input signal and hence spectral regrowth. In this chapter, we analyze the PAR of the signals from several CDMA systems, namely, the single-carrier IS-95 system, direct-spread CDMA2000 system, UMTS WCDMA system, multicarrier IS-95 system, and multicarrier CDMA2000 system.

In Section 5.1, we give a brief introduction on the background of the problem. We formally define the PAR and explain how the PAR is related to spectral regrowth in the power amplifier design. In Section 5.2, we discuss the analysis of PAR. For instance, we discuss and compare the PAR of signals from single-carrier direct-spread CDMA systems (IS-95, CDMA2000, and WCDMA) by simulation in Section 5.1.

In Section 5.2, an analytical model is developed to describe the PAR distribution of multicarrier CDMA systems. Closed-form expressions are obtained. Results are compared with simulations, and a near-exact match is found. This is very useful because simulations of PAR distributions are very costly. The analytical model could provide significant insights into the PAR distribution. In Section 5.3, we discuss the *synthesis methods* of PAR control in both single-carrier and multicarrier CDMA systems. For the single-carrier CDMA, the algorithm and the effectiveness of PAR control by Walsh code selection are examined in both the static scenario and dynamic scenario. For the multicarrier CDMA systems, the effectiveness of PAR control by controlling the combining phase angles is investigated.

5.1 Background and Introduction of PAR

In CDMA systems, a power amplifier is used to boost radiation power out of the antenna. The transfer characteristic of the power amplifier should be linear to avoid distortions [1]. However, a linear class A amplifier is expensive and exhibits poor power efficiency. In practice, a class B amplifier is employed that exhibits a linear transfer characteristic only at a low-input signal level. When the input signal level is too high, nonlinear distortions result, leading to clipping and spectral regrowth. To maintain tight spectral confinement of the transmitted signal, the efficiency of the power amplifier is reduced. Figure 5.1 shows a comparison of power amplifier efficiency versus PAR for both class A and class B amplifiers.

In CDMA systems, the PAR of the aggregate signal for multiple users is usually quite high (above 10 dB). This puts a stringent requirement on the power amplifier and reduces the efficiency in the sense that a higher input backoff factor is needed before the peaks in the signal experience significance distortion[1] due to power amplifier nonlinearity. In other words,

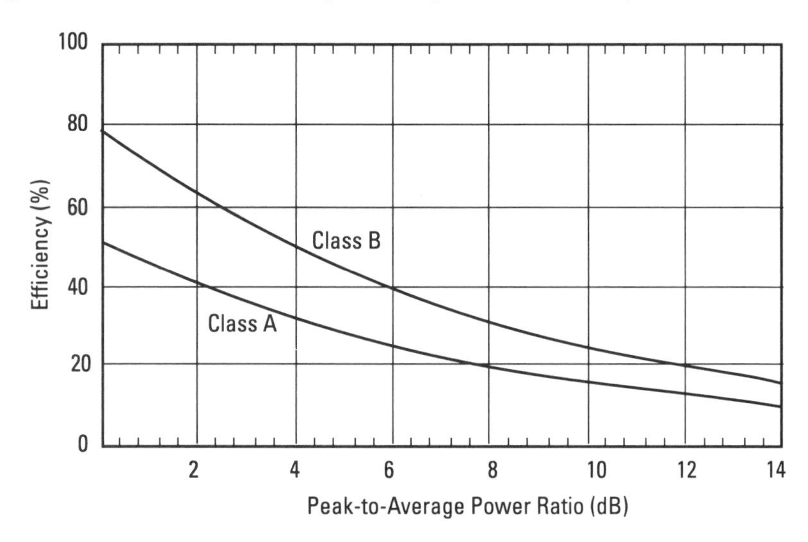

Figure 5.1 Linear amplifier theoretical efficiency limits.

1. Distortion in peaks will cause undesirable spectral regrowth in the adjacent bands.

a more expensive power amplifier is required to cater for the large peak's signal. With the same rating of power amplifier, an input signal with smaller PAR translates into larger average output power. In the base station, this translates into larger cell coverage and capacity of the forward link. In the mobile station, this translates into a higher burst rate in the reverse link.

PAR analysis for *orthogonal frequency division multiplexing* (OFDM) has appeared in [2–7]. However, the PAR behavior of the CDMA signal is quite different from that of OFDM.[2] Before we begin our analysis of the PAR of signals from CDMA systems, we present a general definition and examples of the PAR for various signals.

5.1.1 General PAR Definition

The average power of a signal is well defined to be

$$P_x = \lim_{T \to \infty} \frac{1}{T} \int_{-T/2}^{T/2} |x(t)|^2 dt \tag{5.1}$$

However, there are different interpretations of the meaning of the peak of a signal. Intuitively, a peak of a signal $x(t)$ is given by the maximum of its envelope, $\max_{t \in [-T/2, T/2]} |x(t)|$. However, for a continuous random process, $\max_t |x(t)|$ could reach infinity provided that the observation interval, T, is long enough. Even in a discrete random process where $\max_{t \in [-T/2, T/2]} |x(t)|$ is bounded, it may occur at very low probability, which is not very useful in practice [8]. Therefore, we need a more practical definition of peak in probability terms:

Definition 1

A signal $x(t)$ is said to have a peak at x_p at cutoff probability P_c if

$$\Pr[|x(t)| < x_p] = P_c \tag{5.2}$$

The definition of peak is not unique. The important measure is the level of spectral regrowth as a result of the clipping at the defined "peak"

2. For example, each carrier is modulated by a binary bit in OFDM, whereas each carrier in a multicarrier CDMA system is modulated by a random process of essentially continuous values and the number of carriers is much smaller than the OFDM systems.

level. It has been shown in [9] that the traditional definition of absolute peak could lead to contrary choices of design with respect to the spectral regrowth. For instance, the PAR of O-QPSK modulated signals (according to the tradition definition of absolute peak) is less than the conventional QPSK signals. On the contrary, the O-QPSK signal has a higher adjacent channel power than the regular QPSK signals. In fact, the spectral regrowth depends on the statistical properties of the signal. Hence, the definition of peak should bear a probabilistic structure associated with the signal. Based on the above definition of peak (5.2), the level of spectral regrowth of a Gaussian random process versus the cutoff probability, P_c, is illustrated in Figures 5.2 and 5.3.

Therefore, the PAR of a random process, $x(t)$, could be completely specified by its histogram.[3] If an operation is performed on $x(t)$, the PAR of the output will be changed if and only if its PDF is also changed by the operation. The following gives examples of how the output PAR could be affected by some operations.

Example 1

Let $x_1(t)$ and $x_2(t)$ be two uncorrelated, zero-mean stationary Gaussian processes with variance σ^2, then $y(t) = \frac{1}{\sqrt{2}}[x_1(t) + x_2(t)]$ will have the same PAR because $y(t)$ is still a stationary Gaussian process and the distribution (across the time domain) is unchanged (Gaussian). Therefore, the PARs of $x_1(t)$, $x_2(t)$, and $y(t)$ are the same.

Example 2

In this example, we illustrate the effect of I-Q imbalance to the PAR of signals. Let $x_c(t)$ and $x_s(t)$ be two uncorrelated, stationary Gaussian processes (unit variance, zero mean). Consider the general bandpass (BP) Gaussian process:

$$z_{\mathrm{BP}}(t) = \alpha_c x_c(t)\cos(\omega_0 t) - \alpha_s x_s(t)\sin(\omega_0 t) \tag{5.3}$$

3. The PAR definition in (5.2) refers to the *probability density function* (PDF) (or histogram) generated from its time samples That is, collect N samples of $x(t)$ and plot its histogram. PDF is resulted as N tends to infinity. For an ergodic random process, its PDF in the time domain and ensemble domain is identical. Otherwise, the histogram (or PDF) in the time domain and ensemble domain is, in general, different, and we always refer to the time domain PDF.

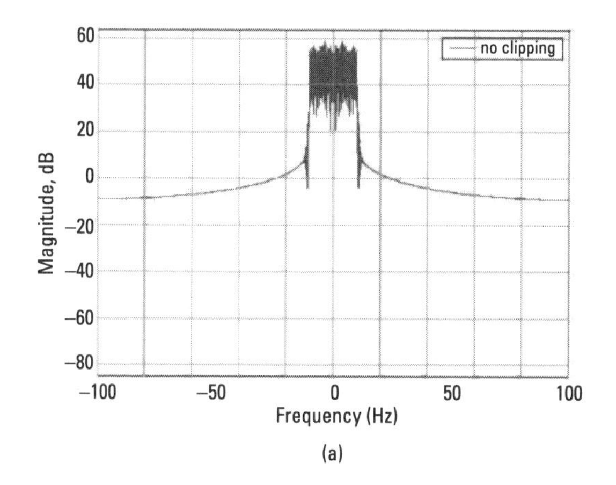

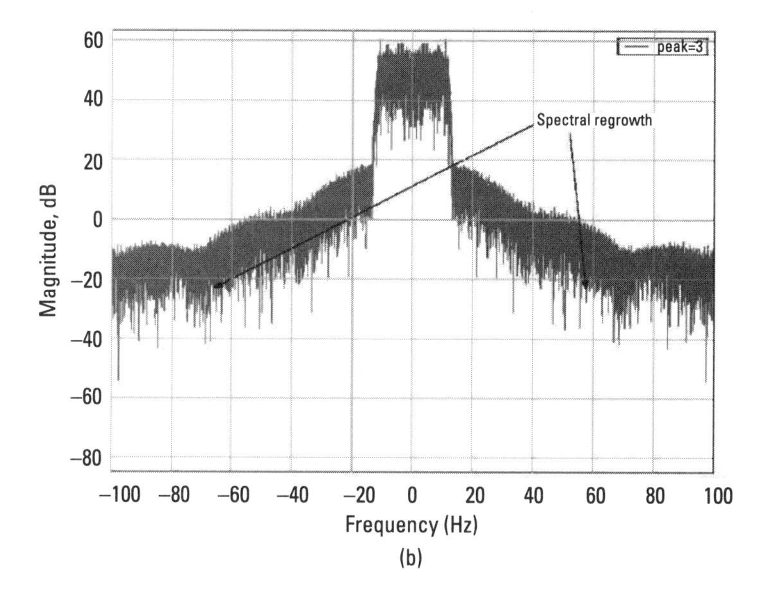

Figure 5.2 Illustration of spectral regrowth versus the cutoff probability of the peak, assuming hard clipping at the defined peak level and a zero-mean unit-variance Gaussian random process: (a) no clipping and (b) clipping at P_c = 99.9%.

where $\alpha_c^2 + \alpha_s^2 = 2$.

Unless $\alpha_c = \alpha_s$, the random process is not stationary and hence is not ergodic. If we collect N samples over an observation time, we could produce a histogram of $\{z_{\mathrm{BP}}(t_i)\}$. Because of the nonergodic nature of $z_{\mathrm{BP}}(t_i)$, the histogram (PDF across the time domain) will be distorted from a Gaussian PDF, resulting in a different PAR (increased) compared with a regular

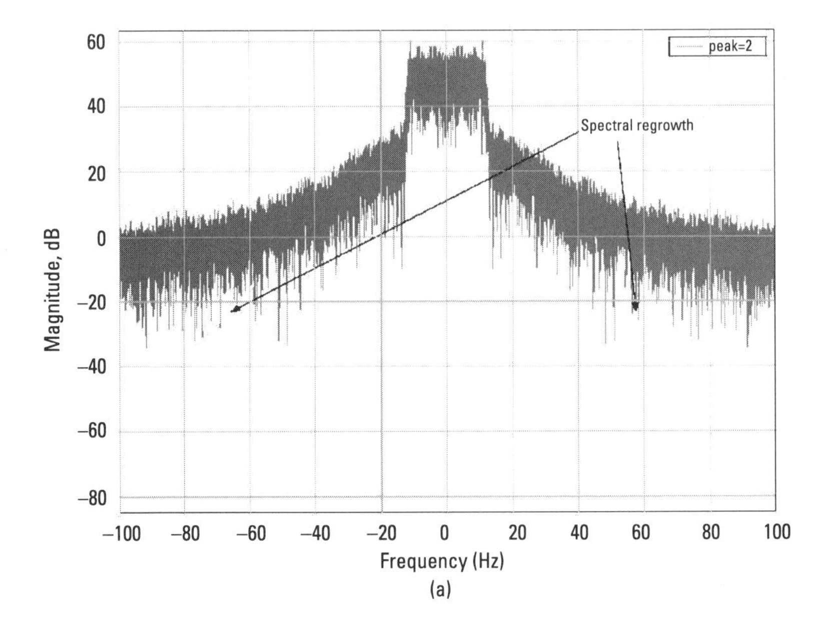

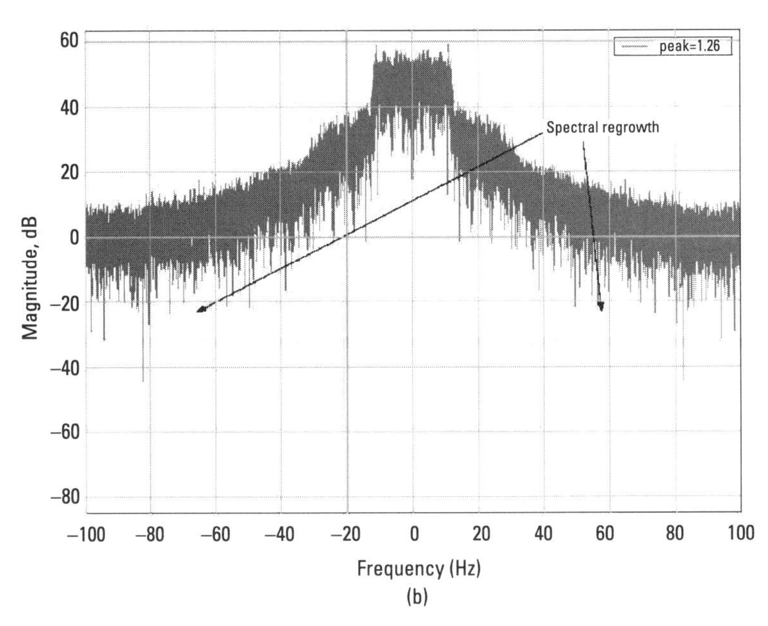

Figure 5.3 Illustration of spectral regrowth versus the cutoff probability of the peak, assuming hard clipping at the defined peak level and a zero-mean unit-variance Gaussian random process: (a) clipping at P_c = 98% and (b) clipping at P_c = 90%.

Gaussian distribution. This is illustrated in Figure 5.4. We see that the distribution of the random process $z_{\mathrm{BP}}(t_i)$ is changed with the higher probability of having the same peak as before. In other words, the PAR of $z_{\mathrm{BP}}(t_i)$ is increased and the amount of PAR increase is determined by the amount of I-Q imbalance α_c / α_s.

5.2 PAR Analysis of CDMA Signals

In this section, we analyze the PAR of CDMA signals for both single-carrier and multicarrier systems. Specifically, we analyze the PAR of single-carrier CDMA systems by detailed simulation and analyze the PAR of multicarrier CDMA systems by both simulation and analytical formulation.

5.2.1 PAR for Single-Carrier DS-CDMA Systems

In this section, we discuss the PAR of signals from single-carrier DS-CDMA systems such as the IS-95, CDMA2000, and WCDMA systems. As mentioned in the introduction, forward link signals of CDMA systems could be approximated by a Gaussian random process. However, this approximation is accurate only when the number of users in the forward link signal is sufficiently large. For practical purposes, we study the mechanism of PAR variations in DS-CDMA signals when the number of users is not very large. For such a case, the traditional Gaussian approximation is not accurate enough to model the PAR dynamics. We evaluate the PAR by detailed link level simulations.

Specifically, we study the dependency of PAR with respect to the *combination of Walsh codes* in the forward link signal, the *modulation format,* and the *spreading format.*

5.2.1.1 Walsh Code Dependency of PAR

In the forward link of IS-95-based CDMA systems, channels corresponding to different users are separated by a different orthogonal code sequence called the Walsh code. We found that the PAR of a DS-CDMA signal actually depends on the Walsh code combination in the forward link signal. Several combinations of Walsh codes will give a large PAR in an aggregate forward link signal. Several combinations of Walsh codes will give a small PAR in the forward link signal. In this section, we investigate the detail dependency of PAR on the Walsh codes. We also suggest an algorithm to optimally select the best combination of Walsh codes in Section 5.3.

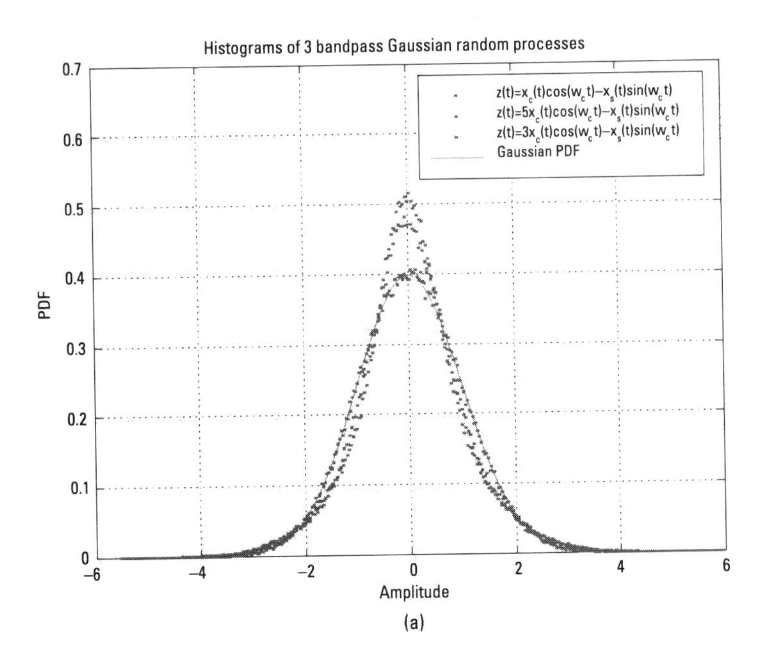

(a)

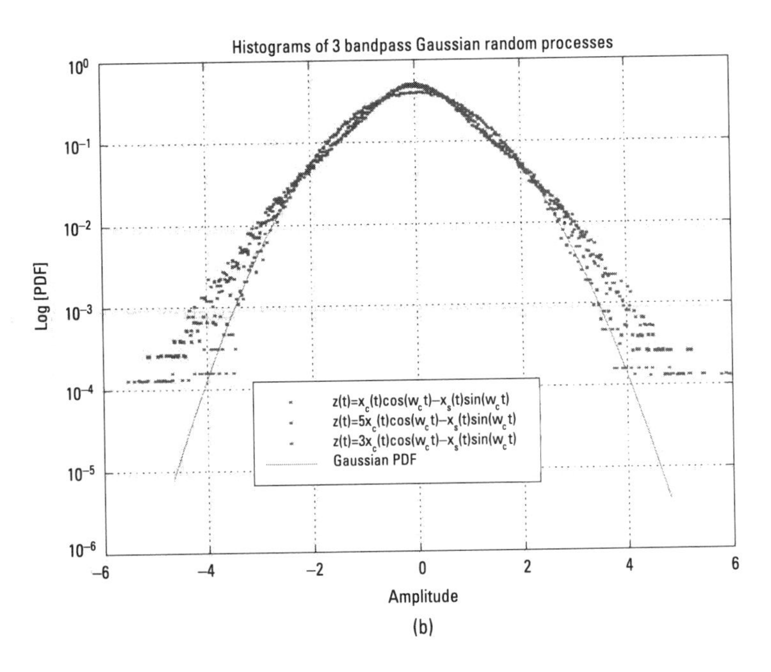

(b)

Figure 5.4 Distortions of PDF due to different degree of I-Q imbalance: (a) linear scale and (b) log scale.

Without loss of generality, consider a coded bit duration T_b. The complex envelope of a DS-CDMA forward link signal could be modeled by:

$$\tilde{s}(t) = \sum_{l=1}^{L} (\lambda_l \mathrm{PN}_I(l) + j\lambda_l \mathrm{PN}_Q(l)) p(t - lT_c) \tag{5.4}$$

where L is the number of chips per T_b, λ_l is defined as the *digital transmission sequence* given by:

$$\lambda_l = \sum_{n=0}^{N_u - 1} d_n w_n(l) \tag{5.5}$$

N_u is the number of users, d_n is the traffic data bit for user n, $w_n(l)$ is the lth chip of Walsh code assigned to user n, PN_I and PN_Q are the pseudorandom short codes, and $p(t)$ is the transmission pulse shape. Index l is used to indicate the chip interval and index n is used to indicate the nth user.

Term λ_l depends only on the choice of Walsh codes and the applied data bits d_n. Due to the band-limited nature of the channel, the pulse width of $p(t)$ will generally span over several chip intervals, causing interchip interference at instances other than the ideal sampling times. Because of the interchip interference and the fact that different users share the same PN sequence, a certain combination of Walsh codes and data bits could result in a particular *digital transmission sequence,* λ_l, that has a higher peak compared with the others. Let's illustrate the above idea with the following cases using the transmit pulse as specified in IS-95.

Consider two different Walsh code sets given by:

- Walsh code set 1 = $\{w_0, w_{32}, w_{16}, w_{48}\}$, $d_n(1) = \{1,1,1,1,1\}$.
- Walsh code set 2 = $\{w_0, w_1, w_{33}, w_{49}\}$, $d_n(2) = \{1,1,1,1,1\}$.

The overall digital transmission sequences, $\lambda_l(1)$ and $\lambda_l(2)$, are shown in Figure 5.5(a) and (b), respectively.

Because of the consecutive peaks at l = 14, 15, and 16, the transmit pulse, $p(t)$, could overlap and add constructively with each other, resulting in a high peak for Walsh set 1 as shown in Figure 5.6(a). However, the previous *bad combination* of consecutive peaks is broken in Walsh set 2 and the resulting peak of the transmitted signal is smaller as shown in Figure 5.6(b). This explains the reason for the Walsh code dependency of PAR.

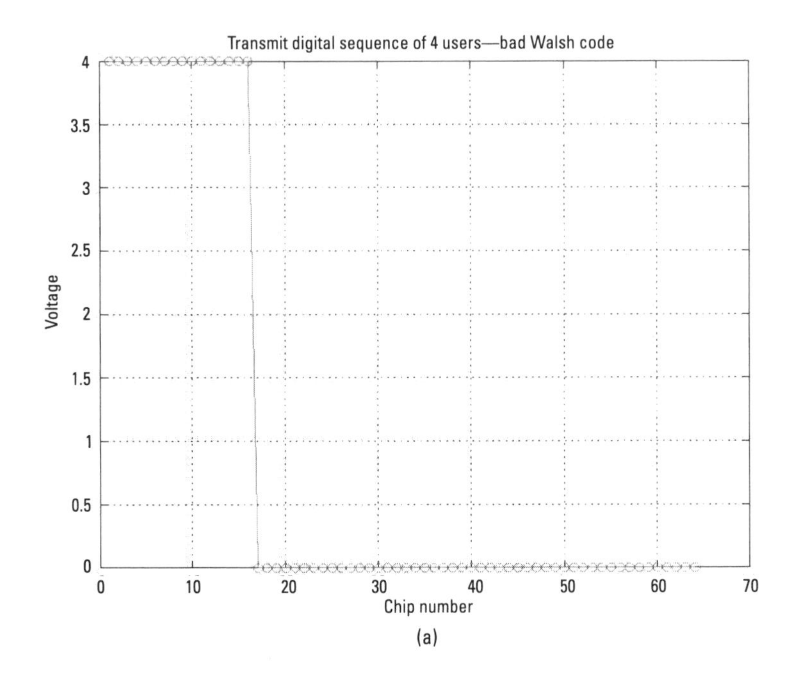

(a)

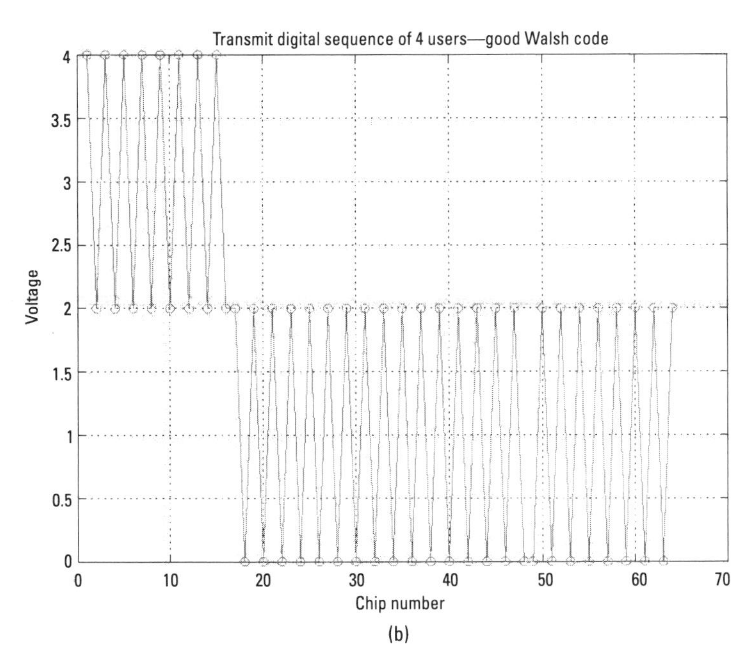

(b)

Figure 5.5 Illustrations of (a) bad and (b) good Walsh code combinations.

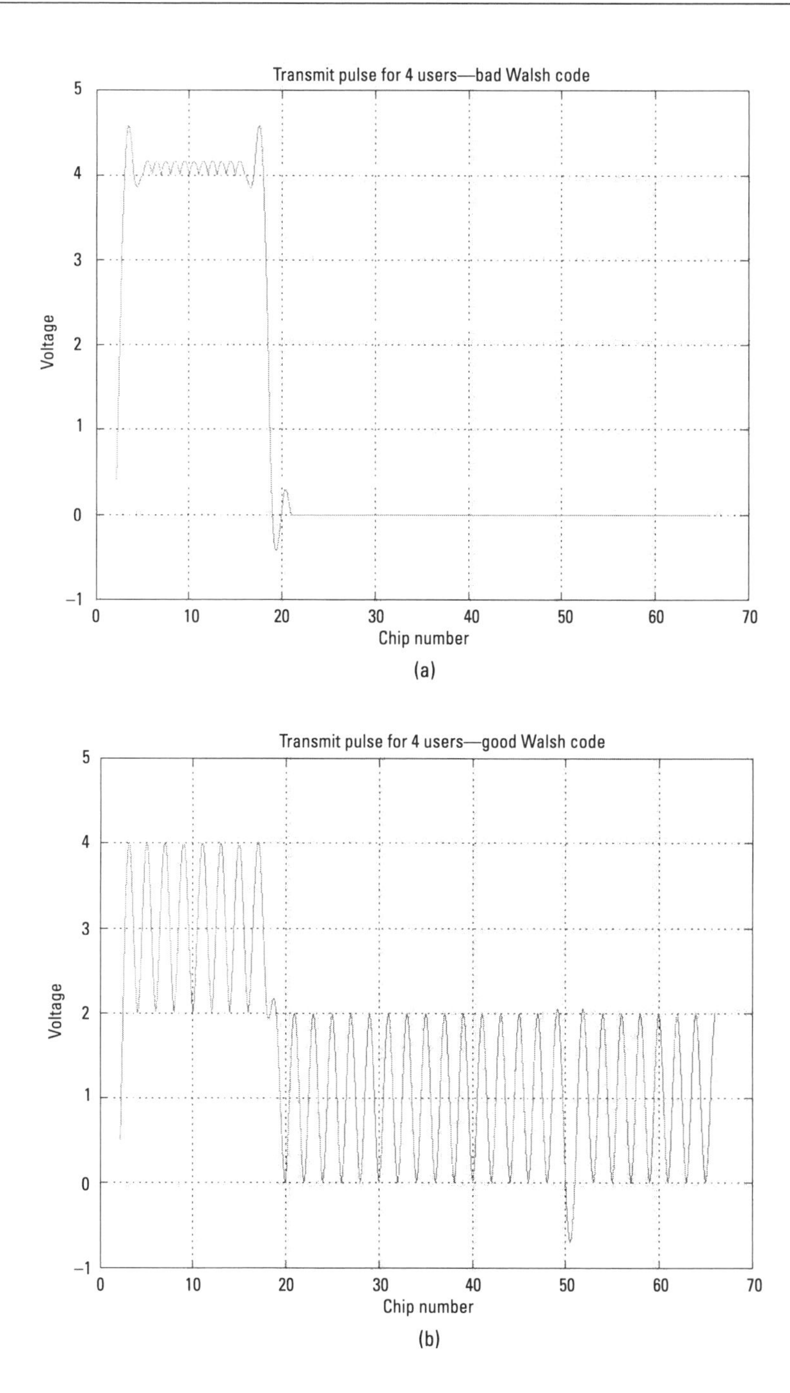

Figure 5.6 Transmitted baseband signal of (a) bad and (b) good Walsh code combinations.

5.2.1.2 Comparison Between IS-95, CDMA2000, and WCDMA Systems

Considering a CDMA2000-1X system[4] or a WCDMA[5] forward link signal, the complex envelope of the signal could be modeled as:

$$\tilde{s}(t) = \sum_{l=1}^{L} (\lambda_l^{(I)} + j\lambda_l^{(Q)})(\mathrm{PN}_I(l) + j\mathrm{PN}_Q(l))p(t - lT_c) \tag{5.6}$$

where $\lambda_l^{(I)}$ and $\lambda_l^{(Q)}$ are given by:

$$\lambda_l^{(I)} = \sum_{n=0}^{N_u-1} d_n^{(I)} w_n(l) \qquad \lambda_l^{(Q)} = \sum_{n=0}^{N_u-1} d_n^{(Q)} w_n(l) \tag{5.7}$$

Comparing with IS-95, both the CDMA2000-1X system and the WCDMA system use QPSK modulation. By QPSK modulation, we mean that the I and Q branches are modulated with independent data symbols. QPSK is used to increase the modulation throughput relative to BPSK modulation as in IS-95. In Section 5.3.1.1 it is shown by simulation that the CDMA2000 system or WCDMA system has a lower PAR value compared with the IS-95 system at any fixed Walsh code selection. We show in the following that the PAR reduction is due to QPSK.

5.2.1.3 PAR Dependency of Modulation Formats: BPSK and QPSK

For an IS-95 system forward link signal, BPSK is employed, and the I and Q signals are given by:

$$\tilde{s}_I(t) = \sum_l \lambda_l \mathrm{PN}_I(l)p(t - lT_c) \qquad \tilde{s}_Q(t) = \sum_l \lambda_l \mathrm{PN}_Q(l)p(t - lT_c) \tag{5.8}$$

Because PN_I and PN_Q are independent pseudorandom binary sequences {+1, −1}, they just toggle the sign of the peaks.

Peak position is totally determined by the *digital transmission sequence.* Hence, for a particular set of data bits d_n that causes λ_l (and hence the I signal) to *peak* at chip position l, the Q signal, being affected by the same λ_l, is very likely to be large in magnitude at the same chip position. In other words, the I and Q signals are highly correlated in magnitude. This

4. CDMA2000-1X system refers to a CDMA2000 system with a 1.25-MHz bandwidth.
5. WCDMA system refers to the WCDMA system specified by the UMTS2000.

could be illustrated by the I-Q plot of the IS-95 CDMA complex envelope in Figure 5.7.

However, QPSK is used for CDMA2000 or WCDMA systems and, hence, $\lambda_l^{(I)}$ and $\lambda_l^{(Q)}$ are independent because they are driven by independent data bits $\{d_n^{(I)}\}$ and $\{d_n^{(Q)}\}$. Therefore, I and Q signals become uncorrelated as illustrated by the spherical shape of the I-Q plot in Figure 5.7(b).

The overall peak of the complex envelope is given by $\sqrt{(\tilde{s}_I^2(t) + \tilde{s}_Q^2(t))}$. A large peak will occur only when both the I and Q signals are large in magnitude. For the IS-95 system, I and Q signals are heavily correlated. For the CDMA2000 or WCDMA systems, I and Q signals are uncorrelated. Hence, the probability of having a large overall peak for CDMA2000 or WCDMA systems is smaller than for the IS-95 system as illustrated in Figure 5.7. Note that the I-Q correlation of the IS-95 signal is due to the fact the different users use the same PN_I and PN_Q sequence. If there is only one single user, the I-Q correlation will disappear.

5.2.1.4 PAR Dependency of Spreading Format: Real Spreading and Complex Spreading

The IS-95 system employs real spreading while CDMA2000 or WCDMA systems employ complex spreading.[6] When complex spreading is applied, the I and Q signals are given by:

$$\tilde{s}_I(t) = \sum_l (\lambda_l^{(I)} \mathrm{PN}_I(l) - \lambda_l^{(Q)} \mathrm{PN}_Q(l)) p(t - lT_c) \tag{5.9}$$

$$\tilde{s}_Q(t) = \sum_l (\lambda_l^{(I)} \mathrm{PN}_Q(l) + \lambda_l^{(Q)} \mathrm{PN}_I(l)) p(t - lT_c) \tag{5.10}$$

In the reverse links of CDMA2000 or WCDMA systems, the gains applied to the I channel data and the Q channel data are different because I channel and Q channel are mapped to the dedicated control channel and the traffic channels, respectively. Without complex spreading, the power of the I branch and the power of the Q branch will be different causing an I-Q imbalance. As illustrated in Example 2 of Section 5.1.1, an I-Q imbalance will induce a large PAR.

When complex spreading is applied, the I-signal power and the Q-signal power are given by:

6. *Real spreading* refers to spreading the I data signal with PN_I and the Q data signal with PN_Q. *Complex spreading* refers to spreading the complex data $(I + jQ)$ with a complex spreading sequence $(\mathrm{PN}_I + j\mathrm{PN}_Q)$.

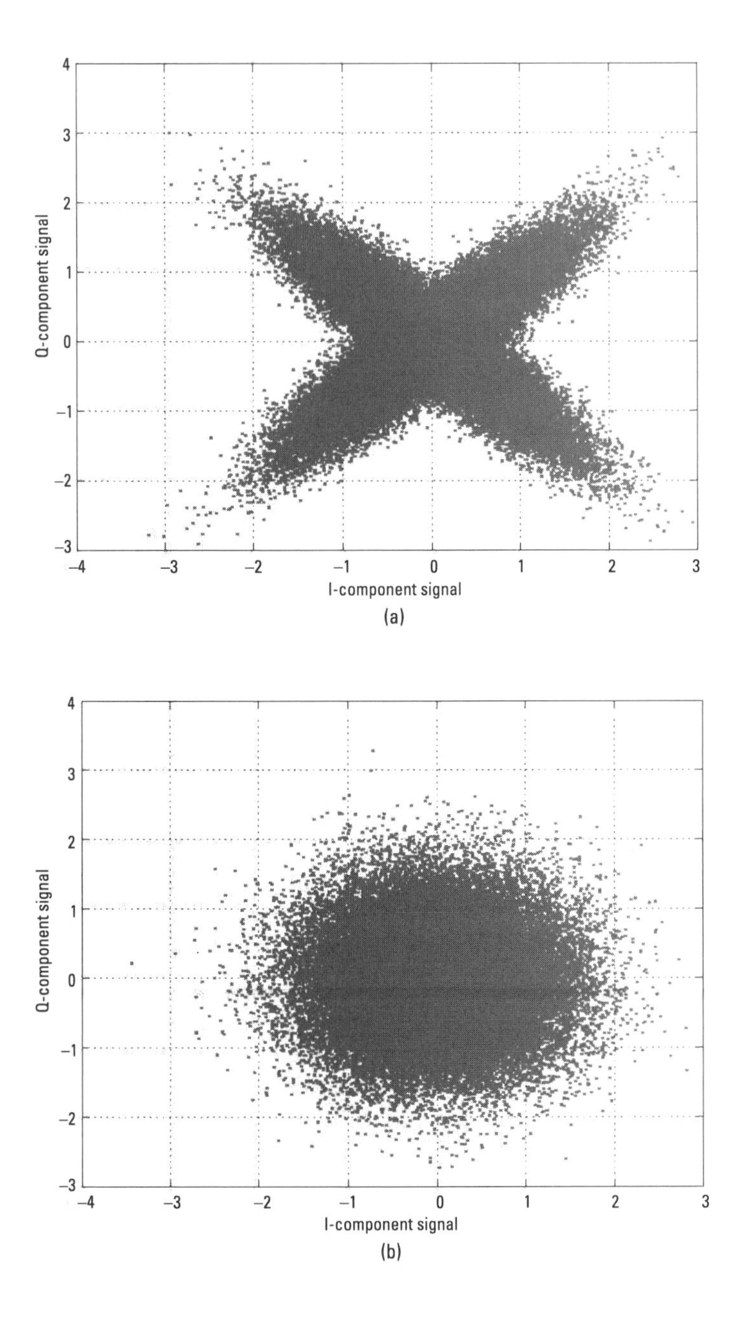

Figure 5.7 Complex envelope of the forward link signal of (a) IS-95 and (b) CDMA2000 or WCDMA systems for N_U = 10. Signals are normalized to have the same root-mean-square power.

$$\sigma_I^2 = \lim_{T \to \infty} \frac{1}{T} \int_{-T/2}^{T/2} \bar{s}_I^2(t)\,dt$$
$$= \mathscr{E}[(\lambda_l^{(I)} \mathrm{PN}_I(l))^2] + \mathscr{E}[(\lambda_l^{(Q)} \mathrm{PN}_Q(l))^2] \tag{5.11}$$
$$- \mathscr{E}[(\lambda_l^{(I)} \lambda_l^{(Q)} \mathrm{PN}_I(l) \mathrm{PN}_Q(l))]$$

and

$$\sigma_Q^2 = \lim_{T \to \infty} \frac{1}{T} \int_{-T/2}^{T/2} \bar{s}_Q^2(t)\,dt$$
$$= \mathscr{E}[(\lambda_l^{(I)} \mathrm{PN}_I(l))^2] + \mathscr{E}[(\lambda_l^{(Q)} \mathrm{PN}_Q(l))^2] \tag{5.12}$$
$$+ \mathscr{E}[(\lambda_l^{(I)} \lambda_l^{(Q)} \mathrm{PN}_I(l) \mathrm{PN}_Q(l))]$$

where $\mathscr{E}(\bullet)$ is the expectation operator. Because PN_I and PN_Q are uncorrelated pseudorandom sequences, the third terms in σ_I^2 and σ_Q^2 are zero. Hence, $\sigma_I^2 = \sigma_Q^2$ and one important benefit of complex spreading is to remove the I-Q unbalance that would otherwise occur. As explained in Section 5.1.1, an I-Q imbalance is undesirable because it could raise the PAR due to the loss of ergodicity.

5.2.2 PAR for Multicarrier CDMA Systems

The PAR for multicarrier systems has received much attention during recent years [10–12]. In this section, we present a novel analytical approach to the PAR problems of multicarrier systems. The analytical approach will help us to gain significant insight into the PAR dynamics and save us expensive simulation efforts.

As mentioned in Section 5.1.1, the necessary and sufficient condition for a change in PAR is the change or *distortion* of the histogram (or PDF) of a signal. In this section, we propose a mathematical model so as to describe the PAR change via analytical formulas. To do that, several assumptions are made to simplify the analysis and the results are compared with simulations. The resulting expression could help us to gain insight into the problem and avoid costly simulations. We first introduce the following theorem.

Theorem 1

Let $\{\tilde{x}_1(t), \tilde{x}_2(t), \ldots, \tilde{x}_{N_c}(t)\}$ be a set of N_c complex lowpass random processes and $\tilde{s}(t)$ be the N_c carriers combined complex signal given by:

$$\tilde{s}(t) = \tilde{x}_1(t)\exp[j(\omega_1 t + \theta_1)] + \tilde{x}_2(t)\exp[j(\omega_2 t + \theta_2)] + \ldots + \tilde{x}_{N_c}(t)\exp[j(\omega_{N_c} t + \theta_{N_c})] \tag{5.13}$$

where ω_k and θ_k are the kth combining frequency and phase. Let f_s be the sampling frequency such that

$$\frac{f_1}{f_s} = \frac{P_1}{Q_1}, \frac{f_2}{f_s} = \frac{P_2}{Q_2} \ldots \frac{f_{N_c}}{f_s} = \frac{P_{N_c}}{Q_{N_c}} \tag{5.14}$$

where P_i and Q_i are relatively prime and $Q = \mathrm{LCM}\{Q_1, Q_2, \ldots, Q_{N_c}\}$. If $\tilde{x}_k(t)$ is ergodic for all k, then the histogram or PDF derived from the time samples of $|\tilde{s}(t)|$ is given by:

$$f_{|\tilde{s}|}(\alpha) = \frac{1}{Q}\left[\sum_{q:|\tilde{s}_q|\neq 0} f_q(\alpha) + \sum_{q:|\tilde{s}_q|=0} \delta(\alpha)\right] \tag{5.15}$$

where $f_q(\alpha)$ is the PDF (in ensemble sense) of $|\tilde{s}_q| = |\tilde{s}(t_{q,n})|$ with $t_{q,n} = nQ + q$, $q \in [1, Q]$.

Proof

Suppose we collect NQ samples from $\tilde{s}(t)$. To derive the histogram for these NQ samples, we divide the samples into Q subsets, each having N samples in such a way that subset q consists of $\tilde{s}(t)$ sampled at $t_{q,n} = nQ + q\{n = [1, N], q = [1, Q]\}$. Note that within the subset q, the samples are given by:

$$\begin{aligned}\tilde{s}_q(n) &= \tilde{x}_1(t_{q,n})\exp\left[j\left(\frac{2\pi P_1}{Q_1}(nQ + q) + \theta_1\right)\right] \\ &\quad + \tilde{x}_2(t_{q,n})\exp\left[j\left(\frac{2\pi P_2}{Q_2}(nQ + q) + \theta_2\right)\right] + \ldots \\ &= \tilde{x}_1(t_{q,n})\exp\left[j\left(\frac{2\pi P_1 q}{Q_1} + \theta_1\right)\right] \\ &\quad + \tilde{x}_2(t_{q,n})\exp\left[j\left(\frac{2\pi P_2 q}{Q_2} + \theta_2\right)\right] + \ldots\end{aligned} \tag{5.16}$$

Observe that all complex sinusoids are independent of n and they are just scaling factors. Hence, all samples within subset q correspond to a complex ergodic random process. This means that the PDF or histogram of the samples in subset q alone is identical to the PDF in the ensemble sense. Let the individual PDF (in the ensemble sense) of subset q be $f_{|\tilde{s}_q|}(\alpha)$ or just $f_q(\alpha)$. The PDF, $f_{|\tilde{s}|}(\alpha)$, is given by:

$$
\begin{aligned}
f_{|\tilde{s}|}(\alpha) &= \lim_{\Delta s \to 0}\left\{\frac{1}{\Delta s}\lim_{N\to\infty}\left[\frac{N_{\{||\tilde{s}|-\alpha|<\Delta s\}}}{NQ}\right]\right\} \\
&= \lim_{\Delta s \to 0}\left\{\frac{1}{\Delta s}\lim_{N\to\infty}\left[\frac{\sum_{q=1}^{Q} N_{\{||\tilde{s}_q|-\alpha|<\Delta s\}}}{NQ}\right]\right\} \\
&= \lim_{\Delta s \to 0}\left\{\frac{1}{Q}\sum_{q}\frac{1}{\delta s}\lim_{N\to\infty}\left[\frac{N_{\{||\tilde{s}_q|-\alpha|<\Delta s\}}}{N}\right]\right\} \\
&= \frac{1}{Q}\left[\sum_{q:|\tilde{s}_q|\neq 0} f_q(\alpha) + \sum_{q:|\tilde{s}_q|=0}\delta(\alpha)\right]
\end{aligned}
\tag{5.17}
$$

5.2.2.1 Application of the Analytical Formulation to CDMA2000 or WCDMA Systems

As described above, $\tilde{x}_i(t)$ is the complex baseband signal of the DS-CDMA signal for carrier i. Because all CDMA carriers share the same common pilot and different carriers use the same PN_I and PN_Q, there would be correlations between baseband signals $\tilde{x}_i(t)$ and $\tilde{x}_j(t)$. In general, since the pilot component is deterministic, it could be taken out and $\tilde{x}_i(t)$ could be expressed as the sum of a deterministic component and a zero-mean random process.

$$
\begin{aligned}
\tilde{x}_1(t) &= \psi_0^{(i)}(\text{PN}_I(t) + j\text{PN}_Q(t)) \\
&\quad + \sum_{n=1}^{N_u}\psi_n^{(i)} w_n^{(i)}(t)\,[d_{n,c}^{(i)}(t) + jd_{n,s}^{(i)}(t)] \times [\text{PN}_I(t) + j\text{PN}_Q(t)] \\
&= \psi_0^{(i)}(\text{PN}_I(t) + j\text{PN}_Q(t)) + \tilde{x}_i'(t)
\end{aligned}
\tag{5.18}
$$

where $\psi_n^{(i)}$ and $\{d_{n,c}^{(i)}, d_{n,s}^{(i)}\}$ are the transmitted digital gain and the [I,Q] data bits for carrier i and user n, respectively. Because of QPSK, $d_{n,c}^{(i)}$ and $d_{n,s}^{(j)}$ are uncorrelated. To include the effect of the correlation between traffic data bits of different carriers, we assume that $\{d_{n,c}^{(i)}, d_{n,c}^{(j)}\}$ and

$\{d_{n,s}^{(i)}, d_{n,s}^{(j)}\}$ are equally correlated [as shown in (5.19)]. At large N_u, the random component, $\tilde{x}_i'(t)$, could be approximated by a zero-mean complex stationary Gaussian process. To simplify the expression, we assume that all carriers are equally loaded and hence, $\mathscr{E}[|\tilde{x}_i'(t)|^2] = \mathscr{E}[|\tilde{x}_j'(t)|^2] = 2\sigma_0^2$. Therefore, we have

$$\begin{aligned}
\mathscr{E}[\tilde{x}'_{i,c}\tilde{x}'_{j,s}] &= 0 \\
\mathscr{E}[(\tilde{x}'_{i,c})^2] = \mathscr{E}[(\tilde{x}'_{i,s})^2] &= \sigma_0^2 = 2\sum_{n=1}^{N_u-1}(\psi_n^{(i)})^2 \\
\mathscr{E}[\tilde{x}'_{i,c}\tilde{x}'_{j,c}] = \mathscr{E}[\tilde{x}'_{i,s}\tilde{x}'_{j,s}] &= \sigma_0^2\rho\forall i \neq j
\end{aligned} \tag{5.19}$$

Consider taking $4NQ$ time samples from $\tilde{s}(t)$ and subdivide the samples into Q subsets with Q defined in Theorem 1. The real part and the imaginary part of the signal samples in subset q are given by:

$$\begin{aligned}
\Re[\tilde{s}_q(n)] = &\sum_{m=1}^{N_c}\left[\psi_0^{(m)}\left(\mathrm{PN}_I(t_{q,n})\cos\left(\frac{2\pi P_m}{Q_m}q + \theta_m\right)\right.\right. \\
&\left.\left.- \mathrm{PN}_Q(t_{q,n})\sin\left(\frac{2\pi P_m}{Q_m}q + \theta_m\right)\right)\right] \\
&+ \sum_{m=1}^{N_c}\left[\tilde{x}'_{m,c}(t_{q,n})\cos\left(\frac{2\pi P_m}{Q_m}q + \theta_m\right)\right. \\
&\left.- \tilde{x}'_{m,s}(t_{q,n})\sin\left(\frac{2\pi P_m}{Q_m}q + \theta_m\right)\right]
\end{aligned} \tag{5.20}$$

$$\begin{aligned}
\Im[\tilde{s}_q(n)] = &\sum_{m=1}^{N_c}\left[\psi_0^{(m)}\left(\mathrm{PN}_I(t_{q,n})\sin\left(\frac{2\pi P_m}{Q_m}q + \theta_m\right)\right.\right. \\
&\left.\left.+ \mathrm{PN}_Q(t_{q,n})\cos\left(\frac{2\pi P_m}{Q_m}q + \theta_m\right)\right)\right] \\
&+ \sum_{m=1}^{N_c}\left[\tilde{x}'_{m,c}(t_{q,n})\sin\left(\frac{2\pi P_m}{Q_m}q + \theta_m\right)\right. \\
&\left.+ \tilde{x}'_{m,s}(t_{q,n})\cos\left(\frac{2\pi P_m}{Q_m}q + \theta_m\right)\right]
\end{aligned} \tag{5.21}$$

Within each subset q, the $4N$ samples could be further subdivided into four smaller subsets of N samples, each corresponding to $\{PN_I, PN_Q\} = \{1,1\}, \{1,-1\}, \{-1,1\}$, and $\{-1,-1\}$. Each of these smaller subsets is labeled by $p \in [1,4]$. For these N samples in each of these subsets, their real and imaginary parts, $\Re[\tilde{s}_{q,p}]$ and $\Im[\tilde{s}_{q,p}]$ could be modeled by two independent Gaussian processes with means, $[\mu_{q,p}(c), \mu_{q,p}(s)]$, given by the first summation terms of (5.20) and (5.21) and variance, $[\Sigma_q^2]$ given by:

$$\Sigma_q^2 \stackrel{\text{def}}{=} \mathscr{E}[\Re(\tilde{s}_{q,p})^2] = \mathscr{E}[\Im(\tilde{s}_{q,p})^2] \tag{5.22}$$

$$= N_c \sigma_0^2 + 2\sigma_0^2 \rho \sum_{i=1}^{N_c} \sum_{\substack{j=1 \\ j>i}}^{N_c} \cos\left[2\pi q\left(\frac{P_i}{Q_i} - \frac{P_j}{Q_j}\right) + (\theta_i - \theta_j)\right]$$

Note that in the ideal case when traffic data bits are uncorrelated between different carriers, $\Sigma_q^2 = N_c \sigma_0^2 \forall q$.

The envelope, $|\tilde{s}_{q,p}(n)|$, is given by $\sqrt{[\Re(\tilde{s}_{q,p})^2 + \Im(\tilde{s}_{q,p})^2]}$, which is the square root of the sum-of-the-square of two *independent* but nonzero mean Gaussian variables. Hence, the PDF of the envelope, $f_{q,p}(\alpha)$, is Rician distributed and is given by:

$$f_{q,p}(\alpha) = \frac{\alpha}{\Sigma_q^2} \exp\left[-\left(\frac{\alpha^2 + \gamma_{q,p}^2}{2\Sigma_q^2}\right)\right] I_0\left(\frac{\alpha \gamma_{q,p}}{\Sigma_q^2}\right) \tag{5.23}$$

where $\gamma_{q,p}$ is given by $\sqrt{\mu_{q,p}^2(c) + \mu_{q,p}^2(s)}$ and Σ_q^2 is given by (5.22). From Theorem 1, the overall PDF (or histogram) of the envelope derived from the $4NQ$ samples of $\tilde{s}(t)$ is given by:

$$f_{|\tilde{s}|}(\alpha) = \frac{1}{4Q} \sum_{q=1}^{Q} \sum_{p=1}^{4} f_{q,p}(\alpha) \tag{5.24}$$

To derive the peak envelope according to the probabilistic definition in (5.2), we consider

$$P_c = \Pr[|\tilde{s}| > \alpha_0] = 1 - \int_0^{\alpha_0} f_{|\tilde{s}_q|}(\alpha)\, d\alpha \tag{5.25}$$

$$= \frac{1}{4Q} \sum_{q=1}^{Q} \sum_{p=1}^{4} \left\{\mathcal{M}_1\left[\frac{\gamma_{q,p}}{\Sigma_q}, \frac{\alpha_0}{\Sigma_q}\right]\right\}$$

where $\mathcal{M}_1(a, b)$ is the Marcum's function given by:

$$\mathcal{M}_1(a, b) = e^{-\frac{a^2 + b^2}{2}} \sum_{k=0}^{\infty} \frac{a^k}{b} I_k(ab) \tag{5.26}$$

Note that since the $\tilde{x}_i(t)$ values are lowpass random processes, the combined signal power, Σ^2, is given by:

$$\begin{aligned} \Sigma^2 &\stackrel{\text{def}}{=} \frac{1}{N} \sum_{n=1}^{N} \left|\tilde{s}[n]\right|^2 \\ &= \frac{1}{4Q} \sum_{q=1}^{Q} \sum_{p=1}^{4} \left[\mu_{q,p}^2(c) + \mu_{q,p}^2(s) + 2\Sigma_q^2 \right] \\ &= 4N_c \psi_0^2 + 2N_c \sigma_0^2 = 4N_c \sum_{m=0}^{N_u - 1} \psi_m^2 \end{aligned} \tag{5.27}$$

Therefore, the PAR of the multicarrier signal is given by:

$$\text{PAR (dB)} = 20 \log_{10} \left\{ \frac{\alpha_0}{\Sigma} \right\} \tag{5.28}$$

To verify the closeness of the analytical results, we considered combining three DS-CDMA baseband signals, each modulated by ω_1, ω_2, ω_3, respectively. The histograms obtained from simulations and (5.28) are compared as shown in Figure 5.8.

The calculated PDF matches that obtained from simulations closely for different combining phases. In Figure 5.9, the simulated and calculated PARs of CDMA2000 three-carrier signals varied with respect to the combining phases $(\Delta\theta)$ and traffic data correlation, ρ, are plotted with different values of pilot transmit gain.

A large pilot transmit gain is used to exaggerate the variations in the PARs and to test the analytical results under extreme conditions. As shown in the figure, individual carriers are essentially uncorrelated at small pilot power where the combined signal, $\tilde{s}(t)$, becomes stationary and the envelope PDF becomes Rayleigh distributed and hence independent of the combining phases. We can see that the simulated PARs fall within 0.05 dB of the calculated PARs for most of pilot transmit gain and $\delta\theta$. Furthermore, the difference in PARs between using the best Walsh codes and the worst Walsh

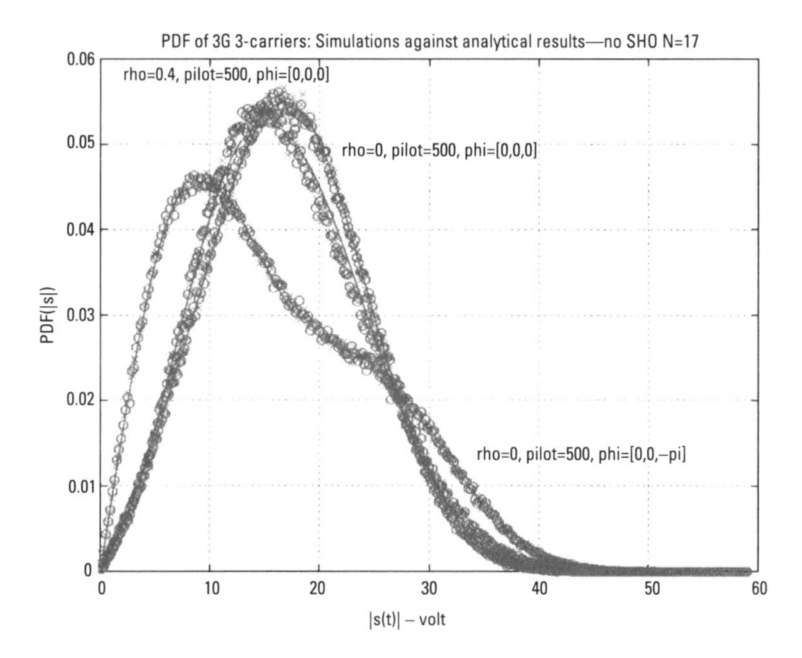

Figure 5.8 Histograms of CDMA2000 three-carrier DS-CDMA signal envelope simulations and analytical results. All pilot dgu = 500. + and o correspond to the simulated histograms using Walsh sets W_{max} and W_{min}, respectively.

codes vanished in the multicarrier situation. This could be explained by the fact that without loss of generality, the combined three-carrier baseband signal, $\tilde{s}(t)$, is given by (5.29) when the middle carrier frequency is factored out:

$$\tilde{s}(t) = \tilde{x}_1(t)\exp[-j(2\pi\Delta ft + \theta_1)] + \tilde{x}_2(t) + \tilde{x}_3(t)\exp[j(2\pi\Delta ft + \theta_3)] \quad (5.29)$$

because Δf is the carrier spacing which is approximately equal to the chip frequency. Any peak variations due to the effect of Walsh code combinations [13] would be smoothed out by the term $\exp[\pm j(2\pi\Delta ft + \theta)]$. A peak will show up only when the subchip interval (corresponding to the peak) aligns with the time when $\exp[\pm j(2\pi\Delta ft + \theta)] = 1$, which is quite unlikely in probability. Therefore, the Walsh code dependency of PAR vanished in multicarrier systems.

Moreover, it is important to distinguish the effect of sharing a common pilot between different carriers and the effect of *traffic data* correlations on the PAR. For the case without traffic data correlations between different

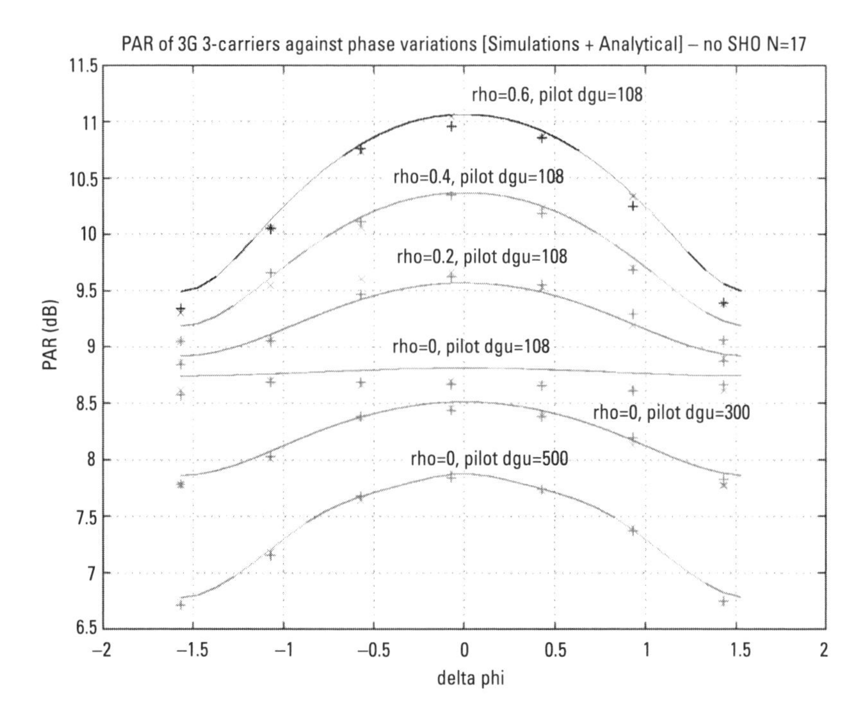

Figure 5.9 Comparison of calculated PARs and simulated PARs for a CDMA2000 three-carrier DS-CDMA signal using pilot dgu = [108,300,500], paging dgu = 64, sync dgu = 34, nominal dgu = 73.4, and rho = [0,0.6] with 17 users. + and x correspond to the simulated PARs for W_{min} and W_{max} Walsh sets, respectively. dgu is the unit corresponding to the amplifier gain applied.

carriers ($\rho = 0$), we observe from Figure 5.9 that the multicarrier combined PAR is actually lower than the PAR before combining. This could be explained by the last paragraph that the sinusoid term actually smooths out the peaks of the baseband signals and $\Sigma_q = \Sigma \forall q$. In the extreme case when we have a very high power pilot, the combined signal is close to a sinusoid and the PAR for a sinusoid is well known to be much lower than the PAR of a Gaussian-like process [14, 15]. However, when we have traffic data correlations ($\rho \neq 0$), the combined PAR is higher than the PAR before multicarrier combining. This is because the total signal power, Σ^2, is unchanged from (5.26) but there exist some $\Sigma_q > \Sigma$ that cause a rise in the tail part of the combined Rician distribution, creating a higher PAR. For example, at $\rho = 0.2$, the rise in PAR after multicarrier combining reaches 1 dB. This suggested the reasons why an increase in PAR of multicarrier CDMA systems has been reported in field test. In [16], it was found that the traffic data bits used to measure the PAR are actually highly correlated and the long code mask for each traffic channel is turned off, causing high traffic data correlations between carriers.

Finally, because of the close match between the analytical and the simulation results, (5.28) could be used to optimize the PAR without doing costly simulations. This is elaborated upon more in Section 5.3.1.

5.2.2.2 Application of the Analytical Formulation to Multicarrier IS-95 DS-CDMA Systems

We apply the analytical model of PAR to IS-95 systems with several carriers deployed in a cell. Following a technique similar to that in Section 5.2.1, the common pilot signal is taken out of each baseband signal. The real and imaginary parts of the N samples in subset (q, p) are given by:

$$\begin{aligned}
\Re[\tilde{s}_{q,p}(n)] &= \psi_0 \sum_{i=1}^{N_c} \left[\mathrm{PN}_I(t_{q,p,n}) \cos\left(\frac{2\pi P_i q}{Q_i} + \theta_i\right) \right. \\
&\quad \left. - \mathrm{PN}_Q(t_{q,p,n}) \sin\left(\frac{2\pi P_i q}{Q_i} + \theta_i\right) \right] \\
&\quad + \sum_{i=1}^{N_c} \left[\tilde{x}'_{i,c}(t_{q,p,n}) \cos\left(\frac{2\pi P_i q}{Q_i} + \theta_i\right) \right. \\
&\quad \left. - \tilde{x}'_{i,s}(t_{q,p,n}) \sin\left(\frac{2\pi P_i q}{Q_i} + \theta_i\right) \right] \\
&= \mu_{q,p}(c) + \Re[\tilde{s}'_{q,p}(t_{q,p,n})]
\end{aligned} \tag{5.30}$$

$$\begin{aligned}
\Im[\tilde{s}_{q,p}(n)] &= \psi_0 \sum_{i=1}^{N_c} \left[\mathrm{PN}_Q(t_{q,p,n}) \cos\left(\frac{2\pi P_i q}{Q_i} + \theta_i\right) \right. \\
&\quad \left. + \mathrm{PN}_I(t_{q,p,n}) \sin\left(\frac{2\pi P_i q}{Q_i} + \theta_i\right) \right] \\
&\quad + \sum_{i=1}^{N_c} \left[\tilde{x}'_{i,s}(t_{q,p,n}) \cos\left(\frac{2\pi P_i q}{Q_i} + \theta_i\right) \right. \\
&\quad \left. + \tilde{x}'_{i,c}(t_{q,p,n}) \sin\left(\frac{2\pi P_i q}{Q_i} + \theta_i\right) \right] \\
&= \mu_{q,p}(s) + \Im[\tilde{s}'_{q,p}(t_{q,p,n})]
\end{aligned} \tag{5.31}$$

Because of BPSK, the I and Q channels share the same data bits for each carrier and hence $\Re[\tilde{x}'_{i,c}]$ and $\Im[\tilde{x}'_{i,s}]$ are heavily correlated. Furthermore, to illustrate the effect of traffic channel bits correlation on PAR, we assume that $\Re[\tilde{x}'_{i,c}]$ and $\Im[\tilde{x}'_{j,s}]$ are correlated as well. Define σ_0^2 and ρ as:

$$\sigma_0^2 \stackrel{\text{def}}{=} \mathcal{E}[(\Re(\tilde{x}'_i))^2] = \mathcal{E}[(\Im(\tilde{x}'_i))^2] = \sum_{m=1}^{N_u-1} \psi_m^2 \tag{5.32}$$

$$\begin{aligned} \rho &\stackrel{\text{def}}{=} \frac{1}{\sigma_0^2}\mathcal{E}[(\Re(\tilde{x}'_i))(\Re(\tilde{x}'_j))] = \mathcal{E}[(\Im(\tilde{x}'_i))(\Im(\tilde{x}'_j))] \\ &= \sum_{m=1}^{N_u-1} \psi_m^2 < w_m^{(i)} w_m^{(j)} >< d_m^{(i)} d_m^{(j)} > \end{aligned} \tag{5.33}$$

The second-order statistics of $\tilde{x}'_i$ are given by:

$$\mathcal{E}[(\Re(\tilde{x}'_i))^2] = \mathcal{E}[(\Im(\tilde{x}'_i))^2] = \sigma_0^2 \tag{5.34}$$

$$\mathcal{E}[(\Re(\tilde{x}'_i))(\Im(\tilde{x}'_i))] = \sigma_0^2 \mathrm{PN}_I \mathrm{PN}_Q \tag{5.35}$$

$$\mathcal{E}[(\Re(\tilde{x}'_i))(\Re(\tilde{x}'_j))] = \mathcal{E}[(\Im(\tilde{x}'_i))(\Im(\tilde{x}'_j))] = \sigma_0^2 \rho \tag{5.36}$$

$$\mathcal{E}[(\Re(\tilde{x}'_i))(\Im(\tilde{x}'_j))] = \mathcal{E}[(\Im(\tilde{x}'_i))(\Re(\tilde{x}'_j))] = \sigma_0^2 \rho \mathrm{PN}_I \mathrm{PN}_Q \tag{5.37}$$

where $\Re[\tilde{s}_{q,p}(n)]$ and $\Im[\tilde{s}_{q,p}(n)]$ are modeled by zero-mean Gaussian processes and their second-order statistics are given by:

$$\begin{aligned} \Sigma_{q,p}^2(c) &\stackrel{\text{def}}{=} \mathcal{E}[\Re(\tilde{s}'_{q,p})^2] \\ &= \sigma_0^2 \sum_{i=1}^{N_c} \left(1 - \mathrm{PN}_I \mathrm{PN}_Q \sin\left(4\pi q \frac{P_i}{Q_i} + 2\theta_i\right)\right) \\ &\quad + \sigma_0^2 \rho \sum_{i=1}^{N_c} \sum_{\substack{j=1 \\ i \neq j}}^{N_c} \left[\cos\left(2\pi q \left(\frac{P_i}{Q_i} - \frac{P_j}{Q_j}\right) + (\theta_i - \theta_j)\right)\right. \\ &\quad \left. - \mathrm{PN}_I \mathrm{PN}_Q \sin\left(2\pi q \left(\frac{P_i}{Q_i} + \frac{P_j}{Q_j}\right) + (\theta_i + \theta_j)\right)\right] \end{aligned} \tag{5.38}$$

$$
\begin{aligned}
\Sigma_{q,p}^2(s) &\stackrel{\text{def}}{=} \mathscr{E}[\Im(\tilde{s}'_{q,p})^2] \\
&= \sigma_0^2 \sum_{i=1}^{N_c} \left(1 + \text{PN}_I \text{PN}_Q \sin\left(4\pi q \frac{P_i}{Q_i} + 2\theta_i\right)\right) \\
&+ \sigma_0^2 \rho \sum_{i=1}^{N_c} \sum_{\substack{j=1 \\ i \neq j}}^{N_c} \left[\cos\left(2\pi q \left(\frac{P_i}{Q_i} - \frac{P_j}{Q_j}\right) + (\theta_i - \theta_j)\right)\right. \\
&\left.+ \text{PN}_I \text{PN}_Q \sin\left(2\pi q \left(\frac{P_i}{Q_i} + \frac{P_j}{Q_j}\right) + (\theta_i + \theta_j)\right)\right]
\end{aligned} \tag{5.39}
$$

$$
\begin{aligned}
\Gamma_{q,p}^2 &\stackrel{\text{def}}{=} \mathscr{E}[\Re(\tilde{s}'_{q,p})\Im(\tilde{s}'_{q,p})] \\
&= \sigma_0^2 \text{PN}_I \text{PN}_Q \sum_{i=1}^{N_c} \cos\left(4\pi q \frac{P_i}{Q_i} + 2\theta_i\right) \\
&+ \sigma_0^2 \rho \sum_{i=1}^{N_c} \sum_{\substack{j=1 \\ i \neq j}}^{N_c} \left[\text{PN}_I \text{PN}_Q \cos\left(2\pi q \left(\frac{P_i}{Q_i} + \frac{P_j}{Q_j}\right) + (\theta_i + \theta_j)\right)\right. \\
&\left.+ \sin\left(2\pi q \left(\frac{P_j}{Q_j} - \frac{P_i}{Q_i}\right) + (\theta_j - \theta_i)\right)\right]
\end{aligned} \tag{5.40}
$$

Since the envelope-square is given by

$$
\left|\tilde{s}_{q,p}\right|^2 = [\mu_{q,p}(c)^2 + \Re(\tilde{s}'_{q,p})]^2 + [\mu_{q,p}(s)^2 + \Im(\tilde{s}'_{q,p})]^2
$$

and $\Re(\tilde{s}'_{q,p})$ and $\Im(\tilde{s}'_{q,p})$ are correlated, the PDF of $\left|\tilde{s}_{q,p}\right|^2$ is no longer a chi-square. It is shown that the PDF is given by [17]:

$$
f_{q,p}(\beta) = \frac{1}{2\pi} \int_{-\infty}^{\infty} \Phi_{q,p}(jv) e^{-jv\beta} dv \tag{5.41}
$$

where $\Phi_{q,p}(\beta)$ is the characteristic function of the PDF. From Theorem 1 again, the peak-square envelope, $\beta_0 = \alpha_0^2$, at probability P_c is given by the solution to the equation:

$$P_c = 1 - \frac{1}{8\pi Q} \sum_{q=1}^{Q} \sum_{p=1}^{4} \int_{-\infty + j\epsilon}^{\infty + j\epsilon} \frac{\Phi_{q,p}(jv)[1 - e^{-jv\beta_0}]}{jv} dv \quad (5.42)$$

Similar to (5.27), the total signal power of $\tilde{s}(t)$ is given by:

$$\Sigma^2 = \frac{1}{4Q} \sum_{q=1}^{Q} \sum_{p=1}^{4} (\mu_{q,p}^2(c) + \mu_{q,p}^2(s) + \Sigma_q^2(c) + \Sigma_q^2(s)) = 2N_c \sum_{m=0}^{N_u - 1} \psi_m^2 \quad (5.43)$$

A comparison of simulated and calculated PARs for IS-95 three-carrier systems at various ρ and pilot transmit gains is shown in Figure 5.10. Similar observations are found with respect to the behavior of PAR.

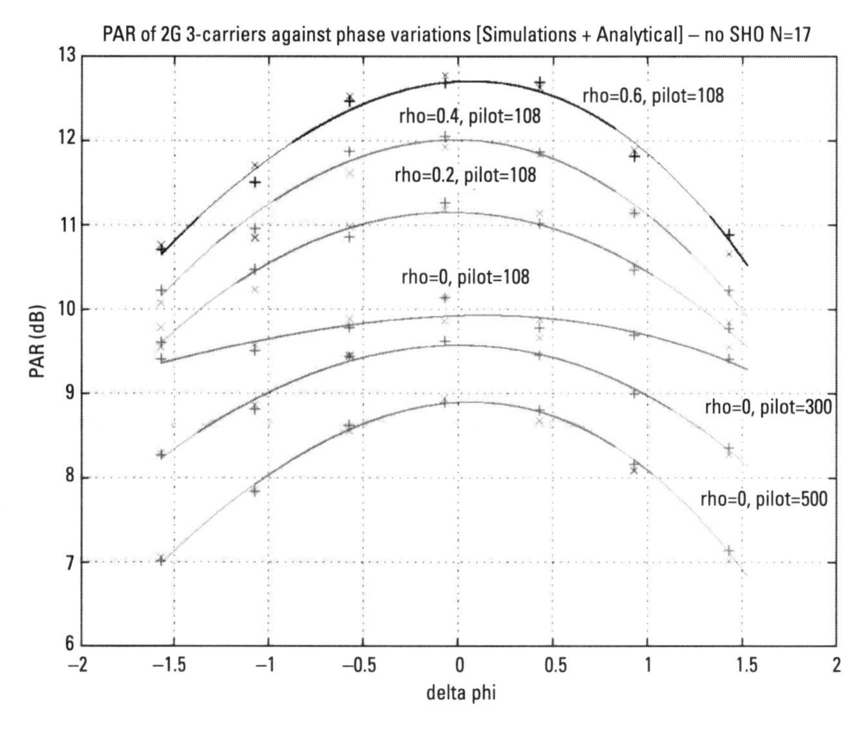

Figure 5.10 Comparison of calculated PARs and simulated PARs for an IS-95 three-carrier DS-CDMA signal using pilot dgu = [108,300,500], paging dgu = 64, sync dgu = 34, nom dgu = 73.4, and rho = [0,0.6] with 17 users. + and x correspond to simulated PARs using W_{max} and W_{min}, respectively.

5.2.2.3 Comparison of PAR Between IS-95 and CDMA2000 Multicarrier Systems

As a conclusion to this section, the change of PAR before multicarrier combining and after multicarrier combining is due to the loss of ergodicity in the combined signal. For a CDMA2000 multicarrier system, the loss of ergodicity is due to the common pilot and traffic data correlation in each carrier. For an IS-95 multicarrier system, the loss of ergodicity is due to the common pilot, the traffic data correlation, and the high correlation (in magnitude) between the I and Q components.[7] For this reason, the PAR for IS-95 systems is higher than the CDMA2000 systems at the same condition in multicarrier environment.

Another important observation is that a multicarrier PAR does not depend on Walsh code combinations of individual carrier anymore in both the IS-95 and CDMA2000 systems (compared to the single-carrier situation in [13]). When traffic data symbols are uncorrelated, the combined PAR is actually lower than the PAR of the baseband signal.

5.3 PAR Control for CDMA Signals

As mentioned in Section 5.1, a large backoff factor is needed to accommodate the forward link CDMA signal with a large PAR. This effectively reduces the allowable average transmitting power for a given linear power amplifier and, therefore, the forward link coverage, capacity, and burst rate are reduced. Hence, it would be desirable to reduce the PAR of the CDMA forward link signal.

In general, PAR control schemes could be classified as *synthesis techniques* and *predistortion techniques*. Synthesis techniques refer to the reduction of PAR by the adjustments or manipulations of the source data or parameters in the process of producing the transmitted signal. Predistortion techniques, in contrast, refer to the reduction of PAR by signal processing techniques on the synthesized transmit signals. Predistortion techniques have the advantage of, in general, better PAR control. In fact, a number of such techniques have been proposed for satellite communication [18, 19]. However, the tradeoff is usually the performance of the physical layer after the signal predistortion, which depends on the required adjacent band power suppression. For example, in [18], a fourth-order Butterworth filter is used for postamplification filtering of MPSK signals. While this may provide adequate

7. The correlation between I and Q signals for an IS-95 system is due to BPSK modulation.

adjacent-channel interference suppression, it will not be sufficient for cellular or PCS applications where the requirement for adjacent band spectral regrowth is quite stringent. However, synthesis techniques have been proposed to tackle the PAR problems of OFDM systems. Much of the work focuses on different coding strategies to avoid a few specific data patterns (that would produce large signal peaks) through data translation. However, such techniques were designed for OFDM signals and, therefore, could not be applied directly to WCDMA systems. Furthermore, data translation means data nontransparency and the requirement to change the specification of CDMA systems. Therefore, in this section, we focus on the possible synthesis techniques for PAR control without requiring any changes in the specifications. The predistortion PAR control techniques specifically for CDMA signaling are still currently under investigation.

5.3.1 PAR Control for Single-Carrier DS-CDMA Systems

Several schemes have been proposed in the literature [4, 20, 21] to deal with the PAR of an OFDM signal by selective mapping or redundancy coding. Such schemes could not be directly applied to DS-CDMA signals. For single-carrier DS-CDMA signals, we introduce a PAR control algorithm by selective assignment of a Walsh code combination in the forward link. This approach has the advantage of being simple and does not require any modification to the existing specifications.

As illustrated in Section 5.2, the PAR of a DS-CDMA forward link signal is determined by the transmit digital sequence, $\lambda_l^{(I)}$ and $\lambda_l^{(Q)}$. From (5.4),

$$\lambda_l^{(I)} = \sum_{n=0}^{N_u-1} d_n^{(I)} w_n(l) \qquad \lambda_l^{(Q)} = \sum_{n=0}^{N_u-1} d_n^{(Q)} w_n(l) \tag{5.44}$$

As shown in (5.44), we could either manipulate the data bits $\{d_n^{(I)}, d_n^{(Q)}\}$ or the Walsh codes $\{w_n(l)\}$ to control $\lambda_l^{(I)}$ and $\lambda_l^{(Q)}$. The simplest way is to selectively assign Walsh codes to each user so as to minimize the PAR because it does not require changes in the MS receiver. This is elaborated in the next section.

5.3.1.1 Static Analysis of Selective Walsh Code Assignment

Due to the limited transmitted bandwidth, *interchip interference* will result due to the transmit pulse width spanning more than one chip interval. Assume that the transmit pulse, $p(t)$, spans $2L + 1$ chip intervals. On each chip interval, we oversample the pulse by K times [i.e., there are K samples (subchips) of transmit pulse, $p(t)$, on each chip]. Without a loss of generality, we consider the peak of either the I signal or the Q signal. Let $S_{N_u}(k, l)$ be the magnitude of the envelope of the I or Q signal at the kth *subchip* between $t = lT_c$ and $t = (l + 1)T_c$. It is given by:

$$S_{N_u}(k, l) = \sum_{l'=-L}^{L} \lambda_{l+l'} \mathrm{PN}_{l+l'} p\left[\frac{(l - l'K)T_c}{K}\right] = \sum_{l'=-L}^{L} \lambda_{l+l'} \mathrm{PN}_{l+l'} p_{k-l'K} \tag{5.45}$$

For any given set of Walsh codes, $\mathcal{W}_{N_u} = \{w_{n_1}, w_{n_2}, \ldots, w_{n_{N_u}}\}$, there exist a unique set of data bits $\mathcal{D}_{N_u} = \{d_1^*, d_2^*, \ldots, d_n^*\}$ such that $\lambda_l = N_u$. Since $\mathrm{PN}_l \in [1, -1]$, the *worst-case peak* of the envelope (5.45) between $t = lT_c$ and $t = (l + 1)T_c$. is given by:

$$S_{N_u}^*(l) = \max_{k \in [0, K-1]} \left[\sum_{l'=-L}^{L} \left|\lambda_{l+l'} p_{k-l'K}\right|\right] \tag{5.46}$$

The problem of Walsh code selection is formulated as follows: To minimize the worst-case peak, $S_{N_u}^* = \max_{l \in [0, M-1]}[S_{N_u}^*(l)]$, over the set of Walsh codes, $\mathcal{W}_{N_u}$, for N_u users with the constraint:

$$\mathcal{W}_{N'} \subset \mathcal{W}_{N'+1} \qquad N' \in [1, 2, \ldots, N_u - 1] \tag{5.47}$$

where M is the number of chips per coded bit (i.e., $M = T_b/T_c$). The constraint is needed because we want to keep the peak to be minimal for all N as new users are added incrementally into the system.

This optimization problem is solved sequentially in [13]. The following is the result of a computer search.

The set of Walsh codes that gives a minimal worst-case peak, $S_{N_u}^*$, for $N_u = 1$–17 is sorted as $\mathcal{W}_{\min} = \{w_0, w_{32}, w_1, w_{27}, w_{25}, w_{13}, w_{26}, w_{20}, w_{14}, w_{15}, \ldots\}$. For the sake of comparison, the set of Walsh codes that gives a maximal worst-case peak is given by $\mathcal{W}_{\max} = \{w_0, w_{32}, w_4, w_{24}, w_{40}, w_{48}, w_8, w_{16}, w_{56}, w_{13}, \ldots\}$. Figure 5.7 shows the difference in PAR (in decibels) of the IS-95 and CDMA2000 signals resulting from $\mathcal{W}_{\min}$ and $\mathcal{W}_{\max}$, respectively.

As shown in Figure 5.11(a), a 1.5- to 2-dB PAR difference exists between the system using $\mathcal{W}_{\min}$ and $\mathcal{W}_{\max}$ Walsh codes for the IS-95 and CDMA2000 systems. Furthermore, given the same Walsh code, there is a

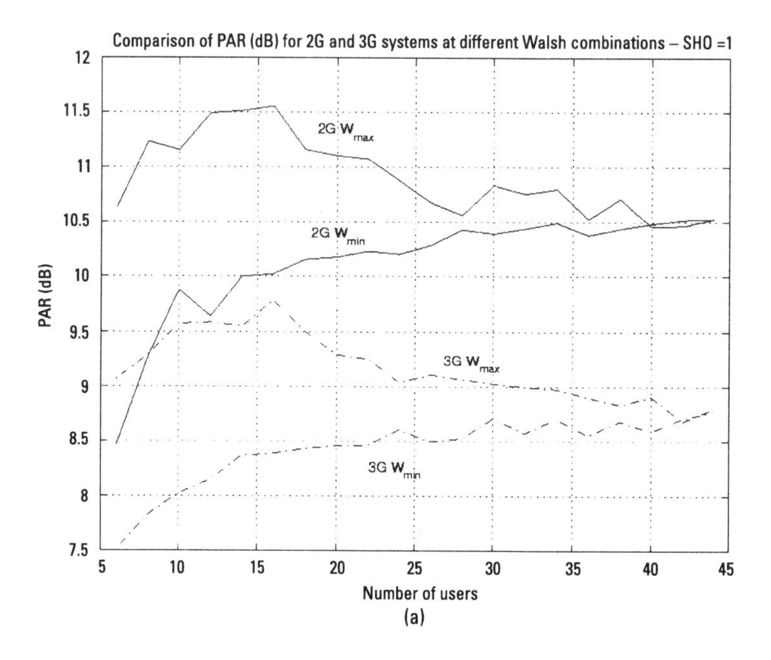

(a)

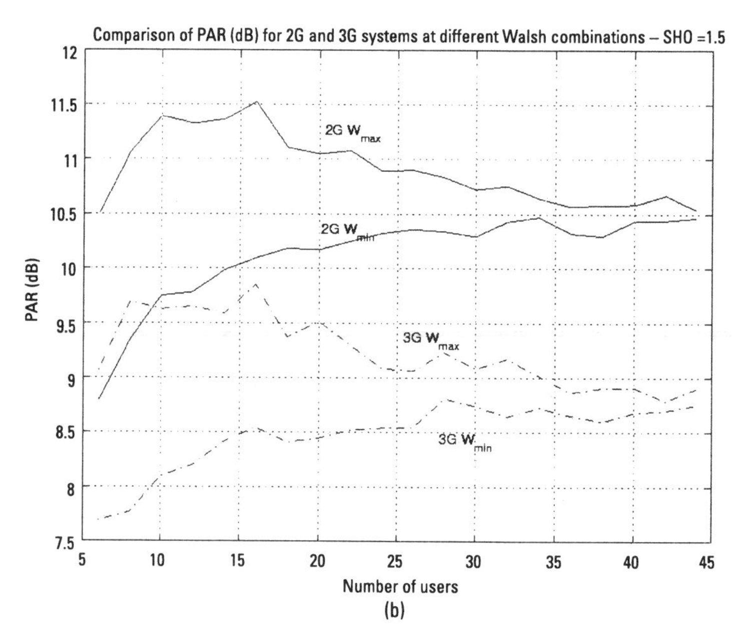

(b)

Figure 5.11 Static comparison of PAR for IS-95 and CDMA2000 systems with $\mathcal{W}_{max}$, $\mathcal{W}_{min}$, and N_u as variables: (a) no SHO and (b) two-way SHO (handoff factor = 1.5).

1.5- to 2.5-dB reduction of PAR for the CDMA2000 system. As the number of users per sector increases, the difference between the PARs of $\mathcal{W}_{max}$ and $\mathcal{W}_{min}$ decreases. This is because the transmitting signal approaches a Gaussian distribution as N_u increases. Because the PDF (histogram) of a Gaussian signal is fixed, the PAR difference tends to vanish.

When an MS is in SHO, the transmitting power of the forward link power control bits is increased by a certain factor (SHO factor) to compensate for the drop in transmitting power of the traffic channel.[8] Hence, SHO will have an adverse effect on the PAR. Figure 5.11(b) shows the difference in PAR when the SHO factor of 1.5 is included. Similar PAR differences are observed.

In current IS-95 systems, the normal loading of a sector is about 14 users. Therefore, the selective Walsh code mapping method could contribute to a 1- to 2-dB reduction in PAR.

5.3.1.2 Dynamic Analysis of Selective Walsh Code Assignment

As shown in the previous section, a considerable PAR improvement results when an appropriately chosen Walsh code combination is used. However, the comparison is static in the sense that we freeze (1) Walsh code sets and (2) individual traffic channel transmit gain (dgu). Therefore, the comparison may be optimistic. In this section, we consider the PAR difference under a more dynamic situation.

Simulations were undertaken to study the PAR difference in a dynamic environment as users enter and leave the system. In the simulation model, the following is included:

1. Call attempts and call completion are modeled as a Poisson process.
2. As a new user enters the system, a Walsh code is assigned in one of three ways: (a) assigned sequentially (implemented in current system); (b) assigned from $\mathcal{W}_{min}$; or (c) assigned from $\mathcal{W}_{max}$. This models the dynamic variation of the Walsh code set as calls come and go.
3. The transmit dgu for each new call is obtained from a truncated log-normal distribution with zero mean and 8-dB variance. This models the dynamic variation of transmit gain for each user due to power control.

8. Power control bits are not combined between SHO legs and hence they need to be transmitted at a higher power to compensate for the reduction in the forward link transmitted power due to power control.

Therefore, the Walsh code set and the composition of the dgu gains are varied dynamically as users come and go. The simulation is executed for 10^5 call attempts. Figure 5.12 shows the results of PAR difference among the three ways of assigning Walsh codes.

As shown in the figure, an approximately 0.5-dB PAR gain is seen between Walsh code assignment schemes (a) and (b) relative to the static situation (1- to 2-dB PAR gain). We could infer from this that it is quite unlikely for a *bad Walsh code combination* to show up by just simply assigning the Walsh code sequentially. When a bad Walsh code combination does show up (artificially by scheme c), the PAR is increased by about 1.3 dB (as shown in the upper curve). Similarly, the comparison for the CDMA2000 system is shown in the same figure.

5.3.2 PAR Reduction Technique for Multicarrier CDMA Systems

For multicarrier CDMA systems, the dependency of PAR on the Walsh code combination is not obvious. For example, as shown in Figure 5.9, the PAR of a three-carrier system is not affected by the selection of Walsh codes. At $\Delta\theta = 0$, the PAR difference between the best Walsh code and the worst Walsh code (in the single-carrier sense) is less than 0.05 dB. In multicarrier systems, another possibility for controlling PAR is to adjust the *combining phases* (θ_1, θ_2, θ_3) of the component signals in multicarrier signals. In [3, 5, 10, 11] this technique has been shown to be effective in reducing the PAR of OFDM systems. However, in OFDM systems, the number of carriers is usually quite large, whereas in multicarrier CDMA signals, the number of carriers is relatively small. This basic difference is important in the effectiveness of the PAR control by phase adjustment. In this section, we study the effectiveness of reducing the PAR by adjusting the combining phases in multicarrier CDMA systems.

For a CDMA2000 multicarrier system, PAR is determined by (5.28). Therefore, the best case and the worst case combining phases are found by optimizing (5.28) with respect to the combining phases. Similar optimization is performed on (5.42) for IS-95 systems.

Figure 5.13 shows the PAR gain[9] of IS-95 and CDMA2000 three-carrier systems with regard to the pilot dgu, ψ_0. As shown in the figure, little PAR gain is seen at low ψ_0. The PAR gains are negligible at low pilot transmit gain and low traffic channel correlation (ρ).

---

9. PAR gain refers to the difference between the best phases and the worst phases.

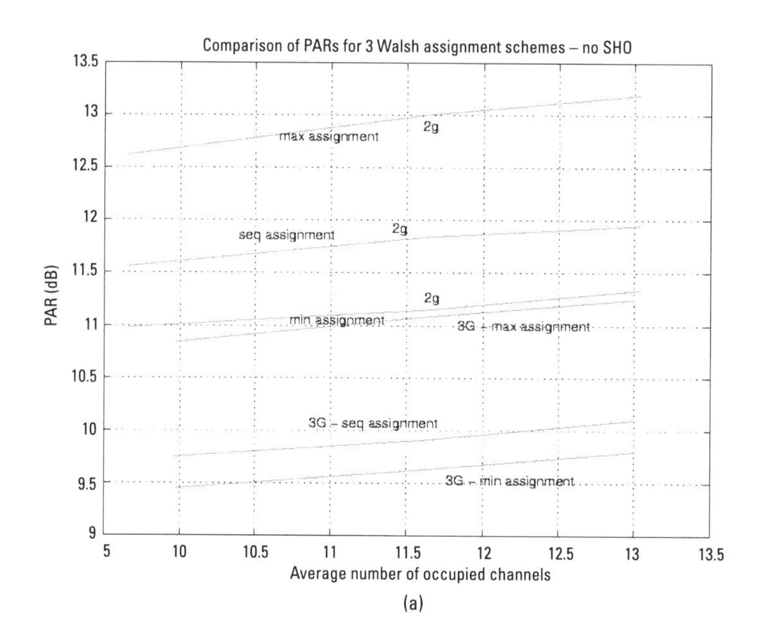

(a)

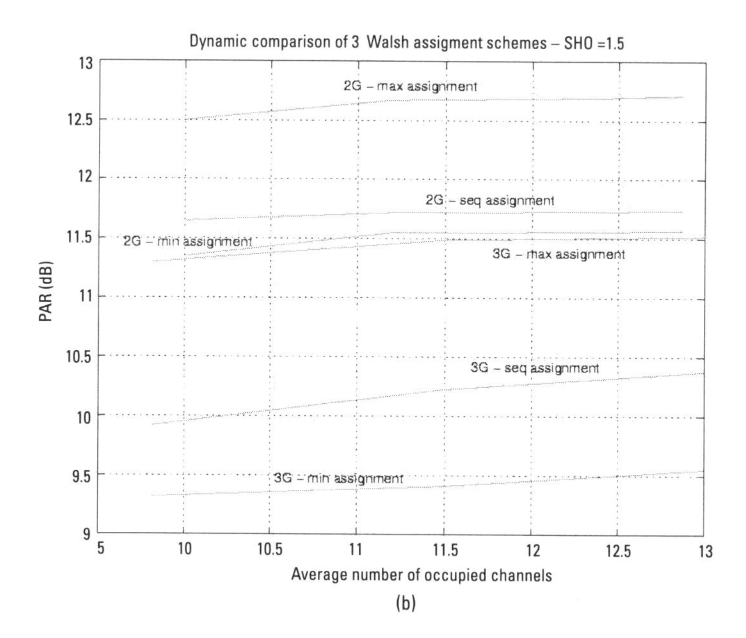

(b)

Figure 5.12 Dynamic comparison of PAR difference for three Walsh code assignment schemes for IS-95 and CDMA2000 systems: (a) no SHO and (b) SHO factor = 1.5.

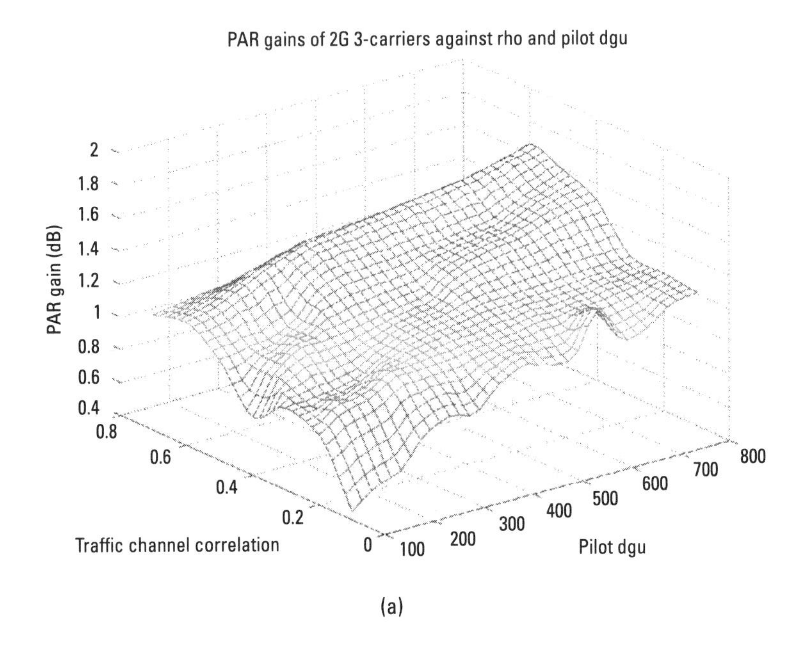

(a)

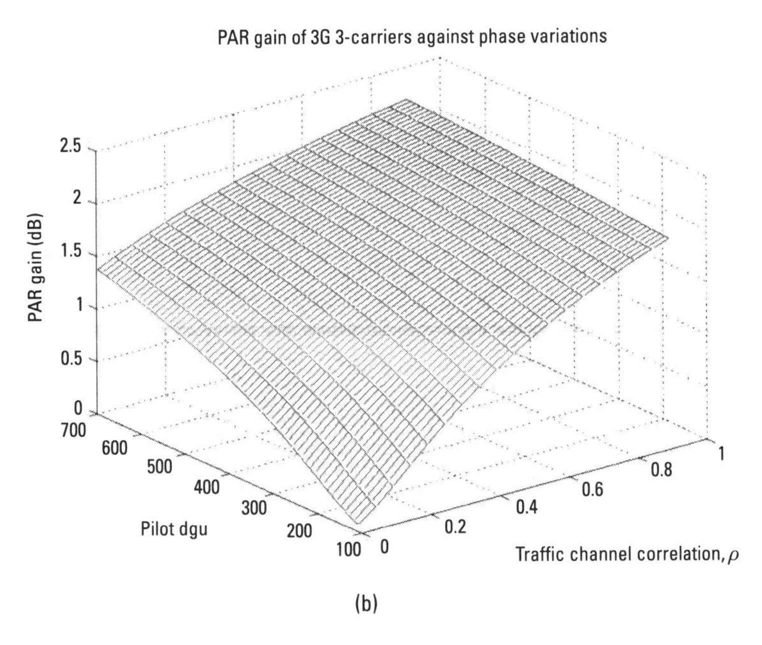

(b)

Figure 5.13 PAR gain for (a) IS-95 and (b) CDMA2000 three-carrier systems with regard to pilot dgu (ψ_0) and traffic channel correlation (ρ).

5.4 Conclusion

In this chapter, we have discussed the PAR issues for both single-carrier and multicarrier DS-CDMA signals. We considered IS-95, CDMA2000, and WCDMA systems as illustrations. For the single-carrier systems, we found that the signal could not be modeled by a Gaussian random process unless the number of users in the forward link is sufficiently large. To accurately study the mechanics of the PAR, we investigate the system by detailed simulation. We found that the PAR depends on (1) the *Walsh code combination* in the forward link, (2) *BPSK or QPSK modulation,* and (3) *real spreading or complex spreading* in the reverse link. An algorithm for controlling the PAR by Walsh code selection is presented. The following are the main conclusions found for the single-carrier situation:

1. The PAR for the IS-95 system is larger than that of CDMA2000 or WCDMA systems.
2. The PAR of the forward link is dependent on the Walsh code combination.
3. The PAR of CDMA2000 or WCDMA systems benefits from QPSK modulation.
4. The PAR of CDMA2000 or WCDMA reverse link signals benefits from complex spreading.

For multicarrier CDMA systems, analytical expressions are formulated to model the histogram (PDF) and the PAR of the signals for IS-95 and CDMA2000 systems. Analytical results matched the simulation results closely. In the process of deriving the expressions, better insight into the reasons for the PAR variations can be obtained. Finally, an algorithm for PAR control that selects the optimal combining phases has been considered and its effect on the PAR is discussed.

The following are the main conclusions found for the multicarrier situation:

1. The PAR of the CDMA2000 system is smaller than that of the IS-95 system in the multicarrier situations.
2. At low pilot transmit gain and traffic channel correlation, the PAR gain of combining phases adjustment is less than 0.3 dB for IS-95 and CDMA2000 multicarrier systems.

3. When traffic data bits (or assigned Walsh codes) are uncorrelated between different carriers, the PAR of the multicarrier combined signal is lower than the baseband PAR. However, when data bits are correlated, the combined PAR is raised.

Therefore, in general, a more active PAR control scheme, namely, a *predistortion technique,* is needed, which is under further investigation.

References

[1] Miller, S. L., and R. J. O'Dea, "Peak Power and Bandwidth Efficient Linear Modulation," *IEEE Trans. on Communications,* Vol. 46, No. 12, Dec. 1998, pp. 1639–1648.

[2] O'Neill, R., and L. B. Lopes, "Envelope Variations and Spectral Splatter in Clipped Multicarrier Signals," *Proceedings of 6th IEEE PIMRC'95,* New York, June 1995, pp. 71–75.

[3] Boyd, S., "Multitone Signals with Low Crest Factor," *IEEE Trans. on Communications,* Vol. 45, No. 10, Oct. 1997, pp. 1338–1344.

[4] Popovic, B. M., "Synthesis of Power Efficient Multitone Signals with Flat Amplitude Spectrum," *IEEE Trans. on Communications,* Vol. 39, No. 7, July 1991, pp. 1031–1033.

[5] Sumasu, A., et al., "A Method to Reduce the Peak Power with Signal Space Expansion (ESPAR) for OFDM System," *IEEE 51st Vehicular Technology Conf. Proc.,* VTC2000, Tokyo, July 2000, pp. 405–409.

[6] Paterson, K. G., and V. Tarokh, "On the Existence and Construction of Good Codes with Low Peak-to-Average Power Ratios," *Proc. IEEE Intl. Symp. Information Theory,* July 2000, p. 271.

[7] Fan, P., and X.-G. Xia, "Block Coded Modulation for the Reduction of the Peak to Average Power Ratio in OFDM Systems," *Proc. Wireless Communications and Networking Conf.—WCNC,* July 1999, pp. 1095–1099.

[8] Freiberg, L., A. Annamalai, and V. K. Bhargava, "Crest Factor Reduction Using Orthogonal Spreading Codes in Multicarrier CDMA Systems," *Proc. IEEE PIMRC'97,* New York, June 1997, pp. 120–124.

[9] Sevic, J. F., and M. B. Steer, "On the Significance of PAR in Estimating the Spectral Regrowth for Microwave RF Power Amplifier," *IEEE Trans. on Microwave Theory and Techniques,* July 2000, pp. 1068–1072.

[10] Tarokh, V., and H. Jafarkhani, "An Algorithm for Reducing the Peak to Average Power Ratio Is a Multicarrier Communications System," *Proc. IEEE 49th Vehicular Technology Conf.,* July 1999, pp. 680–684.

[11] Smith, J. A., J. R. Cruz, and D. Pinckley, "Method for Reducing the Peak-to-Average of a Multicarrier Waveform," *Proc. Vehicular Technology Conf.,* VTC 2000, Tokyo, June 2000, pp. 542–546.

[12] Lau, K. N., "Peak-to-Average Ratio (PAR) of Multicarrier IS-95 and CDMA2000 Signals," *IEEE Trans. on Vehicular Technologies,* Vol. 49, No. 6, Nov. 2000, pp. 2174–2188.

[13] Lau, K. N., "Analysis of Peak-to-Average Ratio (PAR) of Single Carrier IS-95 and CDMA2000 Signals," Internal Technical Memo, Lucent Technologies, Dec. 1998.

[14] Li, X., and L. J. Cimini, Jr., "Effects of Clipping and Filtering on the Performance of OFDM," *IEEE Commun. Lett.,* May 1998, pp. 131–133.

[15] Dinis, R., and A. Gusmao, "Performance Evaluation of a Multicarrier Modulation Technique Allowing Strongly Nonlinear Amplification," *Proc. ICC'98,* Atlanta, GA, June 1998, pp. 791–796.

[16] Ma, Z., H. Wu, and P. Polakos, "On the Peak-to-Average Ratio (PAR) of CDMA Signals," Internal Technical Memo, Lucent Technologies, June 1998.

[17] Johnson, N. L., *Continuous Univariate Distributions,* 2nd ed., New York: Wiley Interscience, 1994.

[18] Feher, H., *Advanced Digital Communications, Systems and Signal Processing Techniques,* Upper Saddle River, NJ: Prentice Hall, 1987.

[19] Simon, M. K., and J. G. Smith, "Hexagonal Multiple Phase-and-Amplitude-Shift-Keyed Signal Sets," *IEEE Trans. on Communications,* Vol. 11, No. 5, June 1971, pp. 1108–1115.

[20] Caswell, A. C., "Multicarrier Transmission in a Mobile Radio Channel," *Electron. Lett.*, Oct. 1996, pp. 1962–1964.

[21] Shepherd, S., J. Orriss, and S. Barton, "Simple Coding Scheme to Reduce Peak Factor in QPSK Multicarrier Modulation," *Electron. Lett.,* July 1995, pp. 1131–1132.

6

IP Mobility Framework for Supporting Wireless and Mobile Internet

Haseeb Akhtar, Emad Abdel-Lateef Qaddoura, Abdel-Ghani Daraiseh, and Russ Coffin

The drive toward high-performance Internet and inventing new ways to deliver content and services efficiently and reliably are the key issues facing the industry in terms of the Internet evolution. The ability to reach the end user, whether that user's computing device is linked to the wired network, a *wireless local area network* (WLAN), or a cellular network, ensures freedom and mobility and may well be the Internet's next killer application. The framework of IP mobility provided in this chapter will enable users to access the Web while roaming freely—and have personalized content follow them—across wireline and wireless networks as they move from building to building or from network to network, using all kinds of handheld devices and laptops. The framework is designed to extend the mobility management (with centralized directory management), end-to-end security, device independence, and application independence that allow users to access Internet applications on any device and any network.

6.1 Introduction

The advent of lightweight portable computers and handheld devices, the spread of wireless networks and services, and the popularity of the Internet have combined to make mobile computing a key requirement for future networks. Due to the dynamic nature of a mobile node's connectivity, providing network support for a mobile node and the users who operate them is a complex task when compared to doing the same thing for stationary nodes and users.

User mobility is an integral part of today's and the future's wireless networks. Mobility management is a key component of the core network in a wireless environment. Today, the wireless arena is made up of different types of access and core network technologies. The heterogeneous nature of today's wireless and wireline networks limits the scope of mobility between these heterogeneous networks. With the convergence of voice and data, networks of the future built to carry multimedia traffic will be based on packet-switched technology, mostly due to inherent advantages offered by the technology (the details of which are beyond the scope of this chapter). The mobility management control messaging needed to support users of these services will also be based on the same packet-switched technology.

The architecture framework defined in this chapter identifies the mobility management components needed to support user mobility for the different types of access networks within an IP centric network. The term *mobility-enabled IP centric networks* is used to define the networks that use IP addressing and routing protocols in concert with a mobility protocol to deliver multimedia traffic to roaming users.

The architecture framework provides a vision on how mobility can be achieved for heterogeneous access networks in an IP centric network. For wireless networks to achieve the vision, they may need to progress through some transitional phases [1].

One of the guidelines used to define the architecture was to make use of existing and/or proposed technologies defined by the Internet Engineering Task Force (IETF) and the ITU, and enhance the technologies as necessary to satisfy the requirements of our mobility-enabled networks.

A few of the key drivers for the current proposed architecture are listed here:

- Mobility includes the user, not just the device used by the user.
- Devices are addressed by IPv4/IPv6 addresses. Local devices' addresses (e.g., IEEE MAC, MIN, and IMSI) are access network specific and transparent to this framework.
- Devices may support their own Layer 2 access methodologies.
- If a network that provides data services to its users wishes to have its users roam, the network needs to have a service level agreement

with all the networks in which its users will roam. These service level agreements provide for a trust relationship between the networks that lays a foundation for user and network authentication, and trustworthiness for users paying their bills.

- In today's networks, when a user chooses a service provider for Internet-based services, for example, AOL, the user has a subscription with that service provider. If the service provider does not own the access network facilities, the user must have another subscription with an access network provider. In the architecture, given the service level agreement, it is not necessary for a user to have a subscription with a specific access network provider just to gain access to the service provider network. User authentication will always be provided by the user's home service provider network, which may or may not be an access network provider. In cases where the access network provider does not have any service level agreement with the user's home service provider network, a service bureau (broker) may be used to provide the user authentication.
- A single security framework should be adopted between network components. IPsec, an IETF security framework, can be used to ensure this security.
- Today's telecommunications networks have their own networking protocols that provide mobility. The wireless networks, for example, use IS-41 for the North American Cellular (NAC), and the GSM-MAP. The wireline networks, however, use the Ethernet protocol for the same purpose. A single network protocol must be established to perform the mobility functions and must be extended to cover wireline access networks. The architecture proposes an enhanced version of the *Authentication, Authorization and Accounting* (AAA) Protocol—an IETF protocol—to provide the network mobility functions independent of the wireless access type.
- IP packet data are end to end, that is, the mobile nodes all have IP stacks and all service applications are data oriented.
- The architecture should avoid having an anchor point to which all data will initially be sent while the user roams within the network unless the home network supports a policy for *hiding* the location of its users from other users in the general IP networks. (Having an anchor point is more commonly referred to as *triangle routing.*)

Some of the components provided by the architecture are as follows:

- The protocol supported by the AAA[1] working group is extended for mobility management, that is, location tracking, handoffs, and routing.
- Security gateways (firewalls) implement IPsec for secure (encrypted) communication links between networks.
- Directory services contain unified subscriber profiles, network usage policies, security policies, and information related to availability of services and their location.
- Policy management services are available for configuration management, security, and QoS issues.
- Domain name services provide address resolution.
- Dynamic host configuration services are used for allocating network resources.

This chapter provides an overview of the architectural framework for mobility-enabled IP centric networks. The framework identifies the functional components and the building blocks required to achieve IP mobility. The organization of this chapter is as follows: Section 6.2 provides the main challenges for implementing IP mobility; Section 6.3 contains the framework for IP mobility architecture; and Section 6.4 provides the challenges for implementing IP mobility using IPv6. The conclusion is presented in Section 6.5.

6.2 Challenges of IP Mobility

User mobility in IP networks introduces some unique problems to the network. Because the user devices may be either tethered or wireless, the issues consist of managing the IP addresses, providing AAA functions in real time, setting up a dynamic *business-to-business* (B2B) relationship, securing the access network from malicious nodes, and so forth. The following sections provide a brief list of challenges that are posed by mobility.

6.2.1 User Identity

A standardized method that can uniquely identify users is desirable. Numerous proposals have been made, each with its own pros and cons. All seem

1. The AAA group is currently a working group of the IETF. The working group recently adopted Diameter as the next-generation AAA Protocol and is currently in the process of finalizing the details.

to have one common attribute: The globally unique user identity must be resident in some "home database" that is accessible by all.

With the Internet being the model for the IP-based network, it makes sense to incorporate into the mobility architecture a user name space that is consistent with what already exists within the Internet. Current Internet naming is based on domain names.

With this in mind, the IP mobility framework supports unique identifiers as specified in Internet RFC 2486, "The Network Access Identifier" [2]. The network access identifier defined in this document is based on Internet domain names. The format of the identifiers is user@realm. An example is John.Doe@ISPxyz. The network access identifier can be used to identify users or to identify devices, such as routers.

The network access identifier is not an e-mail address. However, in the most limited sense, a network access identifier may be a user's actual e-mail address.

When a user accesses a visited network, the mobile node sends the user's network access identifier in the system access message [3]. The network access identifier is used to access the user's profile in a directory service and to help perform other system functions.

A user's device can supply the network access identifier in any of numerous ways. The network access identifier may be:

- On the user's subscriber identification module card (similar to the subscriber identification module card in GSM);
- Configured in the mobile node;
- Input by users when they want to access the system.

The network access identifier relates all of the user's devices that are *powered on* by a particular user at any given time. In addition to managing the user's presence in the network, the network access identifier also helps to identify the domain name of the user's service provider (from the "realm" portion).

6.2.2 Address Management

Managing the IP address of the user device (or devices) is of paramount importance in providing mobility-based services. When a device is mobile, the packets destined to that particular device are routed through a special mechanism. Mobile IP uses IP-in-IP encapsulation (as shown in Figure 6.1) at the home network and host-based routing at the visited network to address

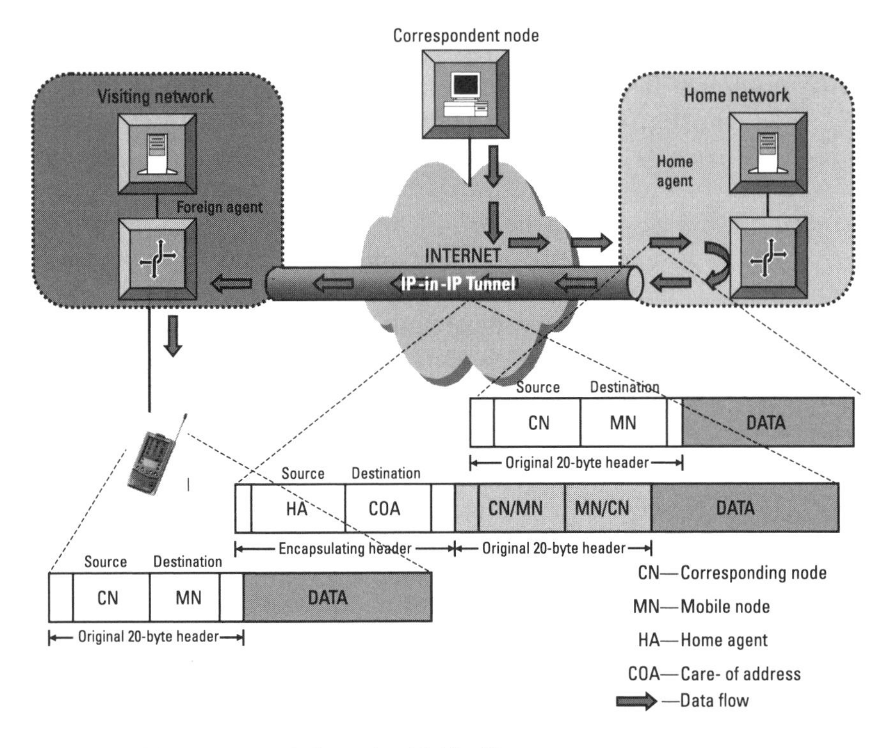

Figure 6.1 IP-in-IP encapsulation using Mobile IP.

this problem [4–6]. The data destined for the mobile node are first intercepted by the home agent and then routed to its care-of address at the visited network. The care-of-address is assigned and managed by the foreign agent. This, however, creates a *triangular routing,* which may not suit the needs of delay-sensitive applications. Assigning a *local address*—an address that has the same topology as the visited network—is one way to solve this problem by forcing the data path to avoid the home network altogether. The IPv6 solution for the Mobile IP [7] uses the same *local address* approach that eliminates the *triangular routing* as well as the foreign agent functionality at the visited network.

Elimination of *triangular routing* as suggested in [8], however, may create another problem. Some users may not feel comfortable about giving out their care-of address as they move across the networks because this, argues the proponent of *triangular routing,* can reveal location information. People, generally speaking, have yet to trust the Internet when it comes to putting private information on the wire.

What about owning multiple devices by a single user? A typical user could utilize a variety of network access devices through which messages can be sent to or received from others, depending on the location and convenience of the device to the user. The address management for all of these devices, to say the least, becomes an onerous task. Several stages are involved in performing this task:

1. First, the network (either the home or the visited network) needs to assign an IP address to all of these devices.
2. The home network needs to be informed of these addresses so that the user's network access identifier can be mapped against them in a database.
3. The network must have the knowledge of the user's preference as it relates to these devices. For example, if a certain user has both of his or her devices [*personal digital assistant* (PDA) and *personal computer* (PC)] on at the same time, the network has to know whether to send packets to the user's PDA only, or to both the PDA and PC simultaneously, or to the PC only and so forth. The user, most likely, would like to have a Web-based interface to a profile manager so that he or she can manage the profile in real time.
4. The network (home or visited) needs to have the intelligence to route packets to the appropriate device(s) accordingly.

The main objective, of course, is to provide the user with the benefit of always on connectivity of the IP mobility and, at the same time, allow maximum flexibility.

6.2.3 Dynamic Host Configuration Protocol and Domain Name Server Interfaces

One of the motivations of using the network access identifier to is to allow the dynamic allocation of IP addresses [3]. In addition to ease of management, the dynamic allocation of IP addresses also relaxes the precious supply of limited IPv4 addresses. The primary issue here is managing the Dynamic Host Configuration Protocol and domain name server interfaces to the mobility components. Here is a list of tasks that are required in this area:

1. The mobility manager at the visited or home network needs to initiate a Dynamic Host Configuration Protocol transaction to temporarily acquire an IP address.
2. The mobility manager also needs to manage the lease of the IP address throughout the entire session.

3. The domain name server that serves the user (either at the visited or home network) may need to be updated. That is, the user's network access identifier needs to be mapped against the mobile node's IP address.
4. The mobility manager needs to be informed as soon as the session is terminated—either by the mobile node or by employing a time-out mechanism.

As mentioned in Section 6.2.2, the complication arises when the user starts to use multiple devices simultaneously. Should that happen, the domain name server may need to carry multiple addresses per user network access identifier.

6.2.4 Mobility-Related Information

Mobility introduces some unique challenges and opportunities to the directory services in terms of information management. The information in this context is the data that are associated with the mobile nature of the user. The following is a brief description of some of these parameters that may be of interest.

- The location information can be of great interest to the content providers as the user continues to move across the globe. A directory that can provide this information in real time can easily be utilized to push location-based services to the user.
- The list of devices that the user is currently using is another piece of information that can be useful as well. This information enables the content providers to push content to the devices that are actually turned on.
- The capability of the user's device (along with its IP address) is yet another important piece of information that can be quite helpful. The content providers can push selective information based on the device capability, for example, graphics only for cellular phones, but full color for laptop PCs.
- The capability of access type—whether Ethernet (802.3), wireless LAN (802.11), or GPRS—is another useful piece of information. This allows the push services to be appraised about the bandwidth limitation of a particular access. A compact disc quality audio can easily be sent if the user is connected at a high-bandwidth access type

(say, 802.11) as opposed to a low-bandwidth access type (cellular, for example).

While all of this information provides new revenue opportunities for the service providers, adding them to the existing directory services is not a trivial task. The mobility manager needs to update a unified directory service as soon as the user experiences a mobility-related change. Some of these events include the following:

- Initial powering up of the user's device;
- Changing the subnetwork point of attachment;
- Moving to a different type of access;
- Powering off the user's device.

Because the unified directory service keeps the mobility-related information of the user, any push-service application can access this information using a standard interface, such as the Lightweight Data Access Protocol. Moreover, using the *persistent search* method, an application can register with the unified directory service to be notified as soon as a change occurs to the user's mobility-related information.

One of the challenges of dynamically updating a directory is to provide the ability to *write* information into a database in real time without causing any significant delay. For a large number of subscribers, this challenge becomes even tougher. The availability of the user's mobility-related data in real time certainly is a big motivator to overcome this challenge.

6.2.5 Security

The challenges of providing a secure system in a packet-based network are numerous [9, 10]. This task gets even more complicated if the user is mobile and is attempting to establish a multimedia session. The following is a brief list of requirements that are needed to secure a mobility-enabled network:

- The network (both home and visited) needs to protect itself from unauthorized users.
- The network (both home and visited) needs to provide access to authenticated users only, except for emergency calls.
- Confidentiality of traffic between the user and the security gateways that reside at the user's home subnetwork is to be protected.

- Confidentiality of traffic between the user and the correspondent node needs to be maintained.
- All nodes within the network need to have reasonable protection against a denial-of-service attack.
- The traffic exchanged between all nodes needs to be nonrepudiated, if necessary.
- The traffic exchanged between all nodes needs to have protection against any replay attacks.
- The visited network needs to be authenticated by the user before initiating any sessions. This is a major paradigm shift from the authentication scheme of today's wireless service providers. According to the current paradigm, the user does not have any need to authenticate the visited network. However, in a packet-based mobile environment, it is relatively easy for a malicious node to pose as a visited network. A WLAN access point can easily impersonate as the point of attachment to a visited network.
- Confidentiality of the user's personality while roaming between subnetwork points of attachments has to be ensured.
- Two or more nodes within the network should to be able to negotiate a security association in real time using the preconfigured security association between the mobile node and the home network. An AAA infrastructure may be required to effectively perform this task. This is especially important for the case when the security association between the visited and the home administrative domains needs to be set up in real time.
- The security credentials shared between two or more nodes need to be renewed automatically. That is, the nodes within the network should have the option to renew their security credentials frequently. The frequency of this change should be user configurable.
- Whenever possible, the user should be served a similar level of security (with respect to active and passive attacks) in a visited network as the user enjoys in his or her home network.

Security must be inherently provided by the network for all types of communication and transactions, and it is an important requirement to be considered for the core network. The degree of security afforded in the network is usually based on policies defined for that network.

6.2.6 User Privacy

User privacy is one of the most important requirements of any mobility-based solution. As the user moves across the networks, it is imperative that the user's current point of attachment not be revealed. It is also equally important to ensure that none of the user's personal information is compromised. The user's network access identifier, IP address, security credentials, or even the device identity that can be related to a particular user is usually considered to be information that can be used to describe the user's personality. As a result, any malicious node that has access to this information can easily detect and monitor the user's movement.

6.2.7 Dynamic Profile Distribution

Subscriber information needs to be maintained in the home network. The subscriber's profile may contain information related to the services to which he or she subscribes—the class of service, bandwidth needs, and security policies. These profiles need to be maintained in a directory and be accessible by the networks visited by the user in order to determine network services to be offered. The visited network may have a service agreement with the home network of the user and need to obtain the user's profile in order to provide local services. Profiles at the visited network may or may not be the same as those at the home network. Figure 6.2 provides an example for a dynamic profile distribution. Note that the profile may be location dependent or may have location-dependent rules associated with it, for example, a subscriber may be authorized for different services in different locations.

This becomes more critical when two separate administrative domains own the visited and home networks. The visited network is then required to retrieve the user's profile from the home network in real time. Without this capability, the user's experience across multiple administrative domains may suffer considerably.

6.2.8 Mobility-Based AAA Protocol

The main functionality of the AAA Protocol is fairly obvious from its name, that is, authentication, authorization, and accounting. These functions are needed for providing any kind of network services to the users, remote or otherwise [11]. However, for enabling mobility across multiple administrative domains, the AAA Protocol needs to be extended to include mobility-specific features. The following are a few of the requirements that are needed for a mobility-based AAA Protocol:

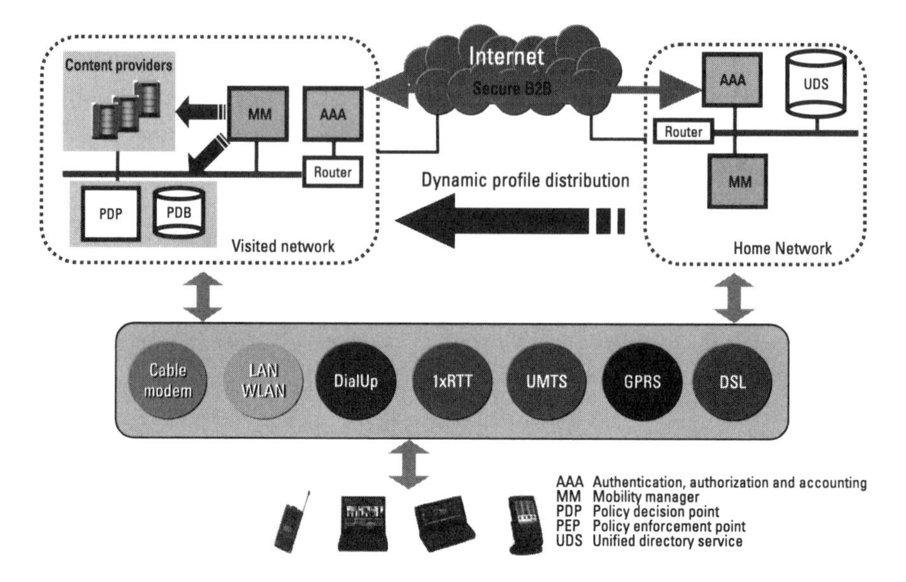

Figure 6.2 An example of dynamic profile distribution.

- The AAA Protocol must be able to encapsulate the control messages that allow a mobile node to register itself (e.g., the registration request message as described in [1]) with both home and visited networks. This allows the user to be authenticated and registered by traversing the Internet only once. This is a key requirement for delay-sensitive applications.
- Whenever possible, the AAA functions need to be done by a single protocol. For example, the control message for registration should not authenticate the same user twice—once by the AAA Protocol and then again by the Mobility Protocol.
- The AAA Protocol must provide enough space to carry *opaque* data for transferring security credentials (e.g., digital signatures and certificates) that may be shared between two or more nodes within the network.
- The AAA Protocol must provide the extensibility to transport the user's profile between multiple administrative domains.
- The AAA Protocol must have the ability to perform selective encryption of its attributes. This is especially required for the brokerage houses that are to arbitrate dynamic service level agreement between multiple administrative domains.

- The AAA Protocol must have the ability to carry the accounting records across multiple administrative domains in real time.

As mentioned earlier, the IETF has already adopted the *Diameter* as the AAA Protocol that can deliver the needs of a mobility-enabled network. The AAA working group is currently in the process of finalizing the Diameter Base Protocol [12].

6.2.9 Handoff

Handoff between networks with different access types poses considerable challenges. The goal here is to minimize the data loss when a mobile node changes its subnetwork point of attachment to the network. The time needed for the mobile node or the network to become cognizant of this change should be minimal. That is, the fact that the mobile node has changed its subnetwork point of attachment ought to be discovered as soon as possible.

Preferably, the network needs to have the intelligence to initiate the handoff process before the mobile node leaves its old subnetwork point of attachment. In the case of movement from one access type to another access type (e.g., from GSM to WLAN), this "make before break" capability would allow the upper layer applications to experience seamless mobility.

Additionally, a mechanism should be in place to handle in-transit packets while the transition is being managed. A mechanism for real-time forwarding as mentioned in [13] may enhance the use of delay-sensitive applications.

6.2.10 Broker Services

As the Internet continues to grow, the number of service providers that provide Internet-based services will continue to increase as well. This landscape is fundamentally different from that of the traditional communication carriers. Although it is commonplace in 2002 to have roaming agreements among the wireless carriers, having similar roaming agreements that can cover all Internet-based service providers—including the traditional wireless and wireline carriers, the Internet service providers, the content providers, and many more—is not practically feasible; it would be a network administration nightmare. The users, however, would like to get the same degree of services whether they are on their home network or on a visited network. This, in our opinion, will force the issue of setting up dynamic service level agreements between two or more administrative domains. As a result of this business pressure, broker services will have to be created. As a first step, the broker servers and the services they support will have service level agreements with their clients. A broker server, if and when requested, will then be able to dynamically and securely negotiate a service level agreement between two or more of its clients.

6.2.11 B2B Service Agreements

One of the attractive services enabled by IP mobility is the ability to seamlessly move between the carrier and the enterprise domains without losing the session. That is, the user can start a TCP/IP application (e.g., downloading a multimedia stream from the Internet) at the office (e.g., using Ethernet access) and continue to have the application running while driving home (e.g., using GPRS access). This, however, requires a B2B agreement between the carriers' and the user's enterprise networks to allow the user's data to be forwarded appropriately. In the absence of any agreement, the user in the example mentioned above will have to exit the application, disconnect the Ethernet access, connect the GPRS access, relaunch the application, and then continue to receive the data.

6.3 IP Mobility Architecture Framework

The primary function of the IP mobility architecture is to provide the capability of mobility to IP-based devices. The IP mobility architecture is functionally equivalent to the Mobile IP architecture [1] and interoperates with it for no additional investment. The framework [14] is designed such that specialized and distributed functional blocks handle all changes in the access medium, the mobility characteristics, or the user profile. This enables the architecture to adapt to changes in these arenas and to scale independently as the service base grows.

6.3.1 IP Mobility Architecture Components

The IP mobility architecture can be broadly sectioned into a framework of four abstract entities:

1. x (any) access network;
2. Local serving function;
3. Network serving function;
4. Mobile node client.

The x access network is built up of entities that directly interface with the mobile node. The local serving function, in association with the x access network, provides ubiquitous reachability to the mobile node. The network serving function is the primary entity that manages and dispenses security and other subscriber-related information about the mobile node or its user.

6.3.1.1 x Access Network

The x access network is media independent and reflects the technology used by the mobile node to obtain connectivity to the network. The media used for achieving physical connectivity to the network may be *local area network* (LAN), token ring, or even wireless, as shown in Figure 6.3. The air interface is the ideal transition medium for delivering uninterrupted services to mobile nodes. It is ubiquitous and requires virtually no setup time at the physical layer. As such, mobile nodes that transition between two physical networks can be provided continuous connectivity to the Internet through an intermediary wireless access network with cell-site coverage that overlaps both networks. Furthermore, from an engineering point of view, a given wireless subnet serviced by an IP router may be built up of multiple cell sites, with

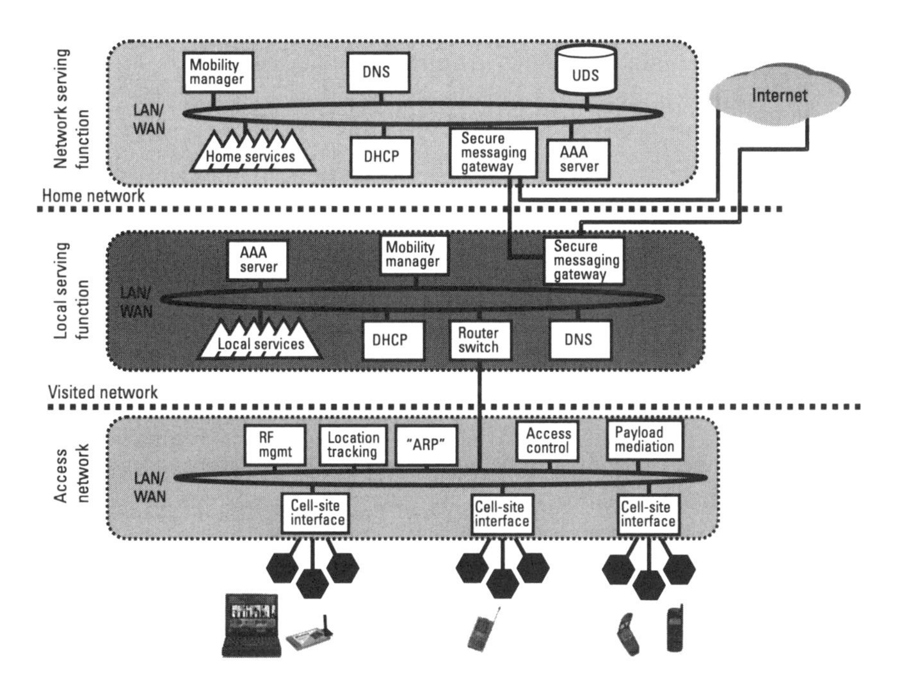

Figure 6.3 IP mobility architecture with abstract functions.

inherent wireless handoff capability that is transparent to upper layer network functionality.

6.3.1.2 Local Serving Function

The local serving function is the serving area network for sets of access networks. It is responsible for managing and dispensing/denying resources to all mobile nodes that visit it. The following is a brief list of its functions:

- The local serving function can support multiple x access networks where each x access network is associated with a different access technology, for example, one x access network may be North American cellular wireless network, another may be the GSM wireless network, and yet another may be Ethernet enterprise network.
- The local serving function supports the local mobility management functions. This is provided by the mobility manager entity of the local serving function. In addition to providing the tunnel management functions as outlined in [1], the mobility manager entity also manages the mobility-related functions across multiple networks that it serves.
- The local serving function is capable of supporting multiple types of access media through a single x access network-local serving function interface, allowing the mobile node to switch media without loss of connectivity. This is accomplished by providing IP layer connectivity between the mobile node and the local serving function.
- The local serving function provides the AAA function to all the users that are served by it. The AAA server located at the local serving function is primarily responsible for authorizing the user by contacting the user's home network. It is also responsible for generating accounting data.
- The local serving function may also provide the address management services to the users by providing the Dynamic Host Configuration Protocol and the domain name server services.
- The local serving function may be protected from the public IP network by a secure messaging gateway that provides a secure back end to all functions located within the local serving function. It also provides a single secure tunnel for all B2B control messages.

A local serving function may provide services to multiple network serving functions (either owned by the single administrative domain or by multiple administrative domains) at the same time.

6.3.1.3 Network Serving Function

The network serving function is the home network of the user and it *owns* the user's subscription and associated profile. The following major functions are performed by the network serving function:

- The network serving function provides mobility to the users on a larger scale. The user can roam in any visited network and the handling of this mobility is done by the mobility manager entity at the network serving function. The mobility manager is responsible for maintaining the current location care-of address of the user. The network serving function accommodates the tunnel management service [1] that functions as the gateway to the current point of attachment of the mobile node. Because the network serving function is associated with a mobile node primarily through its IP address, all packets bound for a mobile node are first routed to the gateway at the network serving function. As long as a mobile node is attached to a subnetwork that is managed by the network serving function, none of the mobility functionality is exercised and datagrams are directly delivered to the mobile node. If, however, the mobile node is attached to a subnetwork that is not managed by the network serving function, the datagrams are then tunneled to that particular subnetwork using the mobile node's care-of address as outlined in [1].
- The network serving function provides the AAA function to all the users that are native to it. The AAA server located at the network serving function is primarily responsible for authenticating and validating the mobile node, in addition to generating accounting data.
- The network serving function maintains the subscriber profile in the unified directory service and makes it available to all other entities—such as the visited network or push-service applications—that require it. The primary responsibility of the unified directory service is to store, maintain, and dispense subscriber profiles. The IP mobility architecture provides a one-stop centralized repository for all subscriber-related information, accessible through a standardized interface (e.g., Lightweight Data Access Protocol).
- The network serving function supports subscriber services and mobility management functions independent of the type of access the subscribers use to connect to the network. It also provides address management via Dynamic Host Configuration Protocol.

- In most topologies, a specific network serving function will be accessible through the Internet. The network serving function offers domain name server service for looking up the IP address currently bound to a mobile node's network access identifier.
- The network serving function may be protected from the public IP network by a secure messaging gateway that provides a secure front end to all functions located within the network serving function. The secure messaging gateway works in association with the AAA to provide impervious security. It also provides a single secure tunnel for all B2B control messages.

In an integrated network serving function/local serving function there is only one set of components that performs the appropriate role based on their association with the user. A service provider's network may logically be composed of a single network serving function with multiple local serving functions to which it provides home network functionality. These local serving functions may be physically (geographically) located elsewhere. Mobile users that are homed in this network serving function can roam in any of the local serving functions that are associated with this network serving function and be considered to be in their home network.

6.3.1.4 Mobile Node Client

The mobile node client is responsible for originating registrations and managing its interface to the local serving function. It implements all safeguards required to maintain a secure control and data path into the network.

6.4 IPv6 Challenges for IP Mobility

The enhanced features of IPv6 clearly eliminate the need for a foreign agent router at the serving network. The Mobile IPv6 standard [7], however, does not address several essential functions needed for providing service at the serving network. The following is a brief list of these functions:

- An AAA infrastructure and the mechanism for authorizing mobile node services are required.
- The mobile node needs to acquire a care-of address from the Dynamic Host Configuration Protocol while initiating a session. No mechanism existed as of this writing for managing these addresses.

- In the multidomain model (where the serving network and the home network do not belong to the same administrative domain), it may be necessary to transfer the user's profile from the home network to the serving network. The current Mobile IPv6 standard does not support this feature.
- While it is understood that IPv6 uses IPsec for ensuring security, the Internet Key Exchange Protocol, however, may not be suitable for delivering keys in real time. Additionally, the public key infrastructure may not be as readily available. A mechanism for distributing keys in real time may still be needed. A detailed description of this proposal can be found in a recent Nortel Networks submission to IETF [15].
- Although IPv6 relaxes the address space of the IPv4 standard, the service providers would most likely use the network access identifier as opposed to the user's home address for identifying the user.
- Performing smooth handoffs (make before break) is an important aspect of IP mobility architecture. Wherever possible, an interface between Layer 2 and Layer 3 would be necessary to gain accurate movement detection of the mobile node from the old local serving function to the new local serving function. The current mobility support for IPv6 does not have this interface.

While the basic Mobile IP for IPv6 provides a good protocol to provide a point of attachment to the network, it falls short of providing a system-wide solution as mentioned above. To address all aspects of mobility, the IPv6 solution requires the following two basic layers: (1) a resource management layer and (2) a routing layer. The resource management layer allocates and manages all the network resources needed for serving the mobile node, while the routing layer controls the mobility bindings. Section 6.4.1 briefly describes these layers.

6.4.1 Resource Management Layer

The resource management layer consists of control signals that reside above the transport (TCP/UDP) layer. The following functions are provided by this layer:

- The resource management layer provides the mechanism for allocating and managing the IPv6 addresses to be used by the mobile node as its care-of address. The unified directory service at home network is also updated to reflect this so that other applications (such as location-based services) can use this information in real time.

- The resource management layer also provides access to the AAA infrastructure. The mobility manager entity at the local serving function interacts with the AAA infrastructure, which may include an interface to the broker AAA as represented in Figure 6.4. In the case where the visited network does not have any service level agreement with the user's home network, the broker AAA can arbitrate the AAA process.
- The resource management layer also provides a mechanism for transferring the user's profile from the home network to the serving network. This information is retrieved from the unified directory service during the initial registration and shipped back to the visited network using the AAA Protocol. The mobility manager at the local serving function can then update any policy decision point or any policy enforcement point located at the visited network with the relevant information.
- Dynamic distribution of security associations is another essential function performed by the resource management layer. The Key Exchange for Network Architectures, as described in [16], can be implemented to do just this by using the mobility-based AAA Proto-

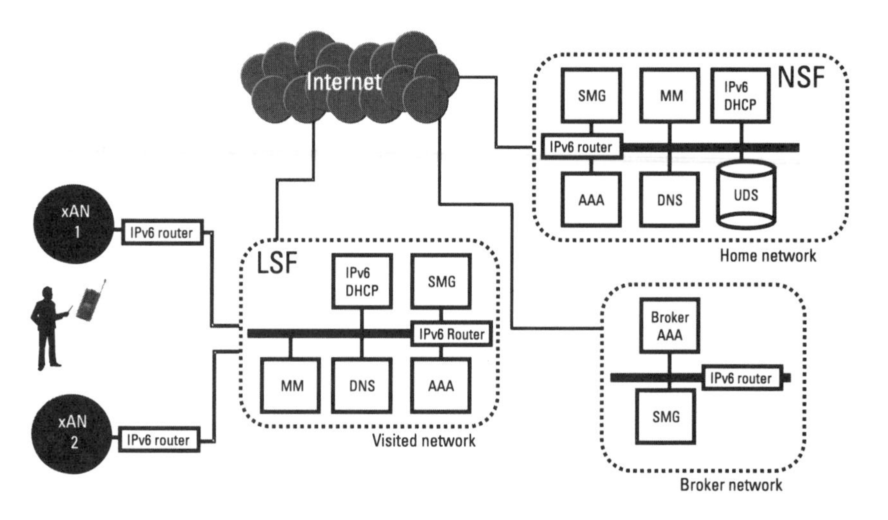

Figure 6.4 Broker AAA between the local serving function and the network serving function.

col. Using the preconfigured security association between the mobile node and the home network, the Key Exchange for Network Architectures can deliver a security association (per session per user) between the mobile node and the visited network. An IPsec session can be initiated later using this security association without resorting to the Internet Key Exchange Protocol. Eliminating the Internet Key Exchange may reduce the delay substantially.

- The resource management layer facilitates user management by allowing the network access identifier to be used. Identifying the user's home network, providing a built-in security parameter index, and retrieving the user's profile from the unified directory service are better served by the use of an network access identifier versus the IPv6 address.
- The resource management layer also provides a way to exchange link layer information between local serving functions during the handoff. Because Mobile IPv6 does not specify a standard interface between all the link layers and the network (IP) layer, this would allow this information to be carried across several domains using this resource management layer. This becomes especially important if the mobile node client uses multiple accesses and needs to arbitrate among those accesses simultaneously.

6.4.2 Routing Layer

This layer provides the basic data path between the correspondent node and the mobile node. It manages the binding between the mobile node and the IPv6 router at the visited network. Although the Mobile IPv6 protocol can be used in this layer, several control messages are required to secure the resources mentioned in Section 6.4.1. A few of those messages are detailed in [17] and are beyond the scope of this chapter.

6.5 Conclusion

Investigations into enhancing performance and research in terms of adopting the optimal approach for introducing mobility support into an existing network with minimal impact are ongoing [18]. User devices (such as PDAs and Pocket PCs) with multiple access types are expected to be available during the rollout of 3G wireless networks [19]. These small but powerful devices, along with the availability of high bandwidth over the air (e.g., 3G and WLAN) are poised to set the stage toward a wider deployment of IP mobility solutions.

The IP mobility architecture described in this chapter highlights the basic infrastructure required to support a mobility-enabled IP network. Implementing the IP mobility solution, as mentioned in this chapter, introduces quite a few nontrivial challenges. The IP mobility-capable network, however, presents ample growth opportunities to service providers. The ability to roam across multiple accesses without the loss of a session, after all, is a significant technological breakthrough. Using this technology, all upper layer protocols—such as the Session Initiated Protocol [20, 21] and FTP—can instantly become access independent, that is, fully mobile.

However, like any new technology, the fate of IP mobility too depends on the entrepreneur savvy—the ability to generate revenue—of the service providers. Let's hope that the ease of use, access independence, and, above all, the convenience provided by the IP mobility will motivate the market economics in the near future.

Acknowledgments

We give credit to Al Javed who started and continuously supported this innovative program. Special thanks to John Crane and his team for their help in speeding up the release of this document. Also, many thanks to the larger team at Nortel Networks, which continuously provided feedback and support.

References

[1] Akhtar, H., E. Qaddoura, and A. Daraiseh, "IP Mobility Architecture for Supporting Wireless and Mobile Internet," invited paper, *Proc. 4th IEEE CAS Workshop on Emerging Technologies in Circuits and Systems,* Hong Kong, Nov. 2000, pp. 1-5-1–1-5-17.

[2] Aboba, B., and M. Beadles, "Network Access Identifier," RFC 2486, Internet Engineering Task Force, Jan. 1999.

[3] Calhoun, P., and C. Perkins, "Mobile IP Network Access Identifier Extension for IPv4," RFC 2794, Internet Engineering Task Force, March 2000.

[4] Perkins, C., "IP Mobility Support," RFC 2002, Internet Engineering Task Force, Oct. 1996.

[5] Simpson, W., "IP in IP Tunneling," RFC 1853, Internet Engineering Task Force, Oct. 1995.

[6] Perkins, C., "IP Encapsulation Within IP," RFC 2003, Internet Engineering Task Force, Oct. 1996.

[7] Perkins, C., and D. Johnson, "Mobility Support in IPv6," Internet Draft, Internet Engineering Task Force, Nov. 2000.

[8] Perkins, C., and D. Johnson, "Route Optimization in Mobile IP," Internet Draft, Internet Engineering Task Force, Feb. 1999.

[9] Rivest, R., "The MD5 Message-Digest Algorithm," RFC 1321, Internet Engineering Task Force, April 1992.

[10] Harkins, D., and D. Carrel, "The Internet Key Exchange (IKE)," RFC 2409, Internet Engineering Task Force, Nov. 1998.

[11] Calhoun P. (ed.), "AAA Problem Statements," Internet Draft, Internet Engineering Task Force, Nov. 2000.

[12] Calhoun, P., et al., "Diameter Base Protocol," Internet Draft, Internet Engineering Task Force, April 2001.

[13] Khalil, M., et al., "Buffer Management for Mobile IP," Internet Draft, Internet Engineering Task Force, Oct. 1999.

[14] Becker, C., B. Patil, and E. Qaddoura, "IP Mobility Architecture Framework," Internet Draft, Internet Engineering Task Force, Oct. 1999.

[15] Khalil, M., H. Akhtar, and E. Qaddoura, "Dynamic Security Association Establishment Protocol for IPv6," Internet Draft, Internet Engineering Task Force, April 2001.

[16] Khalil, M., et al., "Key Exchange for Network Architectures (KENA)," Internet Draft, Internet Engineering Task Force, March 2000.

[17] Khalil, M., et al., "IPv6 Support in IP Mobility," white paper by IPM Solutions, Nortel Networks internal document, Sept. 2000.

[18] Daraiseh, A.-G., E. Qaddoura, and H. Akhtar, "Performance and Capacity of IP Mobility architecture," white paper by IPM Solutions, Nortel Networks internal document, Dec. 2000.

[19] Daraiseh, A.-G., H. Akhtar, and E. Qaddoura, "Architecture Evolution for IP Mobility in IPv6 and 3G Wireless Networks," white paper by IPM Solutions, Nortel Networks internal document, March 2001.

[20] Handley, M., H. Schulzrinne, and J. Rosenberg, "SIP: Initiation Protocol," RFC 2543, Internet Engineering Task Force, March 1999.

[21] Schulzrinne, H., and J. Rosenberg, "The Session Initiation Protocol: Providing Advanced Telephony Services Across the Internet," *Bell Labs Technical J.*, Vol. 3, Oct.–Dec. 1998, pp. 144–160.

7

Software Radio: A Prospective Technology for Future Broadband Communication Systems

Yik-Chung Wu, Tung-Sang Ng, and Kun-Wah Yip

7.1 Introduction

Around every 10 years a new generation of the mobile communication system appears and gradually replaces the old one for public deployment. Although 3G systems are still in the early stage of public deployment at the time of this writing, it is commonly expected that the forthcoming generation will succeed 3G from 2010 onward. If past experience is any guide, it will take 10 years or more to develop a new generation of the mobile communication system from initial conceptualization to technological maturity and deployment. Thus, research efforts have been initiated to investigate various issues related to the post-3G system and to identify key technologies that are of potential importance in its development [1–5]. It has been envisioned that one of the important requirements in the implementation of future-generation mobile communication systems is to have both network and terminal reconfigurability [1]. Software radios, being able to provide reconfigurability for wireless communication devices, are particularly attractive for future systems.

Fundamentally, a software radio uses an analog-to-digital converter to digitize the received signal when the signal frequency is still at the RF band or has been downconverted to the *intermediate frequency* (IF) band, and processes the digitized signal in the digital domain by programmable processor(s). One can reconfigure the software radio by loading different programs into the programmable processors. Because the signal frequency is very high, sampling at the Nyquist rate is extremely difficult even if technically feasible. Bandpass sampling, which provides a considerable reduction in the required sampling rate, is often used in the analog-to-digital converter process. As a result of the presence of multiple access, the received signal contains both the desired signal and unwanted other-user signals. Selection of the appropriate channel in which the desired signal is transmitted is required. This process is known as *channelization.* The desired signal is then downconverted to the baseband frequency. Afterward, the signal is processed entirely in the baseband. Because the downconverted signal has a sampling rate that is higher than is necessary for subsequent baseband processing, decimation (i.e., taking one sample every several samples) is required, so that the implementation complexity of the baseband processing unit can be minimized. Furthermore, because a software radio is generally required to handle different mobile communication standards, the sampling rate of the signal digitized at the analog-to-digital converter may not be an integer multiple of the target data rate. Thus, decimation is usually performed in two steps: decimation by an integer ratio followed by fractional sampling-rate conversion.

In Section 7.2, we provide a brief overview of the foreseen paradigm of mobile communications in the 2010s, and we establish the relevance of software radio technology.

In Section 7.3, an overview of software radio and its historical aspects are given. Because software radio is a large topic and covers a large number of research areas, it can be analyzed from many different perspectives, including the network level, terminal level, device level, hardware implementation aspects, and software aspects. Here we are concerned with signal processing aspects in receiver design. In particular, we emphasize bandpass sampling, decimation filtering, channelization, and fractional sampling-rate conversion, all of which are important in the implementation of software radios. Baseband signal processing, however, is rather common to all receivers and is not treated here. Sections 7.4 and 7.5 detail bandpass sampling and decimation filtering, respectively. Channelization based on discrete Fourier transform filter banks is discussed in Section 7.6. This channelization method is particularly useful when extracting multiple signals transmitted over different channels. Section 7.7 elaborates on the fractional sampling-rate conversion.

7.2 Mobile Communication Scenario in the 2010s

A major impetus for developing 3G systems in the early 1990s was in response to the ever-increasing demand on wireless data transmission, especially for multimedia data. The need for multimedia communications has been increasing rapidly as a result of the explosion of Internet pervasiveness. With people's growing desire to have Internet access anywhere anytime through wireless communication means and the rising expectation of users to have faster Internet access, it is reasonable to expect that the demand for higher rate wireless data transmission will continue. One of the basic requirements in the implementation of future mobile communication systems is, therefore, to provide higher data rates. One study [1] has forecast that the user data rates that are supported by future mobile communications systems are at least 2 Mbps and can be as high as 600 Mbps depending on the communication environment.

However, supporting data transmission rates greater than those of 3G systems is not the only characteristic of future communication systems. Multimedia contents, such as voice, speech, video, still pictures, graphics, and virtual reality objects, have different requirements for QoS. The cellular approach adopted throughout the first three generations does not scale well to heavy multimedia traffic with different QoS requirements. To supplement the need for heavy multimedia data communications, currently a user can install wireless local-area networks as cellular extensions at hot-spot areas such as in the indoor environment. This approach is envisioned to be incorporated and generalized in future wireless systems [1, 3], in which public cellular systems and private networks coexist. Public cellular systems provide wide-area coverage and mobility to users, whereas private networks, probably operating in an unlicensed spectrum, are targeted to provide high-rate, high-capacity communication links. Overlapping coverage of different networks is expected to become commonplace. A user terminal is able to negotiate with various networks and is empowered to select the most appropriate one for communications. A user terminal is therefore required to reconfigure itself to communicate with a variety of inhomogeneous networks. Software radios are particularly attractive in the realization of user terminals due to their ability to provide terminal reconfigurability.

Another characteristic of future mobile communication systems not common in present-day mobile communication systems is the proliferation of machine-to-machine communications as a result of the advance of low-cost, intelligent embedded radios. These embedded radios exist in the forms of, for example, wireless low-cost sensors, smart devices that can be worn by users, and intelligent home appliances. They assist future communication systems to provide location-based, personalized services to users. The intelligence of embedded radios enables them to interact with the surrounding environment automatically. One of the consequences is that these devices can initiate communications without the intervention of users. The need to communicate with a diversity of embedded radios poses a challenge to future mobile networks. Networks need to reconfigure and optimize themselves from time to time in order to communicate with different embedded radios and to handle different requests from these radios. Software radios provide an effective means for dynamic reconfiguration of future mobile communication networks.

7.3 Overview of Software Radios

7.3.1 Motivation to Develop Software Radios and Their Advantages

The superheterodyne principle has been employed for more than 80 years in the design and realization of radio receivers. These receivers are basically analog in nature. The received signal is first processed by an analog RF front end. The amplified signal is then downconverted into the IF by means of an analog mixer. At the IF stage, only the components of the signal at desired frequencies are retained and the others are filtered out. Finally, the IF signal is further downconverted into the baseband frequency. One can use an analog-to-digital converter to convert the analog baseband signal into the digital domain. Digital signal processing is then used to process the signal in order to estimate the transmitted information data sequence. Advantages of receiver implementation based on the aforementioned principle are that the system architecture is proven, the RF/IF analog technology is mature, and the requirements on analog-to-digital converters and digital signal processors are relatively undemanding.

However, one of the greatest disadvantages of analog receivers is their lack of flexibility. One receiver is designed for one communication system. As more and more mobile communication standards emerge and coexist around the world, a multimode radio receiver would certainly be a significant advantage. One approach for achieving multimode capability is to pack several discrete receivers for different standards, but this approach greatly increases the weight and size of the resultant composite receiver. In addition, the required power consumption may not be favorable to mobile communication applications. An alternative solution is to digitize the received signal not at the baseband frequency but at the IF or even at the RF,[1] and then to process the digitized RF/IF signal by a digital processing unit for signal demodulation and detection. This digital processing unit is programmable, and can be a general-purpose digital signal processor, a customized programmable processor, or a group of them. For different communication standards, different programs (in the form of software) are loaded into the processor(s) to demodulate the received signal. Multimode capability is therefore supported without the need to duplicate hardware. This multimode receiver is a software radio.

Apart from the support of multimode capability, many other advantages accrue from using software radios:

- As more signal processing functions are performed in the digital domain, software radio receivers excel in reliability when compared to conventional analog receivers due to the elimination of common problems encountered in analog receivers, such as temperature drift and aging of components.
- Reconfigurability of receivers can be accomplished very quickly by loading appropriate configuration programs into software radios.
- New algorithms can be tested and implemented more easily.
- If a new communication standard emerges or a standard that is in use is revised and updated, transitions from the old standard to the new/revised one can be accomplished in a short time. In the viewpoint of business competitiveness, communication products involving software radios enjoy a faster-to-market advantage.

7.3.2 Historical Background

Software radios have their origin in military applications. The U.S. SPEAKeasy project was a multiphase, joint service technology program to prove the concept of a programmable-waveform multiband, multimode radio [6]. One of the early survey papers on software radios was written by Mitola [7] in 1992. Interest in software radio technology increased rapidly after the first high-quality tutorial on software radios appeared in the May 1995 issue of the *IEEE Communications Magazine* [8]. In 1996, *Flexible Integrated Radio Systems Technology* (FIRST) was launched in Europe to investigate intelligent multimode terminals [9]. The first European workshop on software radios

1. One of the ultimate goals of software radio research is to digitize the signal as close to the antenna as possible, or directly at the RF without downconverting the received signal.

was organized in 1997 [10]. Shortly after this workshop, two more projects were launched in Europe: *Smart Universal Beamforming* (SUNBEAM) and *Software Radio Technology* (SORT), with one dealing with the development of innovative base station array processing architectures and algorithms [11] and the other aiming at demonstrating flexible and efficient software-programmable radios [12]. In 1999, three follow-up special issues [13–15] on the topic of software radios appeared in the literature, illustrating the progress in recent years. Another special issue [16] was published in 2000. In addition to the aforementioned references, we mention the following publications. Li and Yao [17] and Pereira [18] reported on the progress of software radio research in the United States and Europe, respectively. General overviews on software radios have been given by Mitola [19], Kenington [20], Tuttlebee [21], and Leppanen et al. [22]. Realization of trial software radio systems has been reported by Turletti and Tennenhouse [23], Ellingson and Fitz [24], and Swanchara, Harper, and Athanas [25].

7.3.3 Practical Software Radio Architecture

Ideally, a software radio is best operated by having the received signal sampled and digitized directly at the RF with all demodulation and detection functions implemented in programmable digital processor(s). As a result of the need for exceedingly high rate analog-to-digital converters, this software radio implementation is not economically practical at least at this moment, though considerable research effort has always been given to the development of low-cost, higher-speed analog-to-digital coverters. Practical software radios are usually realized with IF sampling and with subsequent digital signal processing being carried out in a combination of application-specific integrated circuits, field-programmable gate arrays,[2] digital signal processors, and other microprocessors. Software radio architecture is an open one and different researchers use slightly different definitions. Despite this, a practical system, shown in Figure 7.1, usually consists of an analog-to-digital converter, a digital downconverter, a set of digital filters, an interpolator, a demodulator, and a decision unit [27–29].

After the analog RF signal is frequency shifted to the IF by using an analog mixer and a bandpass filter, the IF signal is sampled and analog-to-digital converted, wherein the sampling is performed based on the principle of bandpass sampling [30], which performs downconversion as well. A replica

2. Interested readers may refer to [26] for an overview on the application of field-programmable gate arrays for software radios.

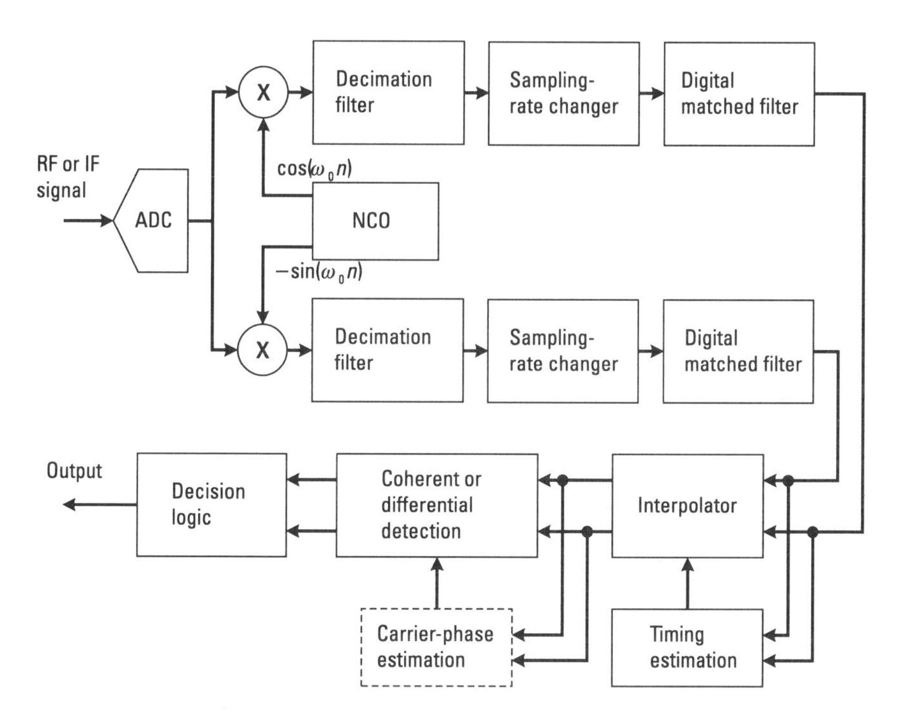

Figure 7.1 A practical software radio architecture.

of the digitized signal is then produced near the zero frequency. Note that at this stage the signal contains the desired-signal component in a given channel as well as other undesired components in other channels. A numerically controlled oscillator, which need not be phase coherent with the incoming signal, shifts (downconverts) the desired signal component from the original frequency band to the baseband. Channelization is thus accomplished. The baseband signal is then processed by a decimation filter.

The purpose of decimation filtering is twofold. It filters out signals of undesired channels without degrading the desired-signal component, and performs decimation on the resultant signal in order to reduce the complexity in the implementation of subsequent signal processing functions. Nonideal effects of decimation filtering (such as the passband of the filter being not flat) can be compensated at a subsequent stage (e.g., in the stage of digital matched filtering) for better performance. Because a software radio is generally required to handle the signals of different mobile communication standards, there is a need to implement a sampling-rate changer in order to adjust the sampling rate of the signal to match with the data transmission rate of the communication standard under consideration.

The resultant baseband signal is then processed by digital matched filters to maximize the signal-to-noise ratio. An interpolator is then used to compute the matched-filter-output samples that are time-offset to the original output samples by a fraction of the sampling interval. The main purpose of using an interpolator is to time-align the local sampling clock with the incoming signal and thereby to eliminate intersymbol interference. Interested readers should refer to [31, 32] for details on interpolation. Coherent or differential detection then follows. Note that carrier-phase synchronization is required for coherent detection. Finally, a decision unit is used to estimate the transmitted data sequence.

7.3.4 Frequency-Domain Illustration of the Receiver Signal Processing

Frequency plots of the signal at various points along the receiver signal processing chain are shown in Figure 7.2. The received signal is first down-converted to the IF and the resultant IF signal is bandpass sampled. Corresponding frequency spectra of the signal at these two stages are depicted, respectively, in Figures 7.2(a) and 7.2(b). The desired-signal component is then frequency shifted to the baseband, producing the spectrum shown in Figure 7.2(c). The signal in the desired channel is extracted by means of digital filtering, and the sampling rate is subsequently reduced. Figures 7.2(d) and 7.2(e) show the spectra of the signals at these two stages. Notice that decimation filtering accomplishes these two steps simultaneously. Afterward, the baseband signal can be manipulated in the baseband.

7.4 Bandpass Sampling

For a bandpass signal, straightforward application of the Nyquist sampling theorem requires that a sampling rate greater than twice the highest frequency component of the signal must be employed, which is generally impractical. For example, digitizing a signal at a frequency of 900 MHz (a GSM signal) requires a sampling rate of at least 1.8-GHz samples per second. By means of bandpass sampling [30], a bandpass signal can be sampled at a much lower rate while the signal can be uniquely reconstructed.

7.4.1 Basic Principle and Implementation Considerations

For a bandpass signal, denoted as BPS in the equations, with a signal bandwidth B and confined to the frequency (f_L, f_U), that is, $f_U - f_L = B$, the allowable sampling rate is given by [30]

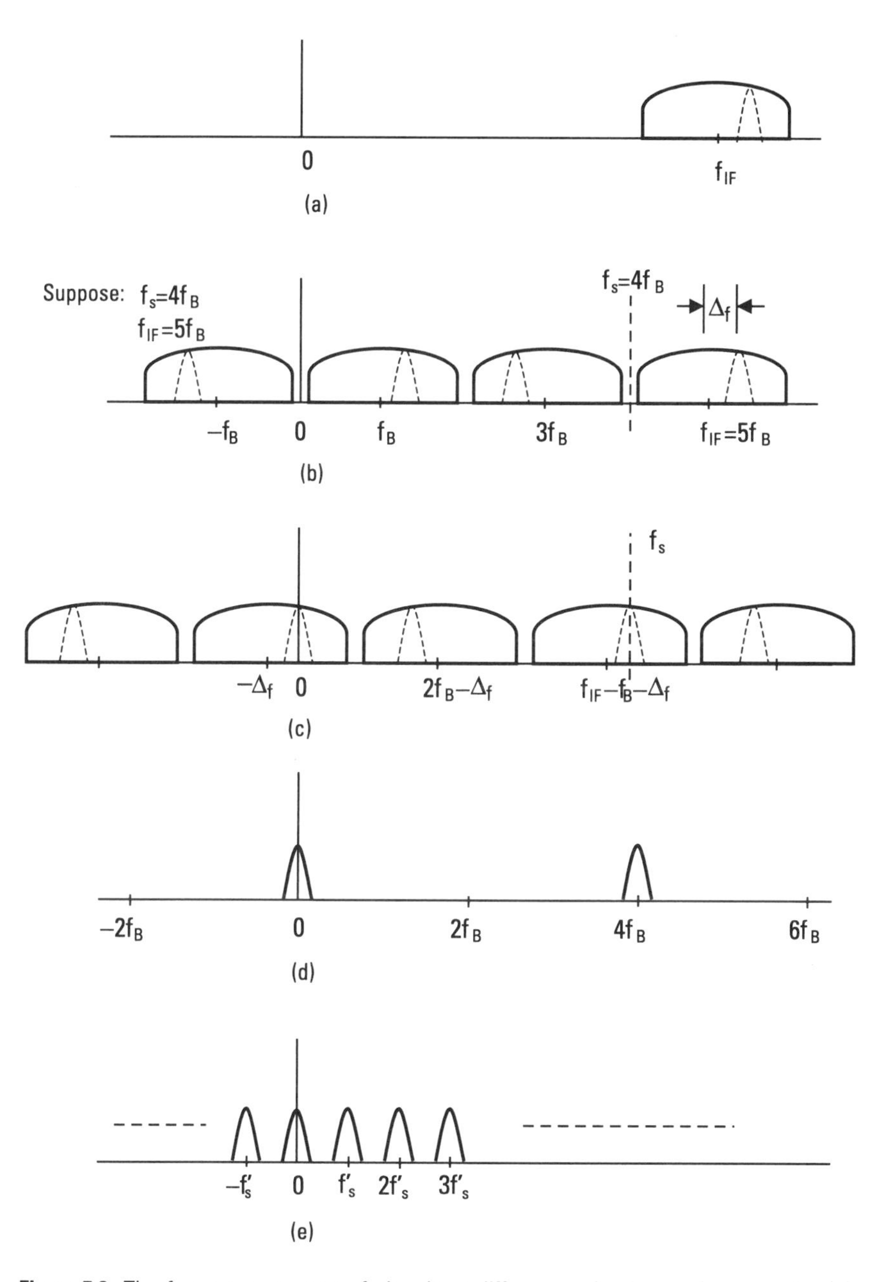

Figure 7.2 The frequency spectra of signals at different nodes in a software radio: (a) bandpass signal at IF, (b) spectrum after bandpass sampling, (c) spectrum after the desired channel is shifted to the baseband, (d) spectrum after lowpass filtering, and (e) spectrum after decimation.

$$\frac{2f_U}{n_{\mathrm{BPS}}} \leq f_s \leq \frac{2f_L}{n_{\mathrm{BPS}} - 1} \tag{7.1}$$

where f_s is the sampling rate, $\lfloor x \rfloor$ denotes the largest integer not exceeding x, and n_{BPS}, satisfying

$$1 \leq n_{\mathrm{BPS}} \leq \lfloor f_U / B \rfloor \tag{7.2}$$

is referred to as the wedge order. This condition can be derived graphically by considering the band position after sampling [33], or it can be derived mathematically [34, 35]. Equation (7.1) can be visualized in graphical form [30] as shown in Figure 7.3. The shaded regions correspond to forbidden sampling frequencies. The center frequency of the resultant signal after bandpass sampling with a sampling frequency f_s can be computed by [36]

$$f_r = \begin{cases} \mathrm{rem}(f_{\mathrm{IF}}, f_s) & \text{if } \left\lfloor \dfrac{f_{\mathrm{IF}}}{f_s/2} \right\rfloor \text{ is even} \\ f_s - \mathrm{rem}(f_{\mathrm{IF}}, f_s) & \text{if } \left\lfloor \dfrac{f_{\mathrm{IF}}}{f_s/2} \right\rfloor \text{ is odd} \end{cases} \tag{7.3}$$

where f_{IF} and f_r are the center frequencies of the signals before and after sampling, respectively, and $\mathrm{rem}(a, b)$ is the remainder when a is divided by b.

We do not recommend sampling the signal at the minimum possible sampling rate because the operating point can easily move into a disallowed region due to sampling and carrier frequency instability, and resulting in aliasing. A simple method to move a "dangerous" operating point to a "safer" one is to use a value of n_{BPS} smaller than the maximum value computed by (7.2). This approach corresponds to, referring to Figure 7.3, moving the operating point vertically from one allowable region to another one. For illustration, consider an example in which a 30-kHz bandwidth signal located within 5.02 to 5.05 MHz is bandpass sampled. The maximum value of n_{BPS} for this case is 168 according to (7.2) so that the allowable sampling frequency is from 60.11904 to 60.11976 kHz, that is, within an extremely narrow range of 0.72 Hz. If $n_{\mathrm{BPS}} = 167$ is used, the allowable sampling frequency is from 60.47904 to 60.481927 kHz, giving a comparatively larger allowable range of 2.88 Hz. Using a smaller value of n_{BPS} yields a wider range of sampling frequency but at the expense of requiring a higher sampling rate.

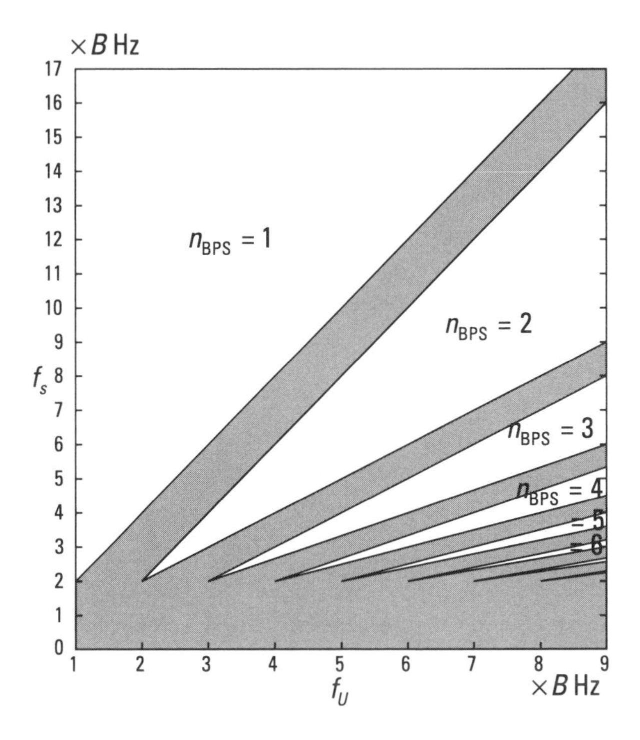

Figure 7.3 Allowable sampling rate f_S (nonshaded region) versus the highest frequency component of the signal, f_U, in bandpass sampling. Note that f_S and f_U are normalized to B, the signal bandwidth.

7.4.2 Extensions of Bandpass Sampling

Qi, Coakley, and Evans [37] have introduced a modified bandpass sampling theorem taking into consideration the sampling frequency uncertainty, Δ_s, and the carrier frequency uncertainty, Δ_c. The minimum sampling rate, $f_s^{(\min)}$, is given by [37]

$$f_s^{(\min)} = \frac{2(f_c + \Delta_c) + B}{I'(1 - p_s)} \tag{7.4}$$

where

$$I' = \left\lfloor \frac{(1 + p_s)[2(f_c + \Delta_c) + B]}{4(f_c p_s + \Delta_c) + 2B} \right\rfloor \tag{7.5}$$

In the last expression, $p_s = \Delta_s / f_s$.

In a multiband receiver, the digitization of a wideband signal having two or more individual signals located in distinct and nonadjacent frequency bands is required. Consider the case for which we desire to digitize a wideband signal composed of N_{SIG} signals in N_{SIG} distinct and nonadjacent frequency bands with bandwidths, B_i, $i = 1, 2, \ldots, N_{SIG}$. Akos et al. [38] have proposed a multiband digitization method wherein the sampling rate f_s is selected such that the following three constraints are satisfied:

$$0 < f_{r_i} - \frac{B_i}{2} \tag{7.6}$$

$$f_{r_i} + \frac{B_i}{2} < \frac{f_s}{2} \tag{7.7}$$

and

$$\left| f_{r_i} - f_{r_j} \right| \geq \frac{B_i + B_j}{2}, \qquad \text{for } i \neq j \tag{7.8}$$

where f_{r_i} and f_{r_j} are f_r values computed by (7.3), $i, j = 1, 2, \ldots, N_{SIG}$. Equations (7.6) and (7.7) ensure that each individual signal does not fold onto itself, that is, that aliasing does not occur. The condition given by (7.8) makes sure that signals of different bands do not overlap after digitization.

The bandpass sampling that we have discussed is a form of uniform sampling because samples are uniformly time spaced. For uniform sampling, note from (7.1) and Figure 7.3 that the minimum sampling rate $2B$ is valid only if f_U is an integer multiple of B and the largest value of n_{BPS} is used, that is, $n_{BPS} = \lfloor f_U/B \rfloor$. Kohlenberg [39] has shown that the second-order nonuniform bandpass sampling enables the use of the minimum sampling rate without the aforesaid restriction. Coulson [40] has generalized this result and developed the theoretical framework for the nth-order nonuniform bandpass sampling. Interested readers may refer to [40] for details.

7.5 Decimation Filtering

After the desired signal is shifted to the baseband by using a numerically controlled oscillator and multiplier(s), it is advantageous to reduce the sampling rate of the resultant signal sequence in order to alleviate processing requirements of subsequent digital circuits. Before this sampling-rate decimation is performed, however, the original signal has to be filtered and band-limited in order to prevent aliasing from occurring after decimation. Filtering also enables extraction of the desired signal from a number of signals in the presence of multiple access. It follows that the whole decimation process comprises both filtering and resampling. Decimation can be accomplished in a single step or through a number of stages. Multistage decimation has an advantage over single-stage decimation in that the computational requirement for filtering can be relaxed. Before elaborating on the details of multistage decimation, we provide a discussion of single-stage decimation and a description of the signal model.

Let $x(n)$ denote the signal to be decimated. We model $x(n)$ as a baseband band-limited signal sampled at a rate f_s, with a frequency spectrum confined to the range $|f| < f_s/2$ or $|\omega| < \pi$. Because $x(n)$ is normally composed of the desired signal and a number of unwanted other-user signals, and because the desired signal has been digitally shifted to the baseband by the numerically controlled oscillator before decimation, we are only interested in the signal components of $x(n)$ that reside within $|f| < f_p$ where f_p is small compared to $f_s/2$.

7.5.1 Single-Stage Decimation

The block diagram of a single-step decimator is shown in Figure 7.4(a). We want to downsample $x(n)$ by a factor $M = f_s/f_s'$, which is the ratio of the original sampling frequency f_s to the new, lower sampling frequency f_s'. To

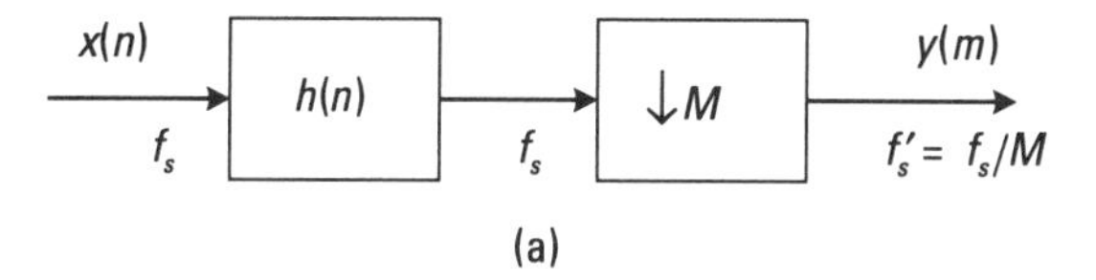

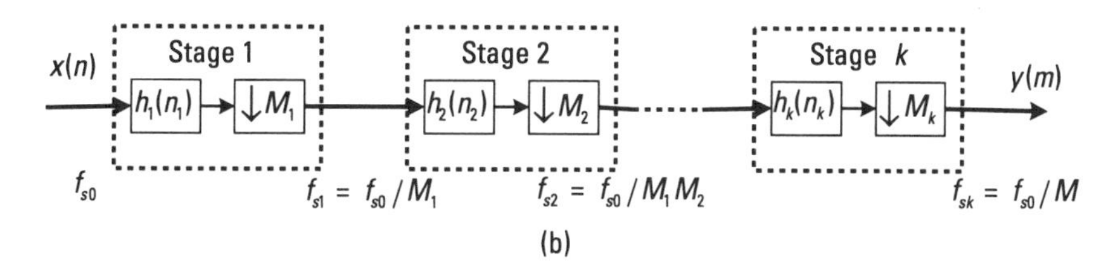

Figure 7.4 Block diagrams of (a) a single-stage decimator and (b) a multistage decimator with k stages.

avoid aliasing after sampling-rate compression, we are required to filter $x(n)$ with a digital lowpass filter. Let $h(n)$ be the impulse response of this filter. Since the sampled signal after decimation has a periodic frequency spectrum with a period f_s' (i.e., f_s/M), and since the desired signal is confined to $\pm f_p$, it follows that the frequency response design specification of the lowpass filter, $\left|H(e^{j2\pi f/f_s})\right|$, can be given by

$$\left|H(e^{j2\pi f/f_s})\right| = \begin{cases} 1 \pm \delta_p & |f| < f_p \\ \text{transition band} & f_p < |f| < f_s/M - f_p \\ \pm \delta_s & f_s/M - f_p < |f| < f_s/2 \end{cases} \tag{7.9}$$

where δ_p is the maximum passband ripple and δ_s is the maximum stopband ripple.

We know that the computational requirement in terms of *multiplications per second* (MPS) for a finite-impulse-response filter is given by [41]

$$R_f = N_f f_s/M \tag{7.10}$$

where N_f is the filter order. Based on the Parks-MacClellan finite-impulse-response design algorithm, N_f can be estimated by [41]

$$N_f = \frac{-10\log_{10}\delta_p\delta_s - 13}{2.324 \times 2\pi \times \Delta f/f_s} \tag{7.11}$$

where Δf is the width of the transition band. Equations (7.10) and (7.11) reveal that if the original sampling rate is very high, the computational requirement is very high too.

We consider an example in which a signal having a 200-kHz bandwidth (e.g., a GSM signal) is required to be extracted from a wideband signal sampled at an original sampling rate f_s = 21 MHz. The Nyquist sampling rate for this lowpass signal (f_p = 100 kHz) is 200 kHz. For convenience, we set f_s' = 210 kHz. Hence, the decimation ratio is given by M = 21 MHz/210 kHz = 100. The design specification of the lowpass filter is shown in Figure 7.5(a) with a transition band Δf = 10 kHz, δ_p = 0.01, and

δ_s = 0.001. Substituting all relevant data into (7.10) and (7.11) yields N_f = 5,320 and $R_f = 1.117 \times 10^9$ MPS, both requirements being too demanding to be practical. To overcome this problem, we can use multistage decimation.

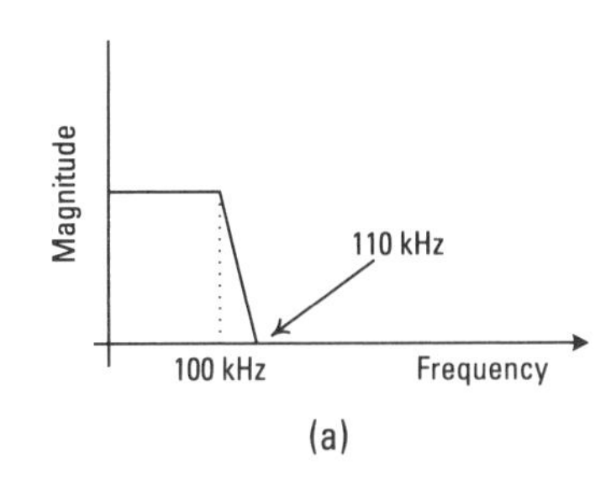

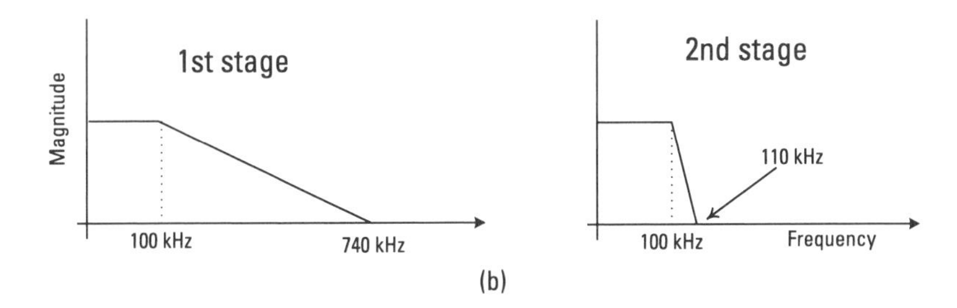

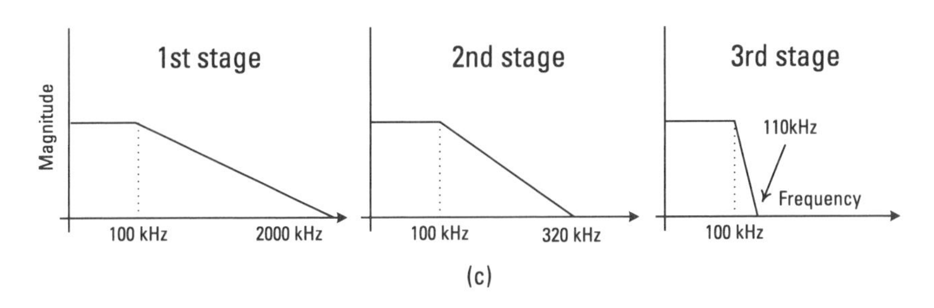

Figure 7.5 Filter specifications for (a) one-stage decimation filter (M = 100; δ_p = 0.001; δ_s = 0.001); (b) two-stage decimation filter (M_1 = 25; M_2 = 4; δ_p = 0.005; δ_s = 0.001); and (c) three-stage decimation filter (M_1 = 10; M_2 = 5; M_3 = 2; δ_p = 0.0033; δ_s = 0.001).

7.5.2 Multistage Decimation

Figure 7.4(b) depicts a block diagram of a multistage decimator. In decimation consisting of k stages, the decimation ratio M is first factored into

$$M = \prod_{i=1}^{k} M_i \tag{7.12}$$

where M_i is the decimation ratio implemented at the ith stage. Let the sampling frequency before and after the ith stage decimation be $f_{s(i-1)}$ and f_{si}, respectively, related by $f_{si} = f_{s(i-1)}/M_i$. The frequency response design specification for the ith-stage anti-aliasing lowpass filter is given by

$$\left|H_i(e^{j2\pi f/f_{s(i-1)}})\right| = \begin{cases} 1 \pm \delta_p & |f| < f_p \\ \text{transition band} & f_p < |f| < f_{si} - f_p \\ \pm \delta_s & f_{si} - f_p < |f| < f_{s(i-1)}/2 \end{cases} \tag{7.13}$$

Note that (7.9) is a special case of (7.13).

For illustration, we consider again the example given in Section 7.5.1 but with two- and three-stage implementations. Because $M = 100$, we assume that $(M_1, M_2) = (25, 4)$ and $(M_1, M_2, M_3) = (10, 5, 2)$ for the two- and three-stage implementations, respectively. Design specifications of the lowpass filters for the two- and the three-stage implementations are depicted in Figures 7.5(b) and 7.5(c), respectively. In the two-stage implementation, the sampling rate at the output of the first stage is 21 MHz/25 = 840 kHz. With a desired signal of a 200-kHz bandwidth, we need to ensure that aliasing does not occur within the frequency range $840n \pm 100$ kHz after decimation, where n is an integer. It follows that the transition band in the first stage can be set from 100 to 740 kHz, consistent with (7.13). We do not care if aliasing occurs in the transition band because it can be removed in the second-stage filtering. The passband ripple specification of the two-stage structure is tightened to be one-half that of the one-stage case since each stage introduces passband ripple. The stopband ripple specification is not changed since the cascade of two lowpass filters reduces the overall stopband ripple. It follows that $\delta_p = 0.005$ and $\delta_s = 0.001$ for both stages. The same approach applies to the three-stage case. Table 7.1 lists the computational requirements for the two- and three-stage implementations together with those for the single-stage implementation for comparison. It is apparent that the complexity can be reduced by 18 times for the three-stage case when compared to the single-stage case.

The reason that multistage decimation provides a reduction in the computational requirement can be deduced from (7.10) and (7.11). In the first few stages of multistage decimation, values of f_s are large but the transition bands are large as well, leading to relatively small computational requirements R_f. In latter stages, the transition bands are small but the sampling rates have already been reduced to small values, again leading to

Table 7.1
Comparison of Computational Requirements for Decimators of Different Numbers of Stages

	Single-Stage	Two-Stage	Three-Stage
First stage	$N_f = 5{,}320$ $R_f = 1.117$ GHz MPS	$N_{f1} = 90$ $R_{f1} = 75.6$M MPS	$N_{f1} = 32$ $R_{f1} = 67.2$M MPS
Second stage	—	$N_{f2} = 230$ $R_{f2} = 48.3$M MPS	$N_{f2} = 28$ $R_{f2} = 11.76$M MPS
Third stage	—	—	$N_{f3} = 120$ $R_{f3} = 25.2$M MPS
Total	1.117 GHz MPS	123.9M MPS	104.16M MPS

small values of R_f. Crochiere and Rabiner have proposed techniques to estimate the optimum number of stages and the decimation ratio in each stage based on computational consideration [42] and storage consideration [43].

In the preceding discussion, we have considered conventional finite-impulse-response filters in the realization of lowpass filters. It is possible to utilize efficient filter structures and some special properties of decimation filtering in order to further reduce the computational requirement, as discussed next.

7.5.3 Reduced-Complexity Implementation Based on the Polyphase Technique

Consider a cascade of a finite-impulse-response filter having a transfer function $H(z) = \sum_{n=-\infty}^{\infty} h(n)z^{-n}$ and a downsampler with a decimation ratio M. Based on the polyphase technique, one can express $H(z)$ as

$$\begin{aligned} H(z) &= \sum_{n=-\infty}^{\infty} h(nM)z^{-nM} + z^{-1}\sum_{n=-\infty}^{\infty} h(nM+1)z^{-nM} + \dots \\ &\quad + z^{-(M-1)}\sum_{n=-\infty}^{\infty} h(nM+M-1)z^{-nM} \qquad (7.14) \\ &= \sum_{\ell=0}^{M-1} z^{-\ell}E_\ell(z^M) \end{aligned}$$

where

$$E_\ell(z) = \sum_{n=-\infty}^{\infty} h(nM + \ell)z^{-n} \qquad (7.15)$$

Therefore, the cascade of the filter and the downsampler can be realized as depicted in Figure 7.6(a). After applying the noble identity [44], the downsampler can be moved to the left and the resultant structure is given

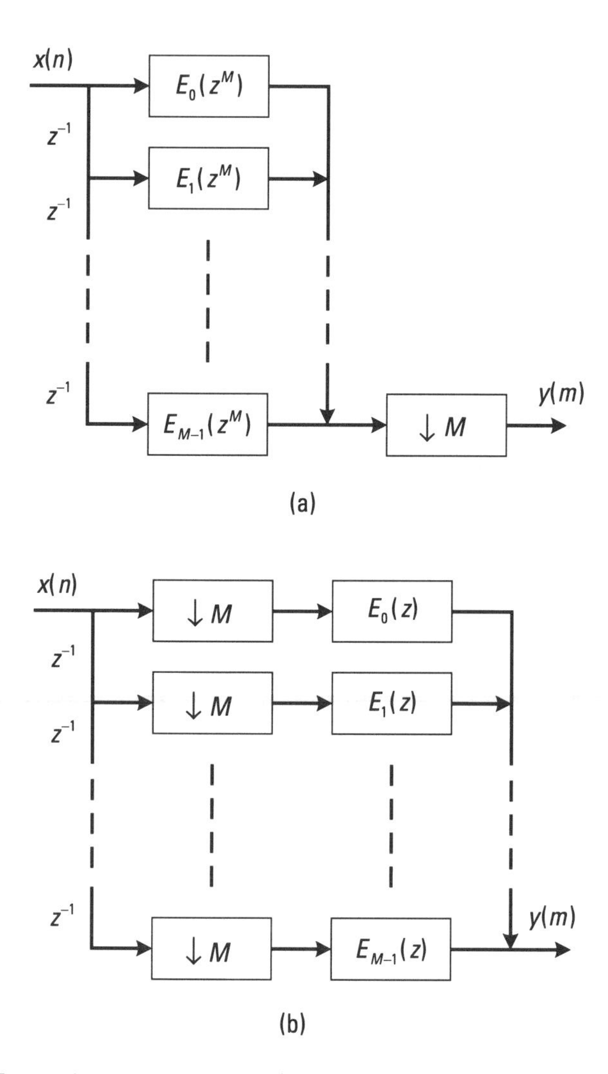

Figure 7.6 Two equivalent realizations of (a) a cascade of a finite-impulse-response filter and (b) a downsampler by the polyphase technique.

in Figure 7.6(b). The main advantage of filter implementation using the structure of Figure 7.6(b) is that all computations can be performed at a rate equal to the output sampling rate rather than the rate of the incoming signal. This gives rise to an approximately M-fold reduction in the number of multiplications and additions.

7.5.4 Reduced-Complexity Implementation by Half-Band Filters

A half-band filter is a special case of finite-impulse-response lowpass filters with frequency response symmetry about $\omega = \pi/2$. Its frequency response, $H(e^{j\omega})$, has the property that

$$H(e^{j\omega}) = 1 - H(e^{j(\pi - \omega)}) \tag{7.16}$$

It can be shown that nearly half of the filter coefficients are zero [45], leading to a reduction in implementation complexity when compared to conventional finite-impulse-response filters. Because of this advantage, half-band filters are attractive for use in decimation by a factor of 2. Multistage half-band filters can be used to achieve a decimation factor 2^p, where p is a positive integer. Notice that half-band filters can also be implemented using the polyphase technique described in Section 7.5.3, resulting in a further reduction of implementation complexity.

7.5.5 Filter Implementation Using Multiple Bandstop Filters

Because we are only interested in preventing aliasing to the desired signal, it is possible to allow aliasing at frequencies that the desired signal does not move into after a change of the sampling rate. In this regard, one may introduce "don't care" bands in the design of filters. Multiple bandstop filters can then be used instead of lowpass filters. With multiple bandstop filters, the filter order can be reduced and so can the computational requirement. The saving is especially large when a "don't care" band is a significant portion of the total frequency band. This situation is commonly encountered in the early stage of decimation.

For a multiple bandstop filter, the frequency-response design specification can be stated as [43]

$$\left|H_i(e^{j2\pi f/f_{s(i-1)}})\right| = \begin{cases} 1 \pm \delta_p & |f| < f_p \\ \pm \delta_s & nf_{si} - f_p < |f| < nf_{si} + f_p,\ n \text{ a positive integer} \\ \text{any value} & \text{otherwise} \end{cases} \tag{7.17}$$

which is also plotted in Figure 7.7. The reduction in filter order by using multiple stopband filters instead of lowpass filters is illustrated by the following example. Consider the first filter of the three-stage decimator described in Section 7.5.2. It is required to implement a lowpass filter with a filter order of 32 (see Table 7.1). The resultant magnitude plot is shown in Figure 7.8(a). For a multiple stopband filter, MATLAB simulation results indicate that a filter of order only 27 is required to provide adequate attenuation within the regions of interest. This leads to a 15% saving in the filter order and hence the complexity. Figure 7.8(b) plots the frequency response of the magnitude of the resultant multiple stopband filter for comparison.

7.5.6 Cascaded Integration-and-Comb Filters

The *cascaded integration-and-comb* (CIC) filter, a special case of multiple stopband filters, was first proposed by Hogenauer [46]. It comprises two building blocks, namely, the integrator section and the comb section, separated by a decimator. A block diagram of a CIC filter is shown in Figure 7.9.

The integrator section consists of N_{CIC} digital integrators, each having a transfer function $H_I(z)$ given by

$$H_I(z) = \frac{1}{1 - z^{-1}} \tag{7.18}$$

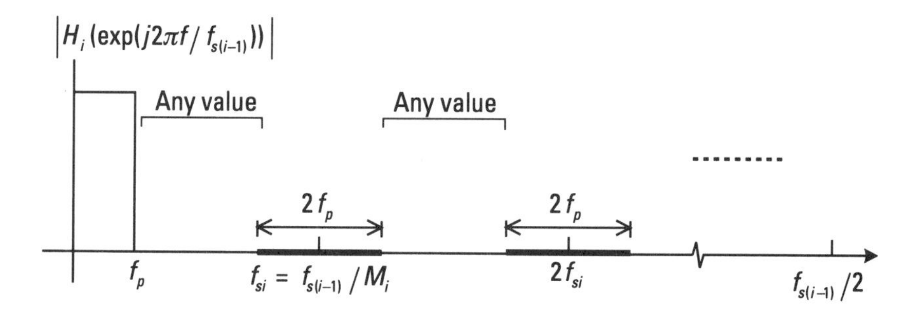

Figure 7.7 Magnitude plot of a multiple bandstop filter for decimation at the *i*th stage.

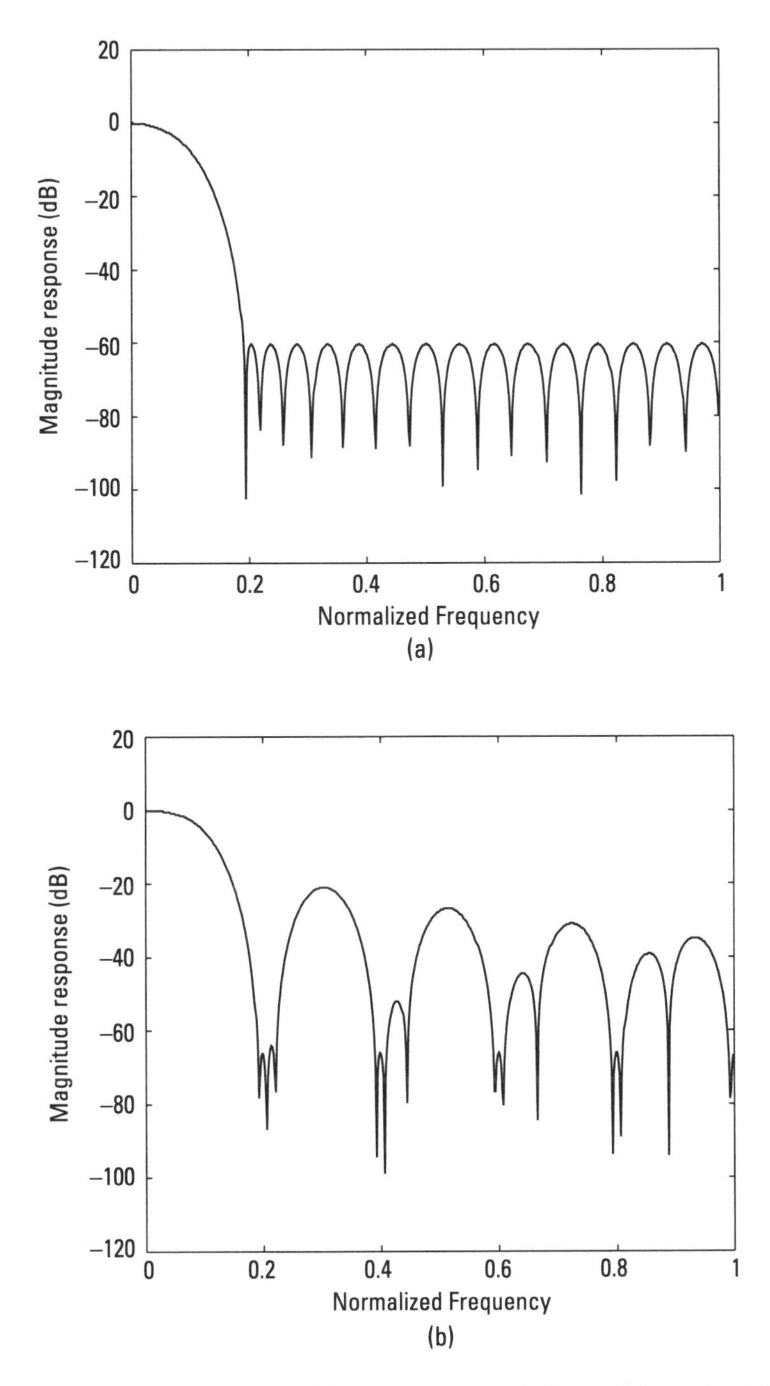

Figure 7.8 Frequency responses of (a) a lowpass filter and (b) a multiple stopband filter.

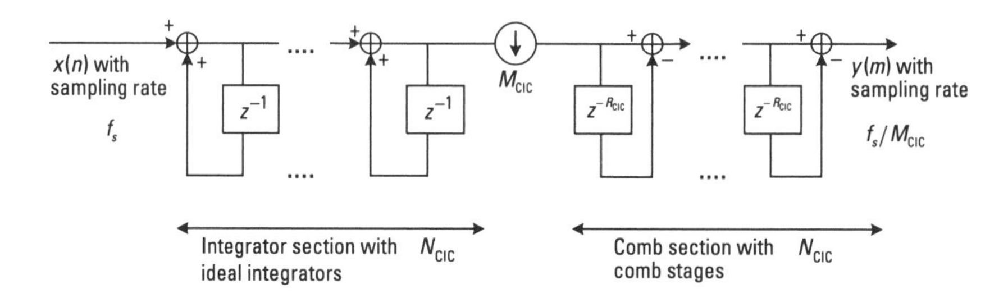

Figure 7.9 Block diagram of a CIC decimator with N_{CIC} cascaded sections.

The comb section operates at a sampling rate decimated by a factor of M_{CIC}, so that the delay element $z^{-R_{CIC}}$ corresponds to a delay of $R_{CIC}M_{CIC}$ input samples. The parameter R_{CIC} is referred to as the differential delay that is used to control the filter's frequency response. The transfer function of a single comb stage, $H_C(z)$, is given by

$$H_C(z) = 1 - z^{-R_{CIC}M_{CIC}} \tag{7.19}$$

For a CIC filter with N_{CIC} cascaded sections, the overall system function is given by

$$H(z) = H_I^{N_{CIC}}(z)H_C^{N_{CIC}}(z) = \frac{(1 - z^{-R_{CIC}M_{CIC}})^{N_{CIC}}}{(1 - z^{-1})^{N_{CIC}}} \tag{7.20}$$

The frequency response is obtained by putting $z = e^{j2\pi f}$ into (7.20). It follows that

$$\left|H(e^{j2\pi f})\right| = \left|\frac{\sin(\pi f R_{CIC}M_{CIC})}{\sin(\pi f)}\right|^{N_{CIC}} \tag{7.21}$$

For illustration, magnitude plots of CIC filters with $f_s = 21$ MHz, $M_{CIC} = 10$, and $R_{CIC} = 1$ are plotted in Figures 7.10(a) and 7.10(b) for $N_{CIC} = 4$ and $N_{CIC} = 6$, respectively.

Three parameters determine the frequency response of a CIC filter, namely, the decimation ratio M_{CIC}, the differential delay R_{CIC}, and the order N_{CIC}. From (7.21), it is apparent that the nulls occur at frequencies that are multiples of $f = (R_{CIC}M_{CIC})^{-1}$. Since M_{CIC} is the sampling-rate compression ratio and is normally fixed, one can adjust R_{CIC} to control the

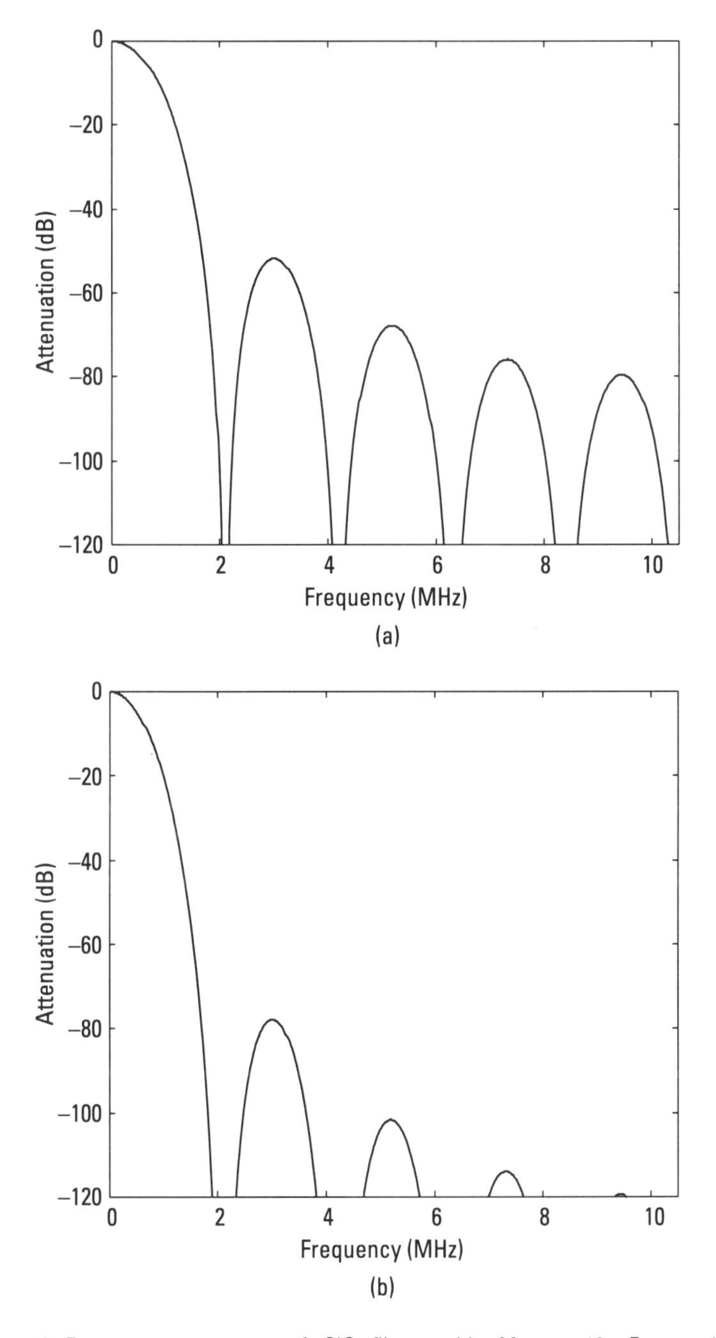

Figure 7.10 Frequency responses of CIC filters with M_{CIC} = 10, R_{CIC} = 1, and (a) N_{CIC} = 4; and (b) N_{CIC} = 6.

null positions. It follows that N_{CIC} determines the width of the allowable aliasing bands at the null positions. A higher value of N_{CIC} yields a wider stopband, as is readily demonstrated from Figure 7.10.

The reason that CIC filters are widely used for decimation filter applications can be attributed to their property that they do not require multipliers and no storage is required for storing filter coefficients. Passband droop, one of the problems of CIC filters, has been addressed by Oh et al. [45] and Kwentus, Jiang, and Wilson [47]. However, if the bandwidth of the desired signal is kept small enough relative to the output sampling rate and if a small differential delay (R_{CIC} = 1 or 2) is used, the passband droop problem is not significant [46].

7.5.7 Choice of Filtering Methods

In practice, a combination of the aforementioned filtering methods is used to implement a multistage decimator. Consider again the example of a three-stage decimation filter given in Section 7.5.2. The first stage can be implemented by a CIC filter since it is multiplierless and can operate at a very high data rate. For the second stage, one can use a simple lowpass filter. Polyphase implementation of the filter provides a reduction of ≈ 5 in the number of MPS. A half-band filter is attractive for the implementation of the third stage because the decimation ratio is 2. For the half-band filter used in this stage, a filter of length 180 is required to satisfy the filter specifications. The computational requirement can be estimated by (7.10) taking into account the facts that (1) because almost half of filter coefficients are zero, the number of multiplications needed is reduced by a factor of ≈ 2; and (2) the polyphase implementation of the half-band filter gives another reduction by a factor of ≈ 2. Hence, the computational requirement for this half-band filter is 9.45M MPS. The total computational requirement for the overall decimation filter is only 0M (as the CIC filter is multiplierless) + 2.35M (data from Table 7.1 plus a reduction by a factor of ≈ 5) + 9.45M = 11.8M MPS. Comparison with the figures of Table 7.1 indicates that this realization of the decimation filter gives an 89% reduction in the multiplication complexity with respect to the case of a three-stage finite-impulse-response lowpass filter, or it yields a multiplication-complexity reduction by a factor of 95 with respect to a single-stage finite-impulse-response lowpass filter.

7.6 Filter-Bank Channelizers

So far, we have considered extraction of only a single signal of interest from a wideband digitized signal that consists of many different signals. This situation arises, for example, in the receivers of MSs. In BSs, a large number of signals arrived from different channels are required to be extracted simultaneously. A direct method is to use multiple discrete channelizers. Consider the case in which we want to extract K channels as shown in Figure 7.11(a). The discrete channelizer implementation is shown in Figure 7.11(b). The incoming signal $x(n)$ is first filtered by a bank of bandpass filters centered at appropriate frequencies. For the kth branch, $0 \le k \le K-1$, let $H_k(z)$ be the transfer function of the kth filter and $x_k(n)$ be the filter output. Let $H_0(z)$ be the transfer function of the causal, prototype lowpass filter having

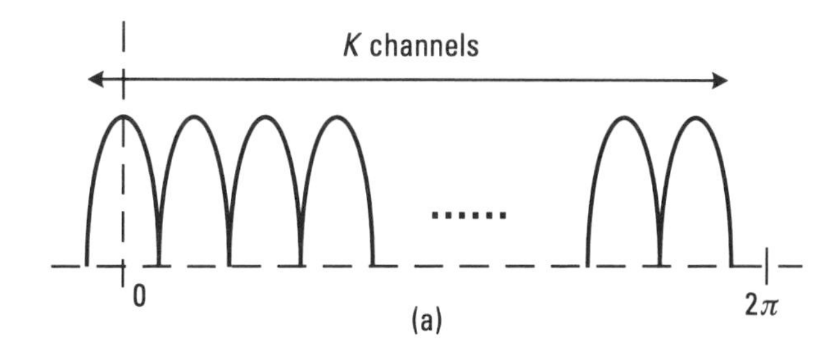

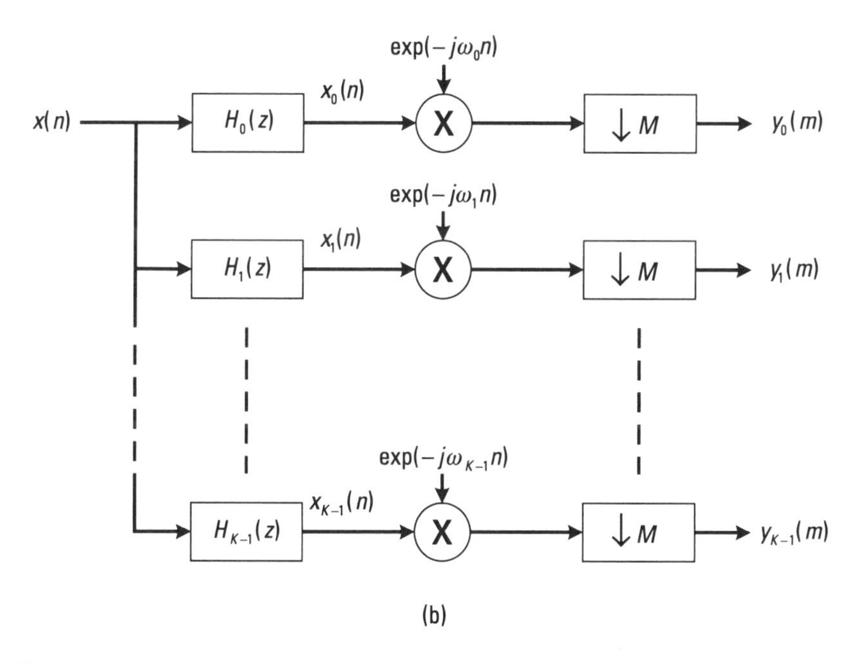

Figure 7.11 Extraction of K signals from a wideband signal: (a) channel arrangement; (b) direct realization by K discrete channelizers; (c) polyphase implementation; and (d) low-complexity polyphase implementation.

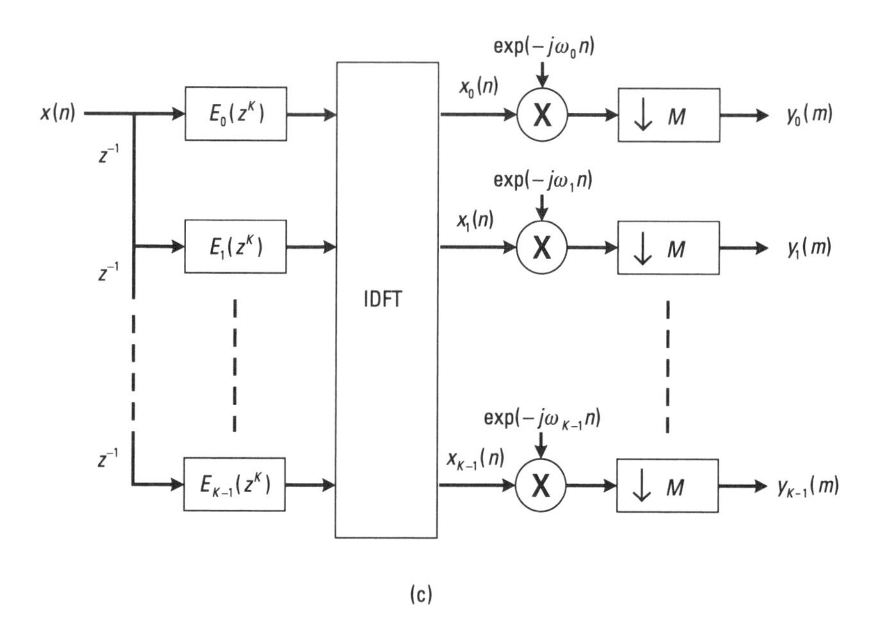

(c)

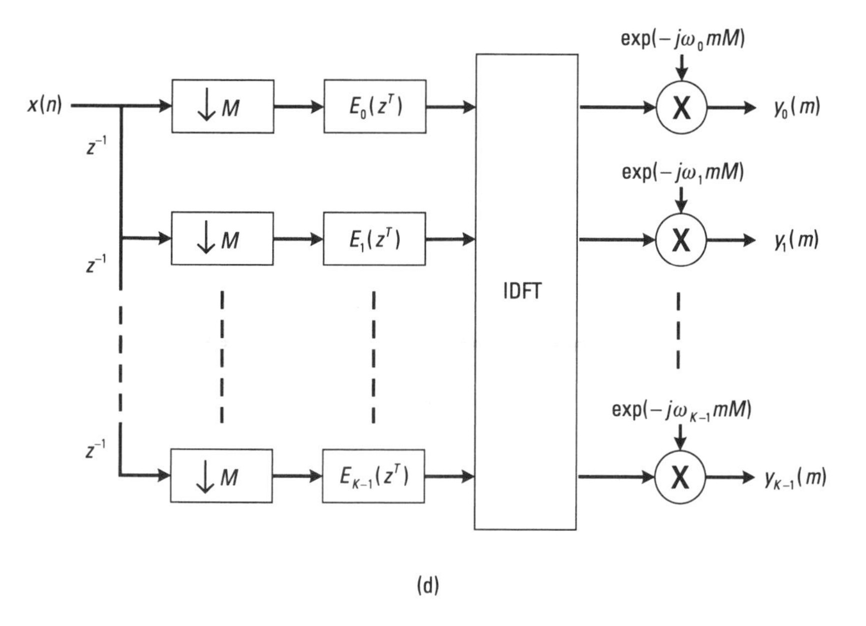

(d)

Figure 7.11 (Continued.)

an impulse response $h_0(n)$. It follows that $H_k(z)$ is obtained by modulating $h_0(n)$ with $e^{j\omega_k n}$ where $\omega_k = 2\pi k/K$. The filtered signal $x_k(n)$ is shifted to the baseband by multiplying with a complex exponential followed by decimating the resultant signal by a factor of M to yield the wanted output $y_k(m)$. This approach of channelization results in an implementation complexity that increases linearly with the number of channels. A more efficient method for performing signal extraction for multiple channels is based on the discrete Fourier transform filter-bank channelizer discussed by Zangi and Koilpillai [48].

Using discrete Fourier transform filter-bank channelizers requires that K and M satisfy

$$K = M \times T \tag{7.22}$$

for some integer T. Then $H_k(z)$ can be expressed as

$$\begin{aligned} H_k(z) &= \sum_{n=-\infty}^{\infty} h_0(n) e^{j\omega_k n} z^{-n} \\ &= \sum_{n=-\infty}^{\infty} h_0(nK) e^{j2\pi knK/K} z^{-nK} \\ &\quad + z^{-1} \sum_{n=-\infty}^{\infty} h_0(nK+1) e^{j2\pi k(nK+1)/K} z^{-nK} + \ldots \\ &\quad + z^{-(K-1)} \sum_{n=-\infty}^{\infty} h_0(nK+K-1) e^{j2\pi k(nK+K-1)/K} z^{-nK} \\ &= \sum_{\ell=0}^{K-1} z^{-\ell} E_\ell(z^K) e^{j2\pi k\ell/K} \end{aligned} \tag{7.23}$$

where $E_\ell(z)$ has the same expression of (7.15). Let $W = e^{-j2\pi/K}$. It follows that

$$H_k(z) = \sum_{\ell=0}^{K-1} z^{-\ell} E_\ell(z^K) W^{-k\ell} \tag{7.24}$$

Since $X_k(z) = H_k(z)X(z)$ where $X_k(z)$ and $X(z)$ are the z transforms of $x_k(n)$ and $x(n)$, respectively, we have that

$$X_k(z) = \sum_{\ell=0}^{K-1} [X(z) \times z^{-\ell} E_\ell(z^K)] W^{-k\ell} \tag{7.25}$$

We can readily tell that $X_k(z)$ is the kth inverse discrete Fourier transform output on $X(z) \times z^{-\ell} E_\ell(z^K)$, $\ell = 0, 1, \ldots, K-1$. Therefore, the discrete bandpass filter bank can be replaced by a cascade of a polyphase filter and an inverse discrete Fourier transform operation, as depicted in Figure 7.11(c).

Alternatively, one can move the downsamplers in Figure 7.11(c) to the left by applying the noble identity [44]. The resultant structure is shown in Figure 7.11(d). Note that K channels can be extracted simultaneously by using one lowpass prototype filter followed by a K-point inverse discrete Fourier transform. Because efficient algorithms for computing inverse discrete Fourier transform are widely available, the implementation cost for inverse discrete Fourier transform is low. Furthermore, the discrete Fourier transform filter-bank channelizer can be operated at the lowest possible frequency, thereby leading to savings in computation and power consumption. Reference [48] reports that when three or more channels are to be extracted, the discrete Fourier transform filter-bank channelizer is more efficient than using multiple discrete channelizers.

7.7 Fractional Sampling-Rate Conversion

Because the sampling rate of the analog-to-digital converter is normally not an integer multiple of the data transmission rate of the system, a fractional sampling-rate converter is often used as the last step of downsampling in order to generate a signal sequence that rate synchronizes[3] with the data transmission rate. Consider the case for which the ratio of the output sampling rate and the input rate of a fractional sampling-rate converter is L/M wherein L and M are relatively prime with $L < M$. Figure 7.12 depicts a direct implementation of a fractional sampling-rate converter, which is a cascade of an interpolator and a decimator. This implementation is very inefficient since the filter $h(k)$ has to be operated at the highest rate within the system. More efficient implementation can be obtained by applying the polyphase technique described in Section 7.5.3 to the decimator or the interpolator. The resultant structures are depicted in Figure 7.13, wherein $E_\ell(z)$ shares the same expression of (7.15) and $R_\ell(z) = \sum_{n=-\infty}^{\infty} h[nL + (L-1-\ell)]z^{-n}$.

3. This means that the rate of signal samples is a multiple of the data transmission rate.

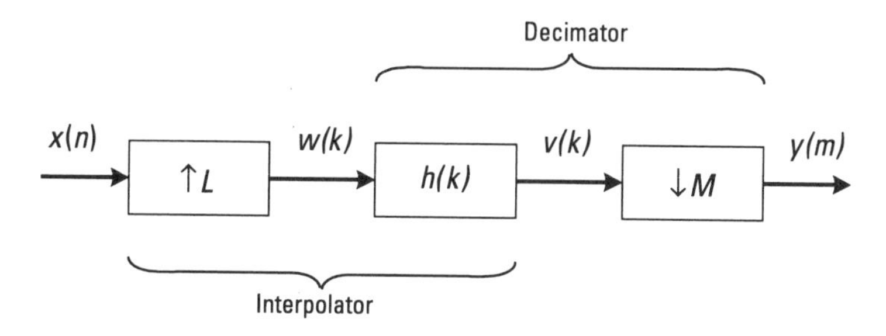

Figure 7.12 Direction implementation of an L/M sampling-rate converter.

Because the two filters only operate at either the input or output sampling rate, which is much lower, they are more efficient than the direct implementation. These two structures are not the most efficient, however. Efficiency can be further improved by moving the sampling-rate expander and downsampler to the rightmost and leftmost positions, respectively, in order that the filters can be operated at the lowest possible sampling rate.

Hsiao [49] has developed a technique that accomplishes this purpose. To illustrate the application of this technique, we consider an example wherein $L = 2$ and $M = 3$. Based on the configuration of Figure 7.13(b), we draw the fractional sampling-rate converter in Figure 7.14(a). Since $z^{-1} = z^{-3}z^{2}$, the converter of Figure 7.14(a) can be equivalently implemented by the one shown in Figure 7.14(b). With the help of the noble identity [44], it is further modified into the one depicted in Figure 7.14(c). Because 2 and 3 are relatively prime, interchanging the downsampler and the expander is possible, yielding Figure 7.14(d). Finally, $R_0(z)$ and $R_1(z)$ are decomposed into their polyphase components as $R_0(z) = R_{00}(z^3) + z^{-1}R_{01}(z^3) + z^{-2}R_{02}(z^3)$ and $R_1(z) = R_{10}(z^3) + z^{-1}R_{11}(z^3) + z^{-2}R_{12}(z^3)$. The resultant structure of the fractional sampling-rate converter is shown in Figure 7.14(e). It is operated at the lowest possible sampling rate and is hence very efficient. In the preceding discussion, we have assumed that the factor of sampling-rate conversion can be expressed as a rational number. For a more general condition that this factor is real, an arbitrary sampling-rate changer is needed. Arbitrary sampling-rate converters using time-varying CIC filters and Farrow structure have been proposed by Henker, Hentschel, and Fettweis [50], and Lundheim and Ramstad [51], respectively.

7.8 Concluding Remarks

In this chapter, we have reviewed the basics of software radio, which is a promising and prospective technology for future broadband wireless systems.

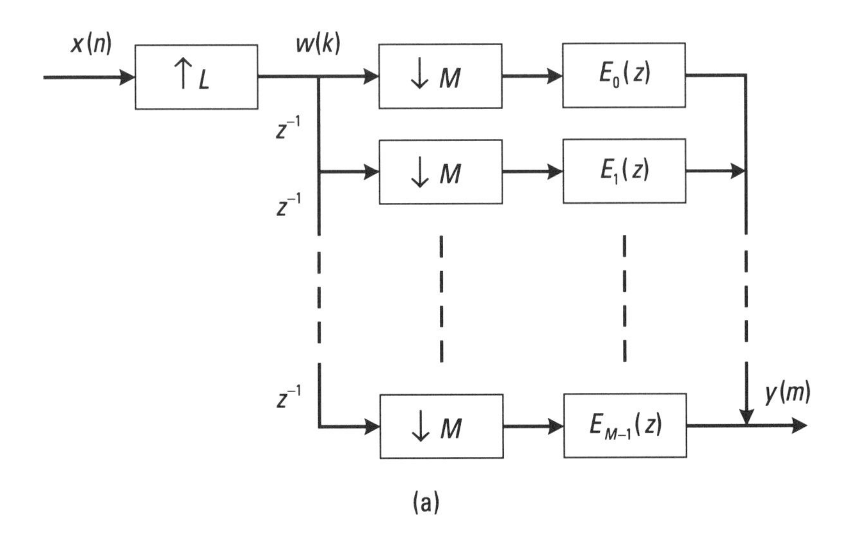

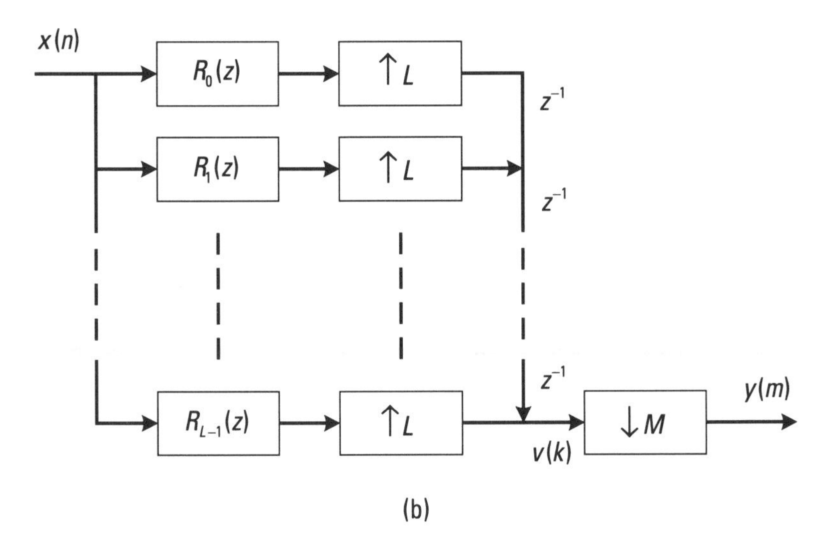

Figure 7.13 Two polyphase implementations of an L/M sampling-rate converter: (a) polyphase technique applied to decimator; and (b) polyphase technique applied to interpolator.

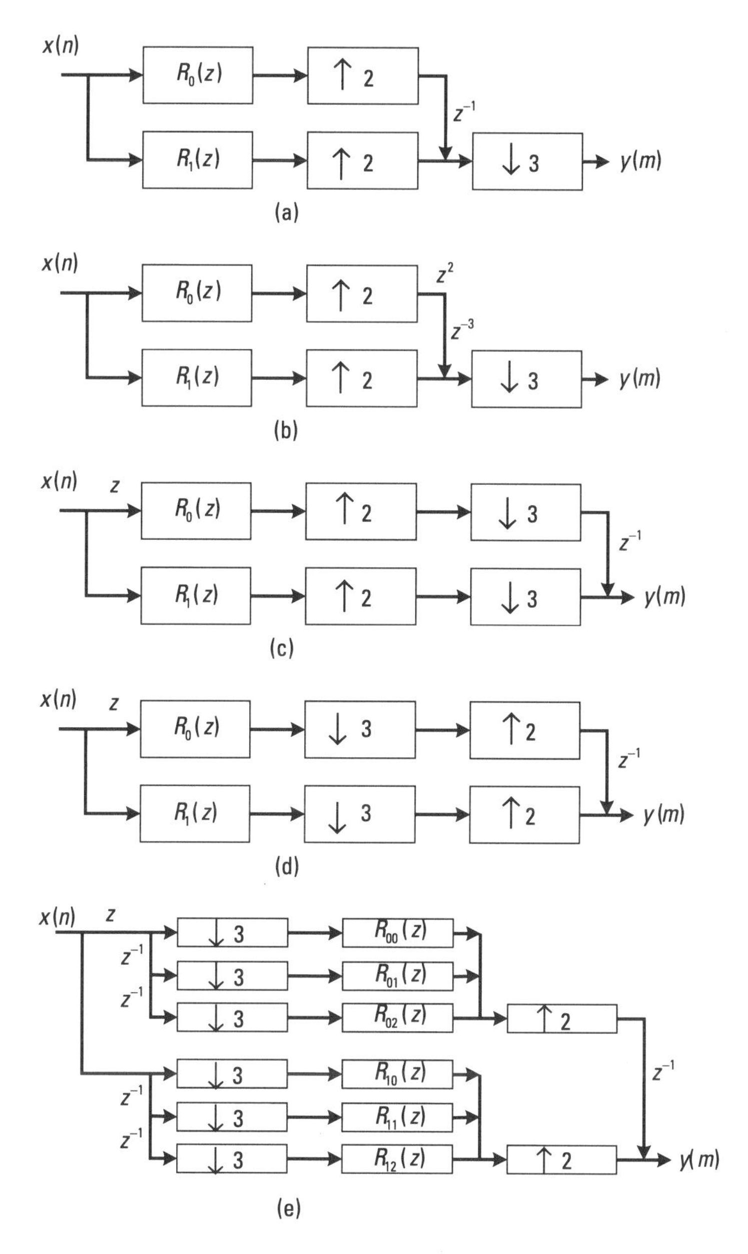

Figure 7.14 Development of an efficient structure of a fractional sampling-rate converter: (a) polyphase technique applied to the interpolator; (b) decomposition of z^{-1} into $z^2 z^{-3}$; (c) noble identity applied to z^2 and z^{-3}; (d) interchanging the downsampler and the expander; and (e) $R_0(z)$ and $R_1(z)$ decomposed into their polyphase components.

In particular, we have covered the theory of bandpass sampling and considered its implementation. We have also discussed the methods of extracting one channel by decimation filtering and multiple channels by discrete Fourier transform filter-bank channelizers, and using fractional sampling rate conversions. The topics presented in this chapter represent only a small part of software radio. We hope that this chapter stimulates more research in this topic.

References

[1] Ohmori, S., Y. Yamao, and N. Nakajima, "The Future Generations of Mobile Communications Based on Broadband Access Technologies," *IEEE Communications Magazine,* Vol. 38, Dec. 2000, pp. 134–142.

[2] Hata, M., "Fourth Generation Mobile Communication Systems Beyond IMT-2000," *Proc. Fifth Asia-Pacific Conference on Communications and Fourth Optoelectronics and Communications Conference (APCC/OECC'99),* Oct. 18–22, 1999, pp. 765–767.

[3] Pereira, J. M., "Fourth Generation: Now, It Is Personal!" *Proc. IEEE PIMRC 2000,* Sept. 18–21, 2000, pp. 1009–1016.

[4] Evans, B. G., and K. Baughan, "Visions of 4G," *IEE Electronics & Communication Engineering J.,* Vol. 12, Dec. 2000, pp. 293–303.

[5] Chuang, J., and N. Sollenberger, "Beyond 3G: Wideband Wireless Data Access Based on OFDM and Dynamic Packet Assignment," *IEEE Communications Magazine,* Vol. 38, July 2000, pp. 78–87.

[6] Lackey, R. J., and D. W. Upmal, "SPEAKeasy: The Military Software Radio," *IEEE Communications Magazine,* Vol. 33, May 1995, pp. 56–61.

[7] Mitola, J., "Software Radios—Survey, Critical Evaluation and Future Directions," *National Telesystems Conf. (NTC-92),* May 19–20, 1992, pp. 13/15–13/23; also in *IEEE Aerospace and Electronics Systems Mag.,* Vol. 8, April 1993, pp. 25–36.

[8] *IEEE Communications Magazine,* Vol. 33, May 1995.

[9] Available at http://www.era.co.uk/first/first.htm.

[10] *Proc. First Workshop on Software Radio,* Brussels: European Commission, May 1997.

[11] Available at http://www.uk.infowin.org/ACTS/RUS/PROJECTS/ac347.htm.

[12] Available at http://www.ifn.et.tu-dresden.de/~sort/.

[13] Feature Topic on the Globalization of the Software Radio, *IEEE Communications Magazine,* Vol. 37, Feb. 1999.

[14] Special Issue on Software Radios, *IEEE J. Selected Areas of Communications,* Vol. 17, April 1999.

[15] *IEEE Personal Communications,* Vol. 6, Aug. 1999.

[16] Special Issue on Software Radios, *IEICE Trans. on Communications,* Vol. E83-B, June 2000.

[17] Li, W., and Y. Yao, "Software Radio: Technology & Implementation," *Proc. Int. Conf. Communications Technology,* Oct. 22–24, 1998, p. 5.

[18] Pereira, J. M., "Beyond Software Radio, Towards Re-Configurability Across the Whole System and Across Networks," *Proc. IEEE VTC'99–Fall,* Sept. 19–22, 1999, pp. 2815–2818.

[19] Mitola, J., "The Software Radio Architecture," *IEEE Communications Magazine,* Vol. 33, May 1995, pp. 26–38.

[20] Kenington, P. B., "Emerging Technologies for Software Radio," *IEE Electronics and Communication Engineering J.,* Vol. 11, April 1999, pp. 69–83.

[21] Tuttlebee, W. H. W., "Software-Defined Radio: Facets of a Developing Technology," *IEEE Personal Communications,* Vol. 6, April 1999, pp. 38–44.

[22] Leppanen, P., et al., "Software Radio—An Alternative for the Future in Wireless Personal and Multimedia Communications," *Proc. IEEE Int. Conf. Personal Wireless Communications,* Feb. 17–19, 1999, pp. 364–368.

[23] Turletti, T., and D. Tennenhouse, "Estimating the Computational Requirements of a Software GSM Base Station," *Proc. IEEE ICC'97,* June 8–12, 1997, pp. 169–175.

[24] Ellingson, S. W., and M. P. Fitz, "A Software Radio-Based System for Experimentation in Wireless Communications," *Proc. IEEE VTC'98,* May 18–21, 1998, pp. 2348–2352.

[25] Swanchara, S., S. Harper, and P. Athanas, "A Stream-Based Configurable Computing Radio Testbed," *Proc. 1998 IEEE Symp. on FPGAs for Custom Computing Machines,* April 15–17, 1998, pp. 40–47.

[26] Cummings, M., and S. Haruyama, "FPGA in the Software Radio," *IEEE Communications Magazine,* Vol. 37, Feb. 1999, pp. 108–112.

[27] Frerking, M. E., *Digital Signal Processing in Communication Systems,* New York: Van Nostrand Reinhold, 1994.

[28] *Intersil Data Sheets: HSP50110, HSP50210, HSP50214,* Intersil Inc.

[29] Joeressen, O., et al., "DIRECS: System Design of a 100 Mbit/s Digital Receiver," *IEE Proc.* Vol. 139, Pt. G, April 1992, pp. 222–230.

[30] Vaughan, R. G., N. L. Scott, and D. R. White, "The Theory of Bandpass Sampling," *IEEE Trans. Signal Processing,* Vol. 39, Sept. 1991, pp. 1973–1984.

[31] Gardner, F. M., "Interpolation in Digital Modems—Part I: Fundamentals," *IEEE Trans. on Communications,* Vol. 41, March 1993, pp. 501–507.

[32] Erup, L., F. M. Gardner, and R. A. Harris, "Interpolation in Digital Modems—Part II: Implementation and Performance," *IEEE Trans. on Communications,* Vol. 41, June 1993, pp. 998–1008.

[33] Hsu, H. P., "Sampling," in J. D. Gibson (ed.), *The Communications Handbook,* Boca Raton, FL: CRC Press, 1997, pp. 13–22.

[34] Proakis, J. G., and D. G. Manolakis, *Digital Signal Processing: Principles, Algorithms, and Applications,* 3rd ed., Upper Saddle River, NJ: Prentice Hall, 1996, pp. 742–746.

[35] Brown, J. L., Jr., "First-Order Sampling of Bandpass Signals—A New Approach," *IEEE Trans. on Information Theory,* Vol. 26, Sept. 1980, pp. 613–615.

[36] Akos, D. M., et al., "Direct Bandpass Sampling of Multiple Distinct RF Signals," *IEEE Trans. on Communications,* Vol. 47, July 1999, pp. 983–988.

[37] Qi, R., F. P. Coakley, and B.G. Evans, "Practical Consideration for Bandpass Sampling," *IEE Electron. Lett.,* Vol. 32, Sept. 1996, pp. 1861–1862.

[38] Akos, D. M., et al., "Direct Bandpass Sampling of Multiple Distinct RF Signals," *IEEE Trans. on Communications,* Vol. 47, July 1999, pp. 983–988.

[39] Kohlenberg, A., "Exact Interpolation of Band-Limited Functions," *J. Appl. Phys.*, Vol. 24, Dec. 1953, pp. 1432–1436.

[40] Coulson, A. J., "A Generalization of Nonuniform Bandpass Sampling," *IEEE Trans. on Signal Processing,* Vol. 43, March 1995, pp. 694–704.

[41] Oppenheim, A. V., and R. W. Schafer, *Discrete-Time Signal Processing,* Upper Saddle River, NJ: Prentice Hall, 1989.

[42] Crochiere, R. E., and L. R. Rabiner, "Optimum FIR Digital Filter Implementations for Decimation, Interpolation, and Narrow-Band Filtering," *IEEE Trans. on Acoustics, Speech, and Signal Processing,* Vol. 23, Oct. 1975, pp. 444–456.

[43] Crochiere, R. E., and L. R. Rabiner, "Further Considerations in the Design of Decimators and Interpolators," *IEEE Trans. on Acoustics, Speech, and Signal Processing,* Vol. 24, Oct. 1976, pp. 296–311.

[44] Vaidyanathan, P. P., *Multirate Systems and Filter Banks,* Upper Saddle River, NJ: Prentice Hall, 1993.

[45] Oh, H. J., et al., "On the Use of Interpolated Second-Order Polynomials for Efficient Filter Design in Programmable Downconversion," *IEEE J. on Selected Areas of Communications,* Vol. 17, April 1999, pp. 551–560.

[46] Hogenauer, E. B., "An Economical Class of Digital Filters for Decimation and Interpolation," *IEEE Trans. on Acoustics, Speech, and Signal Processing,* Vol. 29, April 1981, pp. 155–162.

[47] Kwentus, A. Y., Z. Jiang, and A. N. Wilson, Jr., "Application of Filter Sharpening to Cascaded Integrator-Comb Decimation Filters," *IEEE Trans. on Signal Processing,* Vol. 45, Feb. 1997, pp. 457–467.

[48] Zangi, K. C., and R. D. Koilpillai, "Efficient Filterbank Channelizers for Software Radio Receivers," *Proc. IEEE ICC'98,* June 7–11, 1998, pp. 1566–1570.

[49] Hsiao, C., "Polyphase Filter Matrix for Rational Sampling Rate Conversions," *Proc. IEEE ICASSP'87,* April 6–9, 1987, pp. 2173–2176.

[50] Henker, M., T. Hentschel, and G. Fettweis, "Time-Variant CIC-Filters for Sample Rate Conversion with Arbitrary Rational Factors," *Proc. 6th IEEE Int. Conf. Electronics, Circuits and Systems (ICECS'99),* Sept. 5–8, 1999, pp. 67–70.

[51] Lundheim, L., and T. A. Ramstad, "An Efficient and Flexible Structure for Decimation and Sample Rate Adaptation in Software Radio Receivers," *Proc. ACTS Mobile Communications Submit,* June 1999, pp. 66–668.

Acronyms

2G	Second generation
3G	Third generation
4G	Fourth generation
AAA	Authentication, Authorization, and Accounting
ACK	Acknowledgment
AICH	Acquisition indication channel
AMR	Adaptive multirate codec
AMR-WB	Adaptive multirate wideband codec
AT	Access terminal
ARQ	Automatic repeat request
ASIC	Application-specific integrated circuit
ATM	Asynchronous transfer mode
AWGN	Additional white Gaussian noise
B2B	Business to business
BCCH	Broadcast control channel
BER	Bit error rate
BPSCH	Basic physical subchannel
BPSK	Binary phase-shift keying
BS	Base station
BSC	Base station controller
BSS	Base station subsystem
BSSAP	Base station system application part
BSSGP	Base station system GPRS protocol
BSSMAP	Base station system management application part

BTS	Base transceiver station
CBCH	Cell broadcast channel
CC	Call control
CCCH	Common control channel
CDM	Code-division multiplexed
CDMA	Code-division multiple access
CIC	Cascade integrator-and-comb
CS-i	GPRS coding scheme i
DBPSCH	Dedicated basic physical subchannel
DC	Dedicated control
DCH	Dedicated channel
DGU	Digital gain unit
DL	Data link
DPCH	Dedicated physical channel
DRC	Data rate control
DS-CDMA	Direct-sequence code-division multiple access
DSP	Digital signal processors
DTAP	Direct transfer application part
EDGE	Enhanced data rates for global evolution
EGPRS	Enhanced GPRS
E-TCH	Enhanced TCH
E-FACCH	Enhanced fast associated control channel
FACCH	Fast associated control channel
FBI	Feedback information
FDD	Frequency-division duplex
FIR	Finite-duration impulse response
FPGA	Field programmable gate array
FR	Frame relay
FTP	File Transfer Protocol
GC	General control
GERAN	GSM/EDGE radio access network
GMM	GPRS mobility management
GPRS	General packet radio service
GRA	GERAN registration area
GSM	Global System for Mobile Communications
GSN	GPRS support node
GTP-U	GPRS Tunneling Protocol for the User Plane
HBF	Half-band filter
HHO	Hard handover

HIARQ	Hybrid Type I ARQ
HIIARQ	Hybrid Type II ARQ (a.k.a. incremental redundancy)
ID	Identification number
IEEE	Institute of Electrical and Electronics Engineers
IETF	Internet Engineering Task Force
IF	Intermediate frequency
IMSI	International mobile subscriber identity
IP	Internet Protocol
IR	Incremental redundancy
ITU	International Telecommunication Union
LAN	Local-area network
LDAP	Lightweight Data Access Protocol
LLC	Logical link control
MAC	Medium access control
MAC	Media access control
MAP	Mobile application part
MCS-i	EGPRS modulation and coding scheme i
MIN	Mobile identification number
MPS	Multiplications per second
MRC	Maximum ratio combining
MS	Mobile station
MSC	Mobile switching center
MTP	Message transfer part
NAC	North American cellular
NAK	Negative acknowledgment
NS	Network service
Nt	Notification
O-FACCH	Octal FACCH
OFDM	Orthogonal frequency-division multiplexing
O-TCH	Octal TCH
PACCH	Packet associated control channel
PAR	Peak-to-average ratio
PBCCH	Packet BCCH
PC	Personal computer
PCER	Path collection efficiency ratio
PCH	Paging channel
PCS	Personal communication system
PDA	Personal digital assistant
PDCH	Packet data channel

PDCP	Packet Data Convergence Protocol
PDF	Probability density function
PDP	Packet Data Protocol
PDTCH	Packet data TCH
PDU	Packet data unit
PKI	Public key infrastructure
PLMN	Public Land Mobile Network
PN Sequence	Pseudorandom noise sequence
PS	Packet switched
PSC	Primary synchronization code
PSTN	Public switched telephone network
PTCCH	Packet timing advance control channel
P-TMSI	Packet TMSI
QoS	Quality of service
QPSK	Quadrature phase-shift keying
RAB	Reverse activity bit
RAN	Radio access network
RANAP	Radio access network application part
RF	Radio frequency
RLC	Radio link control
RLP	Radio Link Protocol
RNSAP	Radio network subsystem application part
RR	Radio resource
RRC	Radio resource control
RTP	Real Time Protocol
SACCH	Slow associated control channel
SAP	Service access point
SAPI	Service access point identifier
SBPSCH	Shared basic physical subchannel
SCH	Synchronization channel
SCCP	Signaling connection control part
SCDMA	Synchronous code-division multiple access
SDCCH	Standalone dedicated control channel
SFDR	Spurious free dynamic range
SGSN	Serving GPRS support node
SHO	Soft handoff
SINR	Signal-to-interference plus noise power ratio
SIR	Signal-to-interference power ratio
SM	Session management

SMS	Short messaging service
SNDCP	Sub-Network Dependent Convergence Protocol
SRB	Signaling radio bearer
SSC	Secondary synchronization code
SSDT	Site selection diversity transmission
STTD	Space-time block coded transmit antenna diversity
SSTD	Site selection transmit diversity
TCH	Traffic channel
TCP	Transmission Control Protocol
TDD	Time-division duplex
TDM	Time-division multiplexed
TFI	Temporary flow identity
TMSI	Temporary mobile subscriber identity
TPC	Transmission power control
TRAU	Transcoder and rate adapter unit
TxAA	Transmission antenna array
UDP	User Datagram Protocol
UMTS	Universal Mobile Telecommunication System
URB	User-plane radio bearer
UTRAN	UMTS Terrestrial Radio Access Network
VoIP	Voice-over-IP
WCDMA	Wideband CDMA
WLAN	Wireless local-area network
WWW	World Wide Web
xAN	x (any type) of access network

About the Authors

Jiangzhou Wang received his B.S. and M.S. from Xidian University, Xian, China, in 1983 and 1985, respectively, and his Ph.D. (with Greatest Distinction) from the University of Ghent, Belgium, in 1990, all in electrical engineering.

Dr. Wang was a postdoctoral fellow at the University of California at San Diego, where he participated in the research and development of cellular CDMA systems. He has also worked as a senior system engineer at Rockwell International Corporation, Newport Beach, California, where he participated in the development and system design of wireless communications. He is currently an associate professor at the University of Hong Kong. Dr. Wang has also held a visiting professor position at NTT DoCoMo, Japan. He was a research fellow for 2 years at Southeast University, China, in the 1980s. He was the technical chairman of the IEEE Workshop on 3G Mobile Communications in 2000. Dr. Wang has published more than 100 papers, including more than 20 IEEE transactions or journal papers in the areas of wireless mobile and spread spectrum communications. He is also the editor of *Broadband Wireless Communications: 3G, 4G, and Wireless LAN* (Kluwer, 2001).

Dr. Wang is an editor for the *IEEE Transactions on Communications* and a guest editor on WCDMA for the *IEEE Journal on Selected Areas in Communications.* He holds one U.S. patent for a GSM system. He is a senior member of the IEEE and is listed in the *Marquis Who's Who in the World.*

Tung-Sang Ng received his B.S. Eng. with honors from the University of Hong Kong in 1972, and his M.S. Eng. and Ph.D. from the University of Newcastle, Australia, in 1974 and 1977, respectively, all in electrical engineering.

Dr. Ng has worked for BHP Steel International and the University of Wollongong, Australia. In 1991, he became a professor and the chair of the Electronic Engineering Department at the University of Hong Kong. His current research interests include mobile communication systems, spread spectrum techniques, CDMA, and digital signal processing. He was the general chair of ISCAS'97, vice president, Region 10, of the IEEE CAS Society in 1999 and 2000, and the chair of the Technical Committee on Communications in 2001. He is currently an executive committee member and a board member of the IEE Informatics Divisional Board. He has published more than 170 international journal and conference papers.

Dr. Ng was awarded an honorary doctor of engineering degree by the University of Newcastle, Australia, in August 1997. He received the Senior Croucher Foundation Fellowship in 1999 and the IEEE Millennium medal in 2000. He is a fellow of the IEE and the Hong Kong Institute of Engineers (HKIE), and a senior member of IEEE.

Haseeb Akhtar received his B.S. in electrical engineering from the University of Texas at Arlington in 1989. He has worked as a test engineer at Ericsson Network Systems, where he primarily focused on the area of developing software for AXE10 switches. In 1996, he joined Bell Northern Research (a subsidiary of Nortel Networks) as a software engineer to develop wireless data products. His focus at Nortel has been in the area of designing data communication networks. In 1998, he completed his M.S. in telecommunications engineering from Southern Methodist University. Currently, he is the chief architect for Nortel's IP mobility solution. His present role focuses on providing mobility solutions for IP-based networks.

Russ Coffin has worked in the telecommunications industry since graduating with a B.S. in electrical engineering from the University of Calgary, Alberta, Canada, in 1975. Currently, Mr. Coffin works for Nortel as the director of advanced mobility networks in the Wireless Technology Labs organization focusing on mobility within all IP networks and next-generation information appliances. His previous positions within Nortel include satellite business development, GSM product management, product and market development activities for the U.S. PCS market, North American cellular product management, PCS exploratory development in Richardson, Texas, international product systems engineering in Maidenhead, England, and a variety of positions, including network management systems engineering, ISDN and

SS7 product development, circuit-switched data product development, and custom calling feature development in Ottawa, Canada. Prior to joining Bell Northern Research in 1979, Mr. Coffin worked at Alberta Government Telephones in Edmonton, Alberta, in the Switching Design Engineering Department.

Abdel-Ghani Daraiseh received his B.S. in electrical engineering from the University of Jordan in 1991, and his M.S. and Ph.D. in electrical engineering from Clemson University in 1993 and 1995, respectively. He has worked as a research assistant at the Center for Computer Communication Systems at Clemson University, and as a senior advisor in the Systems Engineering Department of Nortel Networks, where he focused on systems performance, planning, and evolution for CDMA, 1xRTT, and UMTS. Currently, Dr. Daraiseh is an IP network architect in the IP Mobility Department at Nortel Networks in Richardson, Texas. His current role focuses on IP mobility architecture, IPv6, and high-speed Internet. His research interests lie in the areas of TCP/IP over wireless, Internet architecture and mobility, 3G/1xRTT/UMTS, error control coding, and high-speed switching.

Eduardo Esteves received his B.S. from the Military Institute of Engineering, Brazil, in 1989, his M.Sc. from the Catholic University of Rio de Janeiro in 1993, and his Ph.D. from the University of Southern California in 1997, all in electrical engineering. Dr. Esteves has worked for Embratel on the analysis and development of satellite communication systems. He has also been a research assistant at the Communication Sciences Institute in Los Angeles, California. Dr. Esteves joined Qualcomm in 1997 and since then has been working in research and development on the design, standardization, and implementation of the high data rate packet data system (1xEV-DO). Dr. Esteves holds two U.S. patents related to mobile communications.

Hiroshi Furukawa received his B.E. in information engineering from Kyushu Institute of Technology in 1992 and his Ph.D. in electronics engineering from Kyushu University in 1998. He previously worked as a research associate at the Department of Computer Science and Electronics at Kyushu Institute of Technology. He is currently a research scientist at Networking Research Laboratories, NEC Corporation. His research interests cover cellular mobile communication systems.

Vincent Lau obtained a B.Eng. (Distinction First Honors) from the University of Hong Kong in 1992. He worked at HK Telecom for 3 years as a system engineer, responsible for transmission systems design. He obtained the Sir Edward Youde Memorial Fellowship and the Croucher Foundation Fellowship in 1995 and received a Ph.D. in mobile communications from the University of Cambridge. Dr. Lau's Ph.D. thesis was on the practical design and information theoretical aspects of variable rate adaptive channel coding for fast fading wireless communications (TDMA and CDMA). He completed his Ph.D. in 2 years and joined Lucent Technologies/Bell Labs in the United States as member of technical staff in 1997. Dr. Lau was engaged in the algorithm design and standardization of 3G WCDMA systems. He became an assistant professor in the Department of Electrical and Electronic Engineering at Hong Kong University in 1999. There Dr. Lau was appointed as the codirector of the information engineering program as well as the codirector of the 3G Wireless Technology Center responsible for developing an ASIC prototype for CDMA2000. He was the invited session chair of the IEEE WCNC2000 International Conference, the IEEE CAS 3G Workshop 2000, and the SCT2001 International Conference. Dr. Lau was the nonexecutive director of Seamatch Ltd. and the technical advisor of the DAX Group Ltd. His research interests include digital transceiver ASIC architecture design (TDMA and CDMA), adaptive modulation and channel coding, iterative decoding (turbo and LDPC), MIMO systems and BLAST, power control and CREST factor control algorithms, jointly adaptive multiple access protocols as well as short-range wireless ad hoc networking. Dr. Lau has published more than 50 papers in international conference and journals. He has three U.S. provisional patents pending. In 2001 his biography was selected to be included in *Marquis Who's Who in the World.* Dr. Lau joined the Wireless Advanced Technology Labs—Lucent Technology in July 2001 to work on UMTS call processing protocol stack simulation, UMTS one-chip ASIC architecture, and MIMO research. He is currently a senior member of IEEE.

Janne Parantainen received his M.S. in operations research and engineering physics from the Helsinki University of Technology in 1994. He previously worked as a research scientist in the Systems Analysis Laboratory of Helsinki University of Technology. In 1995 he joined Nokia Research Center's Mobile Networks Laboratory and since then he has been working on different fields of telecommunications, including IS-95 CDMA, GSM, and GPRS. Currently, he is involved in the standardization of GERAN and represents Nokia in 3GPP TSG GERAN.

Emad Abdel-Lateef Qaddoura received his B.S. in computer science and engineering from the University of Texas at Arlington in 1986. He began his professional career as a member of the scientific staff at Bell Northern Research in 1986. His work at Bell Northern Research was in the area of data and wireless communications. He received his M.S. in computer science from the University of Texas at Dallas in 1990. In 1991, he joined Fujitsu as a senior firmware and software engineer in the area of fiber optics transmission systems. In 2000, he received his Ph.D. in electrical engineering from the University of Texas at Dallas. In 1993, Dr. Qaddoura rejoined Nortel Networks as a member of the scientific staff, where his focus has been on wireless and data communications. Currently, he is a product director in the area of IP mobility for next-generation communication systems.

Mamoru Sawahashi received his B.S. and M.S. from Tokyo University in 1983 and 1985, respectively, and received his doctorate in engineering from the Nara Institute of Technology in 1998. In 1985 he joined NTT Electrical Communications Laboratories, and in 1992 he transferred to NTT Mobile Communications Network, Inc. (now NTT DoCoMo, Inc.). Since joining NTT, he has been engaged in the research of modulation/demodulation techniques for mobile radio, and research and development of wireless access technologies for WCDMA mobile radio and broadband wireless packet access technologies for beyond IMT-2000. He is now the director of the Wireless Access Laboratory of NTT DoCoMo, Inc.

Benoist Sébire received his M.S. in electronics and computer engineering from the Ecole Nationale Supérieure de Sciences Appliquées et de Technologie, University of Rennes, France, in 1997. He joined Nokia Research Center's Radio Communications Laboratory in 1998. Since then, he has been working on further developments of the GSM/EDGE standard (EDGE Compact and GSM400, for instance). He is currently involved in the standardization of GERAN and represents Nokia in 3GPP TSG GERAN.

Guillaume Sébire received his M.S. in electronics and computer engineering from the Ecole Nationale Supérieure de Sciences Appliquées et de Technologie, University of Rennes, France, in 1997. In 1998, he joined Nokia Research Center's Radio Communications Laboratory in Helsinki, Finland. Since then he has contributed to the development of EDGE, primarily EGPRS, for which he acted as a Nokia delegate in ETSI SMG2 and 3GPP TSG GERAN standardization forums. He is currently involved in the design of the Radio Protocols for GERAN, and represents Nokia in 3GPP TSG GERAN.

Qiang Wu received his B.S. in applied physics in 1991, M.S. in 1994, and Ph.D. in 1997 in electrical engineering, all from the Beijing University of

Posts and Telecommunications in China. In 1997, Dr. Wu joined Qualcomm in the corporate research and development department. Since then, he has been working on the design and implementation of the CDMA2000 high-rate packet data system, also known as HDR, 1xEV-DO, and IS-856. Dr. Wu has several U.S. patents pending related to the 3G high-rate packet data systems.

Yik-Chung Wu received his B.Eng. (honors) and M.Phil. in electronic engineering from the University of Hong Kong in 1998 and 2001, respectively. Currently, he is a research assistant at the same university. His research interests include digital signal processing with applications to communication systems, software radio, and space-time processing.

Kun-Wah Yip received his B.Eng. (honors) and Ph.D. in electrical engineering from the University of Bradford, United Kingdom, in 1991, and the University of Hong Kong in 1995, respectively. He has also worked as a research associate and a postdoctoral fellow at the University of Hong Kong. Currently, he is a research assistant professor at the same university. His research interests include personal wireless communications, communication circuits, spread spectrum techniques, OFDM, and efficient simulation techniques for communication systems.

Index

Recent Titles in the Artech House Mobile Communications Series

John Walker, Series Editor

Advances in Mobile Information Systems, John Walker, editor

Advances in 3G Enhanced Technologies for Wireless Communications, Jiangzhou Wang and Tung-Sang Ng, editors

CDMA for Wireless Personal Communications, Ramjee Prasad

CDMA Mobile Radio Design, John B. Groe and Lawrence E. Larson

CDMA RF System Engineering, Samuel C. Yang

CDMA Systems Engineering Handbook, Jhong S. Lee and Leonard E. Miller

Cell Planning for Wireless Communications, Manuel F. Cátedra and Jesús Pérez-Arriaga

Cellular Communications: Worldwide Market Development, Garry A. Garrard

Cellular Mobile Systems Engineering, Saleh Faruque

The Complete Wireless Communications Professional: A Guide for Engineers and Managers, William Webb

Emerging Public Safety Wireless Communication Systems, Robert I. Desourdis, Jr., et al.

The Future of Wireless Communications, William Webb

GSM and Personal Communications Handbook, Siegmund M. Redl, Matthias K. Weber, and Malcolm W. Oliphant

GSM Networks: Protocols, Terminology, and Implementation, Gunnar Heine

GSM System Engineering, Asha Mehrotra

Handbook of Land-Mobile Radio System Coverage, Garry C. Hess

Handbook of Mobile Radio Networks, Sami Tabbane

High-Speed Wireless ATM and LANs, Benny Bing

Introduction to 3G Wireless Communications, Juha Korhonen

Introduction to GPS: The Global Positioning System, Ahmed El-Rabbany

An Introduction to GSM, Siegmund M. Redl, Matthias K. Weber, and Malcolm W. Oliphant

Introduction to Mobile Communications Engineering, José M. Hernando and F. Pérez-Fontán

Introduction to Radio Propagation for Fixed and Mobile Communications, John Doble

Introduction to Wireless Local Loop, Second Edition: Broadband and Narrowband Systems, William Webb

IS-136 TDMA Technology, Economics, and Services, Lawrence Harte, Adrian Smith, and Charles A. Jacobs

Mobile Data Communications Systems, Peter Wong and David Britland

Mobile IP Technology for M-Business, Mark Norris

Mobile Satellite Communications, Shingo Ohmori, Hiromitsu Wakana, and Seiichiro Kawase

Mobile Telecommunications Standards: GSM, UMTS, TETRA, and ERMES, Rudi Bekkers

Mobile Telecommunications: Standards, Regulation, and Applications, Rudi Bekkers and Jan Smits

Multiantenna Digital Radio Transmission, Massimiliano "Max" Martone

Personal Wireless Communication with DECT and PWT, John Phillips and Gerard Mac Namee

Practical Wireless Data Modem Design, Jonathon Y. C. Cheah

Prime Codes with Applications to CDMA Optical and Wireless Networks, Guu-Chang Yang and Wing C. Kwong

Radio Propagation in Cellular Networks, Nathan Blaunstein

Radio Resource Management for Wireless Networks, Jens Zander and Seong-Lyun Kim

RDS: The Radio Data System, Dietmar Kopitz and Bev Marks

Resource Allocation in Hierarchical Cellular Systems, Lauro Ortigoza-Guerrero and A. Hamid Aghvami

RF and Microwave Circuit Design for Wireless Communications, Lawrence E. Larson, editor

Signal Processing Applications in CDMA Communications, Hui Liu

Spread Spectrum CDMA Systems for Wireless Communications, Savo G. Glisic and Branka Vucetic

Transmission Systems Design Handbook for Wireless Networks, Harvey Lehpamer

Understanding Cellular Radio, William Webb

Understanding Digital PCS: The TDMA Standard, Cameron Kelly Coursey

Understanding GPS: Principles and Applications, Elliott D. Kaplan, editor

Understanding WAP: Wireless Applications, Devices, and Services, Marcel van der Heijden and Marcus Taylor, editors

Universal Wireless Personal Communications, Ramjee Prasad

WCDMA: Towards IP Mobility and Mobile Internet, Tero Ojanperä and Ramjee Prasad, editors

Wireless Communications in Developing Countries: Cellular and Satellite Systems, Rachael E. Schwartz

Wireless Intelligent Networking, Gerry Christensen, Paul G. Florack, and Robert Duncan

Wireless LAN Standards and Applications, Asunción Santamaría and Francisco J. López-Hernández, editors

Wireless Technician's Handbook, Andrew Miceli

For further information on these and other Artech House titles, including previously considered out-of-print books now available through our In-Print-Forever® (IPF®) program, contact:

Artech House
685 Canton Street
Norwood, MA 02062
Phone: 781-769-9750
Fax: 781-769-6334
e-mail: artech@artechhouse.com

Artech House
46 Gillingham Street
London SW1V 1AH UK
Phone: +44 (0)20 7596-8750
Fax: +44 (0)20 7630-0166
e-mail: artech-uk@artechhouse.com

Find us on the World Wide Web at:
www.artechhouse.com